Modern Elementary Mathematics

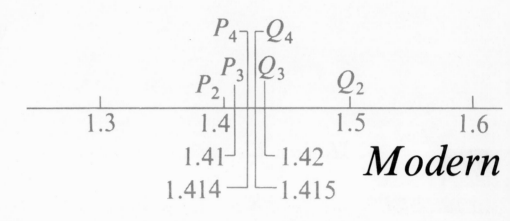

Modern

$I_2:$ |———————————|

$I_3:$ |——|

$I_4:$ |-|

by ANNE E. KENYON

Department of Mathematics
Whittier College

Elementary Mathematics

An Introduction to Its Structure and Meaning

Prentice-Hall, Inc.,
Englewood Cliffs, New Jersey

Modern Elementary Mathematics

ANNE E. KENYON

© 1969 by Prentice-Hall, Inc.,
Englewood Cliffs, New Jersey

Current printing (last digit): 10 9 8 7 6 5 4 3 2 1

13-593251-3

Library of Congress Catalog Card Number 69-19112

Printed in the United States of America

PRENTICE-HALL INTERNATIONAL, INC., *London*
PRENTICE-HALL OF AUSTRALIA, PTY. LTD., *Sydney*
PRENTICE-HALL OF CANADA, LTD., *Toronto*
PRENTICE-HALL OF INDIA PRIVATE LTD., *New Delhi*
PRENTICE-HALL OF JAPAN, INC., *Tokyo*

Foreword
to the Student

> Here and elsewhere we shall not obtain the best insight into things until we actually see them growing from the beginning.
>
> *Aristotle*, Politics

When he penned the words of the quotation above, Aristotle expressed an opinion which has continued to be an article of faith for teachers from the fourth century B.C. to the present. In mathematics, however, it now appears that we have failed to begin at the beginning. Too often, the student emerging from the elementary school has little insight into even the most elementary mathematics. He can usually display some mechanical arithmetical skills, but he understands no more about the structure of the number system which he uses, or its behavior, than he understands about the inner workings of the television set in his home. Far from being an indispensable tool in solving problems, mathematics, for him, is a dreaded discipline that appears to have scarcely any value.

The explosive advances in modern science have necessitated similar advances in "the handmaiden of the sciences," mathematics. This is perhaps the most urgent reason that basic training in mathematics in the elementary school must be improved. Men and women in education, the social sciences, the physical sciences, mathematics, and combinations of these fields, have been studying the elementary school mathematics curriculum and the teacher training program in mathematics with increasing intensity for the past fifteen years or so with the purpose of improving both. At the present time there is an unprecedented concentration of attention on this endeavor by many of the most distinguished professional people in the United States. The upshot of it all has been a radical change in the elementary school mathematics program. At last, it is felt, we are beginning at the "beginning" in mathematics. We are trying to make the structures, the patterns, the processes— the *meaning*—of mathematics explicit, as well as the facts. This textbook is intended for the elementary credential candidate, to give him some of the background necessary to understand and teach from the new elementary school mathematics textbooks. It is also considered appropriate for the liberal arts student who has never come to appreciate the power and beauty of mathematics.

It is hoped that those who have never before enjoyed mathematics, those who have been afraid of its "difficulty," those who have found it dull and unimaginative, will all find reason to change their minds after reading this book.

Credit for the development of this book must be shared with many: with students and student assistants at Whittier College who have tested it; with Amy McHenry, who has spent long hours at the typewriter preparing the final manuscript; and with H. Randolph Pyle, James Holmes, and Bruce Meserve, who have read it at various stages of development and have offered suggestions and encouragement. The author also wishes to express her gratitude to George Allen and Unwin, Ltd., for permission to reprint material from *Mysticism and Logic* by Bertrand Russell.

A. E. K.

Contents

CHAPTER 5

The Whole
Numbers   76

CHAPTER 6

The
Integers   95

CHAPTER 10

Sets of
Points in Space   175

CHAPTER 11

Measurement and
Congruence   205

CHAPTER 12

Areas of
Polygonal Regions   258

Modern Elementary Mathematics

CHAPTER 1

The Language of Sets

1.1 THE SET, ITS MEANING AND ITS MEMBERSHIP

Bertrand Russell has said:

> Pure mathematics consists entirely of such associations as that, if such and such a proposition is true of *anything*, then such and such another proposition is true of that thing. It is essential not to discuss whether the first proposition is really true, and not to mention what the anything is of which it is supposed to be true. Both these points would belong to applied mathematics. We start, in pure mathematics, from certain rules of inference, by which we can infer that *if* one proposition is true, then so is some other proposition. These rules of inference constitute the major part of the principles of formal logic. We then take any hypothesis that seems amusing, and deduce its consequences. *If* our hypothesis is about *anything*, and not about some one or more particular things, then our deductions constitute mathematics. Thus mathematics may be defined as the subject in which we never know what we are talking about, nor whether what we are saying is true.†

† Bertrand Russell, *Mathematics and the Metaphysicians*, The World of Mathematics (New York: Simon and Schuster, 1956), III, 1576–77.

1

This seems a strange statement coming as it does from an eminent mathematician. Nevertheless it tells us about the essential nature of mathematics, and we shall try to come to understand its meaning in the course of our study.

For the present let us investigate a part of one sentence: "It is essential . . . not to mention what the anything is of which it is supposed to be true." Here Russell is saying that in a mathematical system there must be a few terms that we do not define.† If we were to decide that every word must be defined, we would eventually become guilty of circular definition. For instance, a certain dictionary tells us that the noun "interior" means "the interior part of anything." This definition of the noun "interior" makes use of the adjective "interior"; the dictionary tells that the adjective "interior" means "situated within; inner." Suppose the meaning of "inner" in this definition is not clear to us. Referring to the dictionary once more, we find "inner" means "located farther within; interior." We have come full circle!

In a mathematical system, we avoid this process of going around in circles by choosing, as first elements, a few words that are so familiar to everyone—so readily understood in terms of our common experience—that no attempt at definition is necessary. Fairly recently in mathematical history the word *set* has been adopted as such a word. It is the most ambiguous,‡ therefore the most useful, word in the English language, which certainly says something about its familiarity. However in mathematics the word *set* must be used so that its meaning is clear to everyone. We use the noun **set** in precisely the sense that you use the word when you say "a set of dishes," "a set of wrenches" or "a chess set."

The most common sets encountered in mathematics are sets of numbers and sets of points; for example, the set of numbers from one to ten inclusive, and the set of points on a given line. The members of a set are called its **elements**. (The word **element** and its synonym **member**, terms which will be used interchangeably in this book, are two more of the undefined terms in our system.) The elements of the set of counting numbers from one to ten inclusive are the numbers symbolized by the numerals 1, 2, 3, 4, 5, 6, 7, 8, 9, 10.

In mathematics, it is essential that the membership of a set be clearly identified; therefore, we work only with *well-defined sets*. A set is said to be **well-defined** if it is possible to determine whether or not any given thing, or object, is an element of the set. There are two ways to define the membership of a set:

1. *Its elements can be listed*, or
2. *Its membership may be described.*

---

† In a truly *pure* system, these terms would relate to no particular objects, but would be "open" to many possible interpretations; it is this generality which makes a system useful. In our study, however, we shall not attempt to achieve this ultimate abstraction; rather we shall maintain an interchange between the worlds of things and ideas.

‡ This claim can be verified by reference to your dictionary. *A New English Dictionary* (The Clarendon Press, Oxford 1914), for example, contains almost twenty-three full pages, three columns to each page, of meanings for the word "set."

When the elements are to be *listed*, by name (or numeral), it is conventional to place the names within braces and separated by commas; for example, the set $S = \{1, 2, 3, 4, 5, 6, 7, 8, 9, 10\}$. The *order* in which the elements are listed is not important. We could have listed the elements of S by writing $\{3, 6, 5, 9, 1, 2, 10, 8, 4, 7\}$. This listing would indicate no change in the membership of the set S. The same set S may be *described* as follows: the set S of counting numbers from one to ten inclusive. Perhaps it has been noticed that we named the foregoing set "S." This is in keeping with another convention among mathematicians; it is customary to use a capital letter as the "name" of a set.

When we have occasion to indicate that a particular element belongs to a particular set, we use the Greek letter epsilon, $\in$, as a symbol to convey this relationship; for example, $3 \in S$. This is read, "Three is an element of S," or "Three belongs to the set S," or "Three is a member of the set S." If an object, or number, or idea, or thing is *not* a member of a particular set, this can be indicated as $15 \notin S$. The line drawn through the membership symbol $\in$ indicates negation; that is, *not being a member*. The foregoing symbolic sentence is read, "The number fifteen is not an element of the set S," or "Fifteen does not belong to the set S," or "Fifteen is not a member of the set S."

Sets vary widely in the number of elements they contain. Some sets have no elements; for example, the set of all two-headed students in a particular class. A set that contains no elements is called the **empty set**; it is represented by the symbol "$\emptyset$." Its contents may be "listed" by using braces with an empty space between them: $\emptyset = \{ \ \}$. Some sets have so few elements that it is possible and convenient to list all elements in the sets; the set S of counting numbers from one to ten inclusive, used as an example in a preceding paragraph, is such a set. Other sets contain so many elements that, although it is theoretically possible to list all elements in the sets, it is not practical to do so. Sometimes it is best to indicate the contents of such a set by description; for example, the set of all persons present in New York City at a given time. It is possible to use a conventional brace notation to indicate the contents of other very large sets, such as the set of all counting numbers from one to one million inclusive. We can indicate the contents of this set by writing $\{1, 2, 3, \ldots, 1,000,000\}$. Three dots have been used to indicate that some elements have not been listed. Notice that it has been necessary to impose an order on the elements of this set in order to give meaning to the "listing" and in order to indicate that the set has a *last element*.

There are sets that are so large that it is not possible, even theoretically, to list all elements in the sets; for example, the set of all points on a line, or the set of all counting numbers. The elements of the set of all points on a line cannot be indicated by a listing device, but the set of all counting numbers may be indicated as $\{1, 2, 3, \ldots\}$. A brace notation has been used to indicate the elements of this set, but it has again been necessary to list the elements in an order so that the reader can identify the missing elements that have been indicated by the three dots. Since the set of counting numbers has *no last element*, none is indicated in the "listing."

A set may be classified as a **finite set** if all its elements can be listed. This implies that the set *can be ordered in such a way that there is a first and last element.* A set is classified as an **infinite set** if it is not possible to list all its elements, that is, if the set *cannot be ordered so that each element has a next element and there is a last element.*

The reader must understand that an imposition of order on sets such as $\{1, 2, 3, \ldots, 1,000,000\}$ and $\{1, 2, 3, \ldots\}$ in order to make a "listing" of elements meaningful does not constitute a cancellation of the idea that order is irrelevant in listing elements of sets. The membership of elements in a set does not depend in any way upon the order in which the elements are listed. For example, $\{1, 2, 3\}$ and $\{3, 2, 1\}$ are the same sets of elements. Notice that repeated listing also does not affect membership in a set. For example, the set $\{1, 3, 2\}$ may also be expressed as $\{1, 2, 2, 3, 3, 3\}$ since in each case the set consists of the three elements 1, 2, and 3.

EXERCISES 1.1

1. List the members of the following sets using conventional notation:

   Example: The set A of even numbers from 2 to 16 inclusive.

   $$A = \{2, 4, 6, 8, 10, 12, 14, 16\}$$

   (a) The set V of all vowels in the English alphabet.
   (b) The set C of all consonants in the English alphabet.
   (c) The set X of the Great Lakes.
   (d) The set S_1 (read "S sub 1") of odd numbers from 5 to 15 inclusive.
   (e) The set S_2 of integral multiples of 5 from 5 to 15 inclusive.
   (f) The set Y of all inland oceans in Mexico.
   (g) The set X of all positive even numbers.
   (h) The set A of all positive even multiples of 3.

2. Which of the following sets are well-defined?

   (a) The set of books in the King James version of the Bible.
   (b) The set of animals in Wisconsin.
   (c) The set of cities in California whose population as reported in the latest census is over 100,000.
   (d) The set of all wealthy residents of Detroit, Michigan.
   (e) The set of whole numbers from 5 to 25 inclusive.
   (f) The set of all good students in this class.

3. Give two examples of each of the following, first identifying the set by listing its elements and then by describing its contents. (Recall that a list with "..." is still considered to be a listing of the elements of the set.)

   (a) An empty set.
   (b) A finite set.
   (c) An infinite set.

1.2 SUBSETS

Frequently we consider a set whose elements are each elements of another set; for example, the letters of any given word are also letters of the alphabet. Similarly, the set of consonants is a set whose elements are each elements of the set of all letters in the alphabet. The set of consonants is said to be a *subset* of the set of all letters in the alphabet. We define a set X to be a **subset** of a set Y if each element of set X is also an element of set Y. Notice that a set X may be thought of as a subset of a set Y when X is the empty set, when X consists of some but not all elements of Y, and when X contains exactly the same elements as Y. If X is the empty set, then each element of X (there are none) is also an element of Y. As an example of a subset that contains some but not all of the elements of a given set, consider the set X of all equilateral triangles as a subset of the set Y of all isosceles triangles. As an example of a subset that contains exactly the same elements as a given set, consider the set X of all equilateral triangles as a subset of the set Y of all equiangular triangles. Whenever X is a subset of Y, we write $X \subseteq Y$ or $Y \supseteq X$. If X is a subset of Y, but X does not contain exactly the same distinct elements as Y, then X is called a **proper subset** of Y, symbolized $X \subset Y$ or $Y \supset X$. Thus the set of even counting numbers from two to ten inclusive is a proper subset of the counting numbers from one to ten inclusive, that is

$$\{2, 4, 6, 8, 10\} \subset \{1, 2, 3, 4, 5, 6, 7, 8, 9, 10\}$$

If $X = \{t, o\}$ and $Y = \{t, o, t\}$, notice that X and Y contain exactly the same *distinct* elements. Thus X is *not* a proper subset of Y. When sets X and Y do contain exactly the same *distinct* elements, that is, when $X \subseteq Y$ and $Y \subseteq X$, then set X is defined to be **equal** to set Y, and we write $X = Y$. This implies that two sets X and Y are equal if and only if "X" and "Y" are two names for the same set.

To indicate the negations of these statements, we draw lines through the symbols. The expression $X \nsubseteq Y$ is read "X is *not* a subset of Y." The expression $X \not\subset Y$ is read "X is *not* a proper subset of Y." The expression $X \neq Y$ says "X is *not* equal to Y." The following examples indicate correct notation: $\{1, 2, 3\} \nsubseteq \{1, 3, 5, 7\}$, $\{1, 2, 3\} \not\subset \{1, 2, 3\}$, and $\{1, 2, 3,\} \neq \{a, b, c\}$.

It is often convenient to indicate the relationship between sets in a diagrammatic way by the use of drawings called **Euler diagrams**. In these drawings a set is represented by the region that includes the points on and within a closed curve. The set X is represented by the drawing in Figure 1.1.

Fig. 1.1

The drawing in Figure 1.2 indicates that set X and set Y have some elements in common:

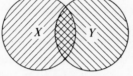

Fig. 1.2

The drawing in Figure 1.3 indicates that set X and set Y have no elements in common:

Fig. 1.3

If two sets X and Y have no elements in common, they are called **disjoint sets**; for example, $X = \{1, 3, 5\}$ and $Y = \{2, 4, 6\}$.

1.3 EQUALITY AND EQUIVALENCE

It is important to understand that the definition of *equal sets*, given in Section 1.2, does *not* say that two sets having the same number of elements are necessarily equal. The sets $A = \{1, 2, 3\}$ and $B = \{a, b, c\}$ are *not* equal even though each has three elements. We use the symbol $n(A)$ to represent the *number of elements* in a set A. Thus we may say, of the sets $A = \{1, 2, 3\}$ and $B = \{a, b, c\}$, $n(A) = n(B)$. Any two sets A and B where $n(A) = n(B)$ are **equivalent sets**. Thus we need a basis for determining whether or not $n(A) = n(B)$. Though the sets A and B defined in this paragraph do not contain identical elements, their elements can be matched to show that *there is exactly one distinct element in A for each distinct element in B*:

$$
\begin{array}{ccc}
1 & 2 & 3 \\
\updownarrow & \updownarrow & \updownarrow \\
a & b & c
\end{array}
$$

Suppose the given sets had been $A = \{1, 2, 2, 3\}$ and $B = \{a, b, c\}$. The number of *distinct* elements in $\{1, 2, 3\}$ is the same as in $\{1, 2, 2, 3\}$. Hence, the matching of distinct elements of sets A and B would not be changed. Also, the matching of the elements in these two sets (or in any other pairs of finite sets with two or more elements) can be done in more than one way; for example, A and B can be matched in at least these two additional ways:

$$
\begin{array}{ccc}
1 & 2 & 3 \\
\updownarrow & \updownarrow & \updownarrow \\
c & b & a
\end{array}
\qquad
\begin{array}{ccc}
1 & 2 & 3 \\
\updownarrow & \updownarrow & \updownarrow \\
a & c & b
\end{array}
$$

A matching that shows exactly one distinct element in the first set for each distinct element in the second set is called a **one-to-one correspondence**. When the distinct elements of two sets X and Y can be put into a one-to-one correspondence, then the set X is defined to be **equivalent** to the set Y; we symbolize this as $n(X) = n(Y)$.

It is not difficult to tell that the sets $\{a, b, c\}$ and $\{1, 2, 3\}$ are equivalent. They are finite sets and the elements in each set can be counted. It is true of any pair of finite sets that they can be tested for equivalence by counting the elements in each. When this is done, however, we are not avoiding the process of showing a one-to-one correspondence between the elements of the sets, because the process of counting elements is a matching process where a one-to-one correspondence is established between the elements of a subset of the counting numbers and the elements of the given set. When two sets have infinitely many elements we cannot "count" their elements and associate a counting number with each set. It is often very difficult to judge equivalence for infinite sets. If we can somehow demonstrate a one-to-one correspondence between the distinct elements of the two sets, then we agree that this would establish equivalence of the sets. Matching the elements of two infinite sets one-to-one can be accomplished in a variety of ways, depending partially on the nature of the sets. The elements of a set can often be indicated by using a *general symbol*, or *variable*, that may represent *any* element of the set. We can, for example, let n represent any counting number; then any even number could be represented by $2n$ for some counting number n, and any odd number could be represented by $2n - 1$ for some counting number n. We employ this symbolism in the following array to illustrate a one-to-one correspondence between the elements of the set of counting numbers and the elements of the set of even counting numbers:

$$
\begin{array}{ccccccc}
1 & 2 & 3 & 4 & \cdots & n & \cdots \\
\updownarrow & \updownarrow & \updownarrow & \updownarrow & & \updownarrow & \\
2 & 4 & 6 & 8 & \cdots & 2n & \cdots
\end{array}
$$

This procedure clearly illustrates that the elements of the set of even counting numbers can be obtained by doubling each element of the set of counting numbers. Thus we conclude that there is a one-to-one correspondence between the elements of the two sets. Therefore the set of even counting numbers is *equivalent* to the set of counting numbers, and the sets have the same *number* of elements even though the sets are infinite sets.

EXERCISES 1.3

1. Copy each of the following sentences and insert any one or more of the relations "is a subset of," "is a proper subset of," "is an element of," "is equal to," "is equivalent to," that will result in a true statement.

(a) The set of even numbers from 2 to 10 inclusive _____ the set of odd numbers from 1 to 9 inclusive.

(b) The number 3 ____ the set of odd numbers.

(c) The set of integral multiples of three from 3 to 15 inclusive ____ the set of integral multiples of three from 3 to 21 inclusive.

(d) The set of integral multiples of three from 3 to 15 inclusive ____ the set containing the numbers 3, 6, 9, 12, and 15.

(e) The empty set ____ the set of Roman numerals from I to VIII inclusive.

(f) The empty set ____ the set of palm trees growing in Antarctica.

(g) The number 34 ____ the set of counting numbers from 1 to 5 inclusive.

(h) The set $\{3, 4\}$ ____ the set of counting numbers from 1 to 5 inclusive.

2. Use a list of elements for each set and a symbol for each relation. Then write each of the sentences in Exercise 1 in symbolic form.

3. Suppose we have a set A of counting numbers from one to five inclusive, and set $B = \{1, 2, 3, 4, 5\}$.

(a) Is A a subset of B?

(b) Is B a subset of A?

(c) Are A and B equal sets?

(d) Are A and B equivalent sets?

4. Answer the questions in Exercise 3 for

(a) $A = \{h, o, t\}$ and $B = \{t, o, o, t, h\}$

(b) $A = \{r, e, s, t\}$ and $B = \{s, t, r, e, e, t\}$

5. Is it true that all equal sets are also equivalent? If not, give an example for which it is not true.

6. Is it true that all equivalent sets are also equal? If not, give an example for which it is not true.

7. Use Euler diagrams and show the following relationships between sets X and Y.

(a) X is a proper subset of Y.

(b) Y is a proper subset of X.

(c) X and Y are disjoint sets.

(d) X is a subset of Y and Y is a subset of X.

(e) X contains some (but not all) elements that belong to Y and Y contains some (but not all) elements that belong to X.

(f) Two different relationships between sets X and Y such that in each relationship X contains some (but not all) elements that belong to Y.

8. For each of the following sets, define each of its subsets by listing the elements.

(a) $A = \{a\}$

(b) $B = \{a, b\}$

(c) $C = \{a, b, c\}$

9. Show that a one-to-one correspondence exists between the elements of the following pairs of sets:

(a) The set of all letters in the word "discourage" and the set of counting numbers from one to ten inclusive.

(b) The set of all counting numbers and the set of all odd counting numbers.
(c) The set of all counting numbers and the set of all unit fractions. (A unit fraction is a fraction whose numerator is 1 and whose denominator is a counting number.)
(d) The set of all counting numbers and the set $S = \{1, 4, 9, 16, \ldots\}$.
(e) The set of all counting numbers and the set $M = \{\frac{1}{3}, \frac{1}{6}, \frac{1}{9}, \frac{1}{12}, \ldots\}$.
(f) The set of all counting numbers and the set $R = \{3, 9, 27, 81, \ldots\}$.

1.4 OPERATIONS ON SETS

During a particular discussion we may be considering several sets that are all subsets of the same set. For example, at Paragon College, the set of students majoring in physics, the set of students majoring in biology, and the set of students majoring in psychology are each subsets of the set of all students at Paragon College. It is customary to call the set of which all other sets in the discussion are subsets, the **universal set**, or simply the **universe**, of that discussion. The symbol commonly used to name the *universe* is U.

Euler diagrams will be used frequently to help clarify set concepts because they distinguish, visually, the several subsets of the universe that may be involved in a discussion. The regions into which the universe is divided, when we use an Euler diagram, represent these subsets. When we use Euler diagrams to represent sets, it is common to let a rectangular region represent the universe U, with circular regions representing the subsets under discussion, as shown in Figure 1.4.

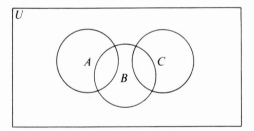

Fig. 1.4

In this diagram, let us suppose that U is the set of all letters in the English alphabet where A represents the set of letters used to name musical notes, B represents the set of letters in the word "clarinet," and C represents the set of all letters in the word "horn." A study of the diagram discloses that A and C have no elements in common, that is, A and C are disjoint; it also shows that some (but not all) elements of A are also elements of B and some (but not all) elements of B are also elements of C.

Suppose we consider the set of elements that are common to sets A and B. When we do this, we are, in a sense, associating a third set with the two original sets. We call this third set the *intersection* of A and B. We use $A \cap B$ to symbolize the intersection of A and B, and we read $A \cap B$ as "the intersection of A and B." The intersection, then, is an operation on two sets that results in a third set. Any operation performed on just two objects at a time is called a **binary** operation; thus intersection of sets is a binary operation. There are three other set operations that we shall consider. We list them, along with intersection, and define all four:

1. **Intersection:**   The intersection of two sets A and B is the set that contains each element of A that is also an element of B, and no other element.

   Symbolized: $A \cap B$.

EXAMPLES: Let $A = \{a, b, c, d, e, f, g\}$, $B = \{c, l, a, r, i, n, e, t\}$ and $C = \{h, o, r, n\}$. Since the elements of a set may be written in any order, we shall, for the sake of convenience, use alphabetical order for sets B and C as well as for set A: $B = \{a, c, e, i, l, n, r, t\}$, $C = \{h, n, o, r\}$. Then $A \cap B = \{a, c, e\}$; $B \cap C = \{n, r\}$. In the diagram in Figure 1.5, the shaded region represents the set $A \cap B$:

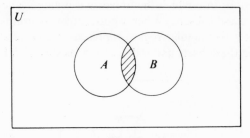

Fig. 1.5

2. **Union:**   The union of two sets A and B is the set that contains each element of A, each element of B, and no other element.

   Symbolized: $A \cup B$ (read the union of A and B).

EXAMPLES: Again, let $A = \{a, b, c, d, e, f, g\}$, $B = \{a, c, e, i, l, n, r, t\}$ and $C = \{h, n, o, r\}$. Then $A \cup B = \{a, b, c, d, e, f, g, i, l, n, r, t\}$; and $B \cup C = \{a, c, e, h, i, l, n, o, r, t\}$. In the diagram in Figure 1.6, the shaded region represents the set $A \cup B$.

3. **Complement:**   The complement of a set A is the set that contains exactly those elements of the universe that are not contained in A.

   Symbolized: A' (read the complement of A).

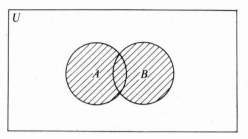

Fig. 1.6

EXAMPLES: Let U be the set of letters in the English alphabet, $A = \{a, b, c, d, e, f, g\}$ and $B = \{a, c, e, i, l, n, r, t\}$. Then $A' = \{h, i, j, k, l, m, n, o, p, q, r, s, t, u, v, w, x, y, z\}$, and $B' = \{b, d, f, g, h, j, k, m, o, p, q, s, u, v, w, x, y, z\}$. In the diagram in Figure 1.7, the shaded region represents the set A':

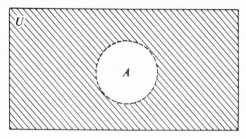

Fig. 1.7

Before defining the fourth set operation, the *Cartesian product*, we need to introduce the idea of an ordered pair. An **ordered pair** is a set of two elements in which one element is designated as the first, and the other element as the second. It is customary to denote an ordered pair by writing the names of the elements, in the proper order, between parentheses and separated by a comma; for example,

$$(1, 2), \qquad (a, b), \qquad \text{(apple, orange)}$$

The important aspect of the ordered pair is that it allows us to distinguish between such pairs as $(1, 2)$ and $(2, 1)$. These are *different*, that is, *distinct*, ordered pairs. In contrast, recall that order is not important in listing the elements of usual sets; for example, $\{1, 2\} = \{2, 1\}$.

We are now ready to define the fourth set operation.

4. **Cartesian product:**   The Cartesian product of two sets A and B is the set of all ordered pairs such that the first entry in each ordered pair is an element of A and the second entry is an element of B.

Symbolized: $A \times B$ (read the Cartesian product of A and B).

EXAMPLE: Let $A = \{a, b, c\}$ and $B = \{d, e\}$. Then $A \times B = \{(a, d), (a, e), (b, d),$ $(b, e), (c, d), (c, e)\}$. As in any set, the order in which the elements are listed is not important. We might have written $A \times B = \{(a, d),$ $(b, d), (c, d), (a, e), (b, e), (c, e)\}$, since the ordered pairs that are the elements of the set are the same as before. The reader will recognize that many other forms of the set $A \times B$, differing only in the order in which the elements are listed, could also be given. Notice that we are *not* saying that $A \times B$ is the same as $B \times A$. Indeed, these sets are usually quite different, as the reader will discover later. A graphical representation of this Cartesian product may be obtained by arranging the elements of A as coordinates on one axis of a rectangular coordinate system and the elements of B as coordinates on the other. The Cartesian product of A and B would be represented by the points determined by the intersection of the "A" lines with the "B" lines, illustrated in the graph in Figure 1.8.

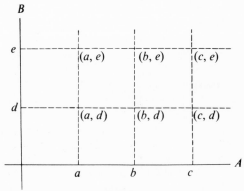

Fig. 1.8

In finding Cartesian products of finite sets, a tabular arrangement is often convenient; for example, given sets $A = \{a, b, c, d, e\}$ and $B = \{f, g, h, i, j\}$, the elements of $A \times B$ can be arranged as shown in Figure 1.9.

$\times$	f	g	h	i	j
a	(a, f)	(a, g)	(a, h)	(a, i)	(a, j)
b	(b, f)	(b, g)	(b, h)	(b, i)	(b, j)
c	(c, f)	(c, g)	(c, h)	(c, i)	$(c\,j)$
d	(d, f)	(d, g)	(d, h)	(d, i)	(d, j)
e	(e, f)	(e, g)	(e, h)	(e, i)	(e, j)

Fig. 1.9

Observe that a display of the elements of a Cartesian product in a graphical representation (Fig. 1.8) or in a tabular arrangement (Fig. 1.9) involves the consideration of the elements in a certain pattern. The imposition of a pattern, arrangement, or order on the elements of a set is strictly for our convenience in thinking about or identifying the elements.

EXERCISES 1.4

1. Choose a universal set for the elements of each of the following collection of sets:

   Example: $A = \{$Ohio, Indiana, Kentucky$\}$, $B = \{$Maine, New Hampshire$\}$, $C = \{$Washington, Oregon, California$\}$. $U = $ the set of all states of the United States of America.
   Note: There is no unique universe for the elements of this collection of sets. We might have chosen U as the set of all continental states of the United States of America, or $U = \{$Ohio, Indiana, Kentucky, Maine, New Hampshire, Washington, Oregon, Maine$\}$, etc.

   (a) $X = \{$Ford, Chevy, Cadillac$\}$, $Y = \{$Plymouth, Chevy, Dodge$\}$, and $Z = \{$Nash, Lincoln$\}$.
   (b) A: the set of chairs in our classroom, B: the set of blackboards in our classroom, C: the set of lighting fixtures in our classroom, and D: the set of all purple floodlights in our classroom.
   (c) S_1: the set of all instruments used in surgery at Presbyterian Hospital, S_2: the set of all implements used in gardening at Paragon College, S_3: the set of all utensils used in preparing meals at Paragon College student dining hall, and S_4: the set of all equipment used to conduct experiments in the Paragon College chemistry laboratories.
   (d) $A = \{$rich man, poor man$\}$, $B = \{$beggarman, thief$\}$, and $C = \{$doctor, lawyer, merchant, chief$\}$.
   (e) $A = \{2, 4, 6, 8\}$, $B = \{4, 8, 12, 16, 20\}$, $C = \{0, 8, 16, 24\}$, and $D = \{2, 6, 18\}$.
   (f) $A = \{b, d, f\}$, $B = \{q, r, s, t\}$, and $C = \{c, s, z\}$.

2. Give three subsets of each of the following universal sets:

   (a) U: the set of all presidents (past and present) of the United States of America.
   (b) U: the set of all cities in the United States of America.
   (c) U: the set of names of all persons qualified to vote in this state in the next election.
   (d) U: the set of all parts of the human anatomy.
   (e) U: the set of all even numbers.
   (f) $U = \{1, 3, 5, 7, 9\}$.

3. Consider set $A = \{a, b, c, d, e, f, g\}$ and set $C = \{h, o, r, n\}$ used in illustrative examples in Section 1.4. What is their union? What is their intersection?

4. The universe $U = \{1, 2, 3\}$ has the following subsets: $U = \{1, 2, 3\}$, $A = \{1, 2\}$, $B = \{1, 3\}$, $C = \{2, 3\}$, $D = \{1\}$, $E = \{2\}$, $F = \{3\}$, and $\emptyset = \{ \ \}$. The table in

Figure 1.10 shows the results of the operation of intersection over the subsets of $\{1, 2, 3\}$:

$\cap$	$\{1, 2, 3\}$	$\{1, 2\}$	$\{1, 3\}$	$\{2, 3\}$	$\{1\}$	$\{2\}$	$\{3\}$	$\{\ \}$
$\{1, 2, 3\}$	$\{1, 2, 3\}$	$\{1, 2\}$	$\{1, 3\}$	$\{2, 3\}$	$\{1\}$	$\{2\}$	$\{3\}$	$\{\ \}$
$\{1, 2\}$	$\{1, 2\}$	$\{1, 2\}$	$\{1\}$	$\{2\}$	$\{1\}$	$\{2\}$	$\{\ \}$	$\{\ \}$
$\{1, 3\}$	$\{1, 3\}$	$\{1\}$	$\{1, 3\}$	$\{3\}$	$\{1\}$	$\{\ \}$	$\{3\}$	$\{\ \}$
$\{2, 3\}$	$\{2, 3\}$	$\{2\}$	$\{3\}$	$\{2, 3\}$	$\{\ \}$	$\{2\}$	$\{3\}$	$\{\ \}$
$\{1\}$	$\{1\}$	$\{1\}$	$\{1\}$	$\{\ \}$	$\{1\}$	$\{\ \}$	$\{\ \}$	$\{\ \}$
$\{2\}$	$\{2\}$	$\{2\}$	$\{\ \}$	$\{2\}$	$\{\ \}$	$\{2\}$	$\{\ \}$	$\{\ \}$
$\{3\}$	$\{3\}$	$\{\ \}$	$\{3\}$	$\{3\}$	$\{\ \}$	$\{\ \}$	$\{3\}$	$\{\ \}$
$\{\ \}$	$\{\ \}$	$\{\ \}$	$\{\ \}$	$\{\ \}$	$\{\ \}$	$\{\ \}$	$\{\ \}$	$\{\ \}$

Fig. 1.10

Make a table showing the results of the operation of union over the subsets of $\{a, b, c\}$.

5. Give all subsets of the universe: $U = \{a, b, c, d\}$.

6. Give the complement of each subset of $\{a, b, c, d\}$.

7. Which of the following are true statements for any given distinct elements a, b, c, d, x, y, z?

(a) $\{a, b\} \in \{a, b, c, d\}$
(b) $(a, b) = \{a, b\}$
(c) $a \in \{a, b, c, d\}$
(d) $\{a, b, c\} \neq \{x, y, z\}$
(e) $(a, b) \notin \{a, b, c\}$
(f) $(a, b) = (b, a)$
(g) $\{a, b\} = \{b, a\}$
(h) $\{a, b\} \subset \{a, b, c, d\}$
(i) $\{a, b\} \subset \{a, b\}$
(j) $\{a, b\} \subseteq \{a, b\}$
(k) $\{a, b\} \supseteq \{b, a\}$
(l) $b \subset \{a, b, c\}$

8. Find the following Cartesian products:

(a) $\{a, b, c\} \times \{a, b, c\}$
(b) $\{a, b, c\} \times \{a, b\}$
(c) $\{a, b\} \times \{a, c\}$
(d) $\{a, c\} \times \{a, b\}$
(e) $\{a, b\} \times \{c\}$
(f) $\{a, b, c\} \times \{\ \}$
(g) $\{a, c\} \times \{\ \}$
(h) $\{\ \} \times \{a, c\}$

9. When we have a problem where more than one operation is involved, we often use grouping symbols to indicate what operation is to be done first, second, etc. For example, $A \cap (B \cap C)$ means that we must find the intersection of B and C first, then find the intersection of A and the set $B \cap C$. Use sets $U = \{1, 2, 3\}$, $A = \{1, 2\}$, $B = \{1, 3\}$, $C = \{2, 3\}$ and perform the indicated operations.

(a) $(A \cap B) \cap C$
(b) $A \cap (B \cap C)$
(c) $A \cup (B \cup C)$
(d) $(A \cup B) \cup C$
(e) $A' \cap B'$
(f) $(B \cap C)'$
(g) $(A \cup B)'$
(h) $B' \cup C'$
(i) $A \times (B \times C)$
(j) $(A \times B) \times C$
(k) $(A \cap B) \cup (A \cap C)$
(l) $A \cap (B \cup C)$
(m) $(A \cup B) \cap (A \cup C)$
(n) $A \cup (B \cap C)$

CHAPTER 2

Elements
of Set Theory

2.1 VENN DIAGRAMS

In Chapter 1, we used Euler diagrams to picture given relationships among sets. In this chapter we shall use *Venn diagrams*. Both sorts of drawing employ circles (and other simple closed curves) and their interior regions to represent sets. Further, both Euler diagrams and Venn diagrams are used to show relationships that may exist among sets. The difference between Euler diagrams and Venn diagrams is this: Each particular relationship that exists among two or more sets requires a separate and different Euler diagram, whereas the regions of one Venn diagram can be interpreted to represent each of the possible relationships that may exist between two sets, and the regions of another Venn diagram can be used for each relationship among three sets. For example, if two sets are being considered, the Venn diagram in Figure 2.1 can be used to represent any of the relationships that may exist between them:

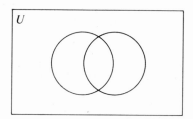

Figure 2.1

15

Notice that in Figure 2.1 we have a picture of the universe divided into four regions. If we let the circular regions (the circles and their interiors) in the diagram represent sets A and B, then these four regions can be identified as $A \cap B$, $A \cap B'$, $A' \cap B$, and $A' \cap B'$, as in Figure 2.2.†

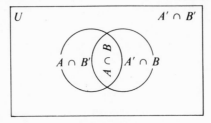

Figure 2.2

There are six relationships that two sets A and B may have with each other. We first illustrate these relationships in Figure 2.3, using Euler diagrams. The reader will observe that each relationship requires an Euler diagram different from the others.

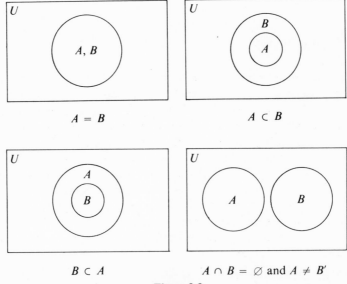

Figure 2.3

---

† A set of points that consists of the points on a closed curve and the points interior to the curve is a *closed set*. A set of points that consists of just the points interior to a closed curve is an *open set*. Notice that the sets of points that represent A, B, and $A \cap B$ in the Venn diagram in Figure 2.2 are closed sets. But the sets of points that represent $A \cap B'$, $A' \cap B$ and $A' \cap B'$ are neither open nor closed sets. The sets of points that represent $A \cap B'$ and $A' \cap B$ include some but not all the points on the closed curves that bound them. The set of points that represents $A' \cap B'$ is not bounded by a closed curve. (See discussion of curves in Section 10.6, Chapter 10.)

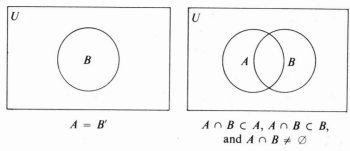

$A = B'$ $A \cap B \subset A, A \cap B \subset B,$
and $A \cap B \neq \varnothing$

Figure 2.3 (continued)

We next illustrate these same six relationships using Venn diagrams. The symbol $\varnothing$ is used to indicate that a region is empty. In each diagram in Figure 2.4, the circular region on the left represents set A and the circular region on the right represents set B

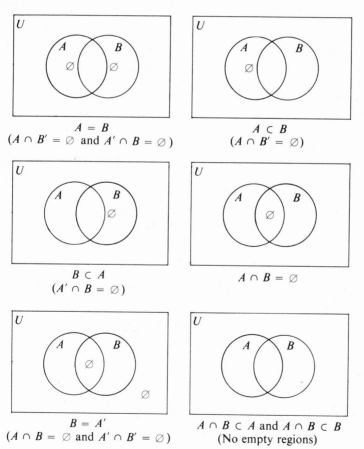

$A = B$
$(A \cap B' = \varnothing$ and $A' \cap B = \varnothing)$

$A \subset B$
$(A \cap B' = \varnothing)$

$B \subset A$
$(A' \cap B = \varnothing)$

$A \cap B = \varnothing$

$B = A'$
$(A \cap B = \varnothing$ and $A' \cap B' = \varnothing)$

$A \cap B \subset A$ and $A \cap B \subset B$
(No empty regions)

Figure 2.4

2.2 PROPERTIES OF SETS

In Chapter 1 we developed a vocabulary to use in considering sets; in this chapter we shall give attention to some elementary theory of the "Algebra of Sets," a very interesting and important mathematical system. We have already explained that every mathematical system begins with some few words that are undefined; these words serve as a basis of definition for all other key words used in the system. We now proceed to discuss other components of a mathematical system. We refer once more to a part of the quotation from Bertrand Russell that appears on the first page of Chapter 1:

> Pure mathematics consists entirely of such associations as that, if such and such a *proposition* [italics not in original] is true of *anything*, then such and such another proposition is true of that thing. It is essential not to discuss whether the *first proposition* [italics not in original] is really true, ... We start ... from certain rules of inference, by which we can infer that if *one proposition* [italics not in original] is true, then so is some other proposition. ...

In this excerpt Russell is saying that, at the beginning of the process of building a mathematical system, we must list some statements that we are willing to accept as true without proof. From these as a basis, by a process of deductive logic,† we develop new statements.

These beginning statements are variously called **assumptions, axioms, postulates, hypotheses.** In our study, the assumptions that we shall accept as true will be found to be quite consistent with intuitive notions we have acquired through extensive experience with many sets. The first group of assumptions involves properties, or characteristics, of sets under the set operations we have defined. They are listed and explained below.

Note: In the examples used to clarify these definitions, we shall use as the universe U the set of all letters in the English alphabet, and subsets $A = \{p, a, r, a, g, o, n\}$, $B = \{c, o, l, l, e, g, e\}$, and $C = \{c, o, e, d, s\}$. Since the order in which elements of a set are listed has no significance and since two sets are equal if they contain exactly the same *distinct* elements, we can arrange the elements of each of these subsets in alphabetical order and eliminate repetition of names for elements. Then

$$A = \{a, g, n, o, p, r\}$$
$$B = \{c, e, g, l, o\}$$
$$C = \{c, d, e, o, s\}$$

1. The commutative property of intersection: A change in the order in which sets are considered when their intersection is found produces no

---

† The axioms and theorems of deductive logic are the "rules of inference" of which Russell speaks.

change in the result.

Symbolized: $A \cap B = B \cap A$.

EXAMPLE: When we use specific sets A and B listed in the preceding paragraph we find that $A \cap B = \{g, o\} = B \cap A$.

In Figure 2.5 the intersection of A and B (or the intersection of B and A) is represented by the shaded region.

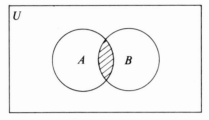

Figure 2.5

2. The commutative property of union: A change in the *order* in which sets are considered when their union is found produces no change in the result.

Symbolized: $A \cup B = B \cup A$.

EXAMPLE: We use the same sets A and B, and we find that $A \cup B = \{a, c, e, g, l, n, o, p, r\} = B \cup A$.

In Figure 2.6 the shaded region represents the union of A and B (or the union of B and A).

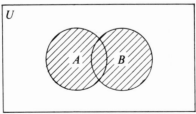

Figure 2.6

3. The associative property of intersection: When an intersection of three sets is to be found, we can find the intersection of only two sets at a time, since intersection is a binary operation. A change in the way three sets are grouped when their intersection is found produces no change in the result.

Symbolized: $(A \cap B) \cap C = A \cap (B \cap C)$.

EXAMPLE: We use the sets A, B, and C specified for these examples and we first determine the elements of $(A \cap B) \cap C$:

$$(A \cap B) \cap C = (\{a, g, n, o, p, r\} \cap \{c, e, g, l, o\}) \cap \{c, d, e, o, s\})$$
$$= \{g, o\} \cap \{c, d, e, o, s\}$$
$$= \{o\}$$

Next, we find the elements of $A \cap (B \cap C)$:

$$A \cap (B \cap C) = \{a, g, n, o, p, r\} \cap (\{c, e, g, l, o\} \cap \{c, d, e, o, s\})$$
$$= \{a, g, n, o, p, r\} \cap \{c, e, o\}$$
$$= \{o\}$$

The double shaded region in each of the Venn diagrams in Figure 2.7 represents the indicated intersection.

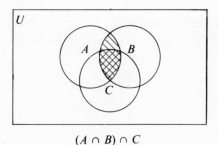

$(A \cap B) \cap C$

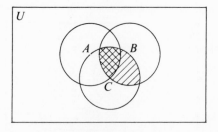

$A \cap (B \cap C)$

Figure 2.7

4. The associative property of union: When a union of three sets is to be found, we can find the union of only two sets at a time, since union is a binary operation. A change in the way three sets are grouped when their union is found produces no change in the result.

Symbolized: $(A \cup B) \cup C = A \cup (B \cup C)$.

EXAMPLE: $(A \cup B) \cup C = (\{a, g, n, o, p, r\} \cup \{c, e, g, l, o\}) \cup \{c, d, e, o, s\}$

$\qquad\qquad\qquad = \{a, c, e, g, l, n, o, p, r\} \cup \{c, d, e, o, s\}$

$\qquad\qquad\qquad = \{a, c, d, e, g, l, n, o, p, r, s\}$

$A \cup (B \cup C) = \{a, g, n, o, p, r\} \cup (\{c, e, g, l, o\} \cup \{c, d, e, o, s\})$

$\qquad\qquad\qquad = \{a, g, n, o, p, r\} \cup \{c, d, e, g, l, o, s\}$

$\qquad\qquad\qquad = \{a, c, d, e, g, l, n, o, p, r, s\}$

The entire shaded region in each of the Venn diagrams in Figure 2.8 represents the indicated union:

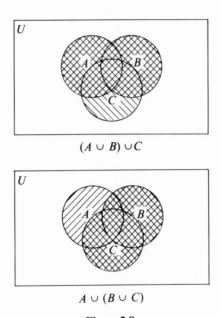

$(A \cup B) \cup C$

$A \cup (B \cup C)$

Figure 2.8

5. The distributive property of intersection over union: Since a verbal statement of this property is very complicated, we shall use only a symbolic statement.

Symbolized: $A \cap (B \cup C) = (A \cap B) \cup (A \cap C)$.

EXAMPLE: $A \cap (B \cup C) = \{a, g, n, o, p, r\} \cap (\{c, e, g, l, o\} \cup \{c, d, e, o, s\})$

$\qquad\qquad\qquad = \{a, g, n, o, p, r\} \cap \{c, d, e, g, l, o, s\}$

$\qquad\qquad\qquad = \{g, o\}$

$$(A \cap B) \cup (A \cap C) = (\{a, g, n, o, p, r\} \cap \{c, e, g, l, o\})$$
$$\cup (\{a, g, n, o, p, r\} \cap \{c, d, e, o, s\})$$
$$= \{g, o\} \cup \{o\}$$
$$= \{g, o\}$$

We illustrate this property by Venn diagrams in Figure 2.9.

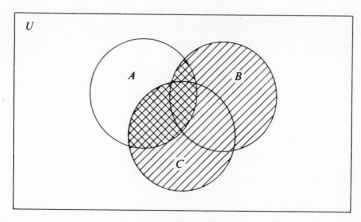

$$A \cap (B \cup C)$$
(Double shaded region)

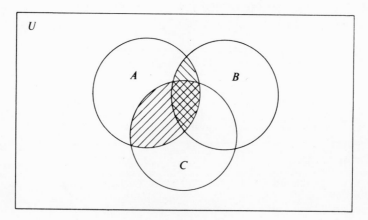

$$(A \cap B) \cup (A \cap C)$$
(Entire shaded region)

Figure 2.9

6. The distributive property of union over intersection: A symbolic statement of this property is: $A \cup (B \cap C) = (A \cup B) \cap (A \cup C)$.

EXAMPLE:

$$A \cup (B \cap C) = \{a, g, n, o, p, r\} \cup (\{c, e, g, l, o\} \cap \{c, d, e, o, s\})$$
$$= \{a, g, n, o, p, r\} \cup \{c, e, o\}$$
$$= \{a, c, e, g, n, o, p, r\}$$

$$(A \cup B) \cap (A \cup C) = (\{a, g, n, o, p, r\} \cup \{c, e, g, l, o\}) \cap (\{a, g, n, o, p, r\}$$
$$\cup \{c, d, e, o, s\})$$
$$= \{a, c, e, g, l, n, o, p, r\} \cap \{a, c, d, e, g, n, o, p, r, s\}$$
$$= \{a, c, e, g, n, o, p, r\}$$

We illustrate this property by Venn diagrams in Figure 2.10.

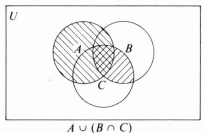

$A \cup (B \cap C)$
(Entire shaded region)

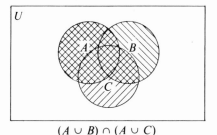

$(A \cup B) \cap (A \cup C)$
(Double shaded region)

Figure 2.10

7. The closure property of intersection: The intersection of two sets A and B is the *set* that contains each element of A that is also an element of B, and no other elements. Since each element of $A \cap B$ is an element of A, each element of $A \cap B$ must be an element of the universe under consideration, and $A \cap B$ must be a subset of that universe. We describe this property of subsets of any given universe by saying that the universe is closed under intersection.

8. The closure property of union: The union of two sets A and B is the *set* that contains each element of A, each element of B, and no other element. Since each element of $A \cup B$ is found in A or in B or in both A and B, each element of $A \cup B$ must be an element of the universe under consideration, and $A \cup B$ must be a subset of that universe. We describe this property of subsets of any given universe by saying that the universe is closed under union.

9. DeMorgan's law for the complement of a union: (Augustus De-Morgan has been given credit for discovering this law and the next one.) The complement of the union of two sets is equal to the intersection of their complements.

Symbolized: $(A \cup B)' = A' \cap B'$.

EXAMPLE: Since $A \cup B = \{a, c, e, g, l, n, o, p, r\}$, its complement is:

$$(A \cup B)' = \{b, d, f, h, i, j, k, m, q, s, t, u, v, w, x, y, z\}$$

The complements of A and B are:

$$A' = \{b, c, d, e, f, h, i, j, k, l, m, q, s, t, u, v, w, x, y, z\}$$

and

$$B' = \{a, b, d, f, h, i, j, k, m, n, p, q, r, s, t, u, v, w, x, y, z\}$$

The intersection of A' and B', then, is

$$A' \cap B' = \{b, d, f, h, i, j, k, m, q, s, t, u, v, w, x, y, z\}.$$

We illustrate this law by Venn diagrams in Figure 2.11:

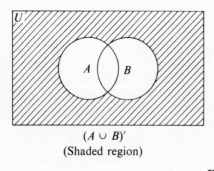

$(A \cup B)'$
(Shaded region)

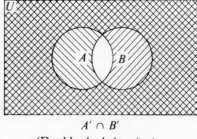

$A' \cap B'$
(Double shaded region)

Figure 2.11

10. DeMorgan's law for the complement of an intersection: The complement of the intersection of two sets is the union of their complements.

Symbolized: $(A \cap B)' = A' \cup B'$.

EXAMPLE: Since $A \cap B = \{g, o\}$, its complement $(A \cap B)'$ contains all the letters in the alphabet except g and o:

$$(A \cap B)' = \{a, b, c, d, e, f, h, i, j, k, l, m, n, p, \ldots, x, y, z\}$$

Referring to the elements of A' and B', we see that their union is also the set of all letters in the English alphabet except g and o:

$$A' \cup B' = \{a, b, c, d, e, f, h, i, j, k, l, m, n, p, \ldots, x, y, z\}$$

We illustrate this law by Venn diagrams in Figure 2.12:

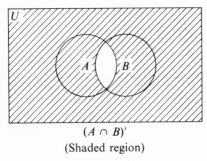

$(A \cap B)'$

(Shaded region)

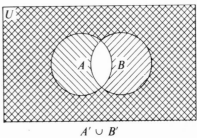

$A' \cup B'$

(Entire shaded region)

Figure 2.12

EXERCISES 2.2

1. The commutative property seems trivial until we realize that not all operations are commutative. Let $A = \{a, b, c\}$, $B = \{1, 2\}$.

   (a) Find $A \times B$ (Cartesian product).
   (b) Find $B \times A$.
   (c) Is the Cartesian product commutative?
   (d) Can you find at least one case in which Cartesian products are commutative? Assuming you have found a case, can you find others?

2. Let the universe be the set of all letters in the English alphabet. Use an Euler diagram first, and then a Venn diagram, to show the relationships that exist between each of these pairs of sets:

   (a) A: the set of consonants; B: the set of vowels.
   (b) $A = \{a, b, c, d, e\}$; $B = \{c, e\}$

(c) $A = \{a, b, c, d, e\}$; $B = \{d, e, f, g\}$
(d) $A = \{a, b, c, d, e\}$; $B = \{a, b, c, d, e, f, g\}$
(e) $A = \{a, b, c, d, e, f, g\}$; B: the set of the first seven letters of the English alphabet
(f) $A = \{a, b, c\}$; $B = \{x, y, z\}$

3. Verify each statement for the given sets:
   $U = \{1, 2, 3, 4, 5, 6, 7, 8, 9, 10\}$, $P = \{1, 2, 3, 4, 5\}$,
   $Q = \{1, 5, 9\}$, $R = \{2, 4, 6, 8, 10\}$.

   (a) $(P \cap Q) \cap R = P \cap (Q \cap R)$
   (b) $(P \cup Q) \cup R = P \cup (Q \cup R)$
   (c) $P \cup (Q \cap R) = (P \cup Q) \cap (P \cup R)$
   (d) $P \cap (Q \cup R) = (P \cap Q) \cup (P \cap R)$
   (e) $(P \cup Q)' = P' \cap Q'$
   (f) $(P \cap Q)' = P' \cup Q'$

4. Illustrate the statements in Exercise 3 by drawing appropriately shaded Euler diagrams, labeled properly.

5. Let the universe be the set of all letters in the English alphabet. Use an Euler diagram first, and then a Venn diagram, to show the relationships that exist among each of these collections of sets:

   (a) $A = \{a, b, c, d, e\}$; $B = \{d, f, g\}$; $C = \{a, c, e, f\}$
   (b) $A = \{a, b, c, d, e\}$; $B = \{d, f, g\}$; $C = \{g, h, i\}$
   (c) $A = \{a, b, c\}$; $B = \{p, q, r\}$; $C = \{x, y, z\}$

6. In a class of freshmen, 37 students have had algebra in high school, 25 students have had geometry, and 19 have had both. If each freshman has had at least one of these two courses, how many students are there in this class of freshmen?

   Suggestion: The Venn diagram is a powerful tool in visualizing the inform-ation given in this sort of problem. Suppose we let A = the set of students who have had algebra and B = the set of those who have had geometry. Then $n(A) = 37$, $n(B) = 25$, and $N(A \cap B) = 19$. The number in $A \cup B$, that is, $n(A \cup B)$ is what we are to find. If we add $n(A) + n(B)$, the Venn diagram helps us to see that we are including $n(A \cap B)$ twice. Can you see that a formula for the solution of our problem would be $n(A \cup B) = n(A) + n(B) - n(A \cap B)$?

7. One day a group of 18 girls came to school wearing either white blouses or black skirts. If 13 girls were wearing white blouses and 17 girls were wearing black skirts, how many girls were wearing white blouses with black skirts?

8. In a senior class, 57 students are taking chemistry, 65 are taking mathematics, and 59 are taking physics. Of these students, 30 take both chemistry and mathe-matics, 29 take both chemistry and physics, and 42 take both mathematics and physics. Also 13 of the students take all three of these courses. If all seniors taking any of these courses are to attend a Science Club picnic, how many seniors will attend? (Use Venn diagrams to "picture" this problem and help you answer the question.)

9. In a survey of 250 students, the numbers studying various languages were found to be: German 151, French 136, Russian 27, French and German 63, Russian and French 7, German and Russian 11, and 4 students were studying all three languages.

(a) How many students were studying Russian only?
(b) How many students were studying German or French or both?
(c) How many students were taking German and French, but not Russian?
(d) How many students were studying no language?
(e) How many students were studying at least two languages?

10. The following problem appeared in an employment advertisement by Nortronics, a division of the Northrop Corporation:

Consider the sets of males, blonds, and children, and label the Venn diagram to identify (a) blond boys, (b) blond women, and (c) women who are not blond.

   The advertisement said, "If you can solve this problem within 60 seconds, and are experienced in advanced solid state circuit design, there's an opportunity waiting for you at Nortronics." Try to solve the problem. Also write a symbolic statement for each of the sets indicated in (a), (b), and (c), using M for the set of males, C for the set of children, and B for the set of blonds, and using symbols for union, intersection, and complement.†

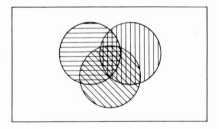

Figure 2.13

2.3   THEOREMS ON PROPERTIES OF
THE UNIVERSE AND THE EMPTY SET

We have listed some statements describing properties of sets under certain set operations. These are statements that we *assume* to be true. At this point we introduce some statements that can be proved deductively, about other properties of sets. Statements that are proved are called **theorems**. The *deductive process* demands that theorems follow logically from the assumptions of the system. Usually a "chain" of reasoning is needed to show this logical progression. This chain consists of statements that can be justified by any of the types of assumed

† Cited by permission of the Northrop Corporation, Beverly Hills, California.

or previously proved statements. These may be

1. Given statements
2. Definitions
3. Assumptions
4. Theorems
5. Previous statements in the chain

To illustrate this process, we have chosen a few theorems involving properties of the universe and the empty set. These have been selected in part for the simplicity of their proofs and in part for their importance in pointing up the unique behavior of the universe and the empty set.

Theorem I: The intersection of any set A with the empty set is the empty set:

$$A \cap \emptyset = \emptyset \cap A = \emptyset$$

PROOF $A \cap \emptyset = \emptyset \cap A$ because intersection is commutative. (See the commutative property of intersection, Section 2.1.) By definition of intersection, $A \cap \emptyset$ contains each element of A that is also an element of $\emptyset$ and no other element. Since, by definition, $\emptyset$ contains no elements, $A \cap \emptyset = \emptyset \cap A = \emptyset$.

Theorem II: The union of any set A with the empty set is the set A:

$$A \cup \emptyset = \emptyset \cup A = A$$

PROOF $A \cup \emptyset = \emptyset \cup A$ because union is commutative. By definition of union, $A \cup \emptyset$ contains each element of A, each element of $\emptyset$, and no other element. Since $\emptyset$, by definition, contains no elements, $A \cup \emptyset = \emptyset \cup A = A$.

The property proved in Theorem II is often described by saying that the empty set is the **identity element for union of sets**.

Theorem III: The complement of the empty set is the universe:

$$\emptyset' = U$$

PROOF   By definition of complement, the complement of the empty set is the set containing just those elements in the universe that are not contained in the empty set. Since, by definition, the empty set contains no elements, all elements in the universe are in $\varnothing'$, and $\varnothing' = U$.

We next list three important theorems on the unique properties of the universe and two theorems on properties of sets in general. Proofs of these theorems will be left to the student as an exercise (Exercise 1).

Theorem IV: The complement of the universe is the empty set:

$$U' = \varnothing$$

Theorem V: The union of any set A with the universe U is the universe:

$$A \cup U = U \cup A = U$$

Theorem VI: The intersection of any set A with the universe U is the set A:

$$A \cap U = U \cap A = A$$

The property stated in Theorem VI is often described by saying that the universe is the **identity element for intersection of sets.**

Theorem VII.   The intersection of any set A with itself is the set A:

$$A \cap A = A$$

The property stated in Theorem VII is called the **idempotent property of intersection of sets.**

Theorem VIII: The union of any set A with itself is the set A:

$$A \cup A = A$$

The property stated in Theorem VIII is called the **idempotent property of union of sets.**

While many theorems, such as these eight theorems, can be proved by a simple statement or two quoting a definition or an assumption (or a proved theorem), the proofs of many other theorems are more complicated, requiring "chains" of justifying statements. In these cases a formal pattern for the proofs is usually desirable.

In Algebra of Sets, most of our theorems are statements that two sets are identical (equal). The first set, expressed as the result of a series of operations on two or more given sets, is to be proved equal to a second set expressed as the result of a different series of set operations. For example, we may wish to prove that $S_1 = (A \cup B) \cup (A \cup C)$ is identical with $S_7 = A \cup (B \cup C)$. Then our theorem could be stated symbolically: Prove that $S_1 = S_7$. The pattern might be of this form:

PROOF

Statement	Reason
1. $S_1 = S_2$	(A single statement of definition
2. $S_2 = S_3$	or property would be given for
3. $S_3 = S_4$	each of the statements of equality
4. $S_4 = S_5$	of two sets.)
5. $S_5 = S_6$	
6. $S_6 = S_7$	

The statements $S_1 = S_2$ and $S_2 = S_3$, considered together, seem to imply to us that $S_1 = S_3$; the statements $S_1 = S_3$ and $S_3 = S_4$ imply that $S_1 = S_4$; the statements $S_1 = S_4$ and $S_4 = S_5$ imply that $S_1 = S_5$, etc. Thus, at the end of this chain of statements, by virtue of the way our minds work, we conlude that $S_1 = S_7$. The assumption that $S_1 = S_7$, if all of the preceding statements are ·justified, is an example of the **extended transitive property of equality.**† Thus, a seventh and final statement in the proof would be:

Statement	Reason
7. $S_1 = S_7$	7. The extended transitive property of equality.

Now we return to the specific sets $S_1 = (A \cup B) \cup (A \cup C)$ and $S_7 = A \cup (B \cup C)$ and use the pattern that we have described to prove the statement $S_1 = S_7$.

Theorem: $(A \cup B) \cup (A \cup C) = A \cup (B \cup C)$

PROOF

Statement	Reason
1. $(A \cup B) \cup (A \cup C) =$ $A \cup [B \cup (A \cup C)]$	1. Associative property of union of sets.
2. $A \cup [B \cup (A \cup C)] =$ $A \cup [(B \cup A) \cup C]$	2. Associative property of union of sets.

† This property will be considered again, more carefully, later in our study.

3. $A \cup [(B \cup A) \cup C] =$ $A \cup [(A \cup B) \cup C]$

3. Commutative property of union of sets.

4. $A \cup [(A \cup B) \cup C] =$ $A \cup [A \cup (B \cup C)]$

4. Associative property of union of sets.

5. $A \cup [A \cup (B \cup C)] =$ $(A \cup A) \cup (B \cup C)$

5. Associative property of union of sets.

6. $(A \cup A) \cup (B \cup C) =$ $A \cup (B \cup C)$

6. The idempotent property of union of sets.

7. $(A \cup B) \cup (A \cup C) =$ $A \cup (B \cup C)$

7. The extended transitive property of equality.

The structure of Algebra of Sets and, indeed, the structure of *any* mathematical system is made up of:

1. Some undefined terms
2. Some defined terms
3. Some assumptions
4. Some theorems

(Reread Lord Russell's statement, Chapter 1, Section 1.1.)

It will be recognized that this structure hangs together because of our confidence in *deductive reasoning*. Furthermore, since all formal mathematical systems have this general structure, we can say that mathematics is *essentially deductive*.

In Exercise 2, most students find a formal proof, such as that demonstrated in this section, necessary only for the last two theorems. It is recommended that the student attempt to do these proofs, even though a sufficient basis of study and practice for proficiency in such proofs has not been provided. The formal deductive proof has been introduced only because it reveals something of the nature of mathematics. An attempt to do such proofs, even if it is unsuccessful, will be an aid to your understanding of what mathematics is.

EXERCISES 2.3

1. Prove Theorems IV, V, VI, VII, and VIII listed in Section 2.3.

2. Prove each of these theorems, where A, B, and C may be any sets. Illustrate each theorem by appropriately shaded Venn diagrams, properly labeled.

   (a) If $A \subseteq B$, then $A \cup B = B$.
   (b) $(A \cap B) \subseteq A$.
   (c) $(A \cup B) \supseteq A$.
   (d) $A \cup (A \cap B) = A$.
   (e) $A \cap (B \cap C) = (A \cap B) \cap (A \cap C)$.

3. List for the Algebra of Sets several (a) undefined terms, (b) defined terms, (c) assumptions, and (d) theorems.

CHAPTER 3

An Algebra of Propositions

3.1 . LOGIC AND MATHEMATICS

We know that mathematics and logic are, in some way, related. It has been argued that logic, as a discipline, is dependent on mathematics.† On the other hand, it has been stated that logic contains mathematics as a subdivision. Bertrand Russell mediates these views when he declares that "... formal logic ... is identical with mathematics."‡ Without deciding which of the three arguments is correct, we shall be content to accept the statement that mathematical systems are logical systems, essentially deductive; and, in this chapter, we shall take a brief look at a mathematical system that is often called "Mathematical Logic": an Algebra of Propositions.

---

† Charles Sanders Peirce, in *The Essence of Mathematics*, The World of Mathematics (New York: Simon and Schuster, 1956), III, 1773, says "... I am persuaded that logic cannot possibly attain the solution of its problems without great use of mathematics."

‡ Bertrand Russell, *Mathematics and the Metaphysicians*, The World of Mathematics (New York: Simon and Schuster, 1956) III, 1577.

3.2 PROPOSITIONS

As in the field of mathematics, so in other fields, we reason, saying "... if such and such a *proposition* is true ..., then such and such another *proposition* is true ... "† [italics not in original]. Propositions, as a part of a logical argument, occupy the place of sets in the Algebra of Sets. We did not attempt to define the word *set*; we shall not attempt to define the word *proposition*. We shall simply characterize a **proposition** by saying it is a sentence, or statement, that is capable of being classified as either true or false. This means that sentences such as "Please come home," "Why are you late for class?" and "Wish me luck!" are not propositions; we cannot assign them **truth values**, that is, we can not say whether they are true or false.

3.3 CLASSIFICATION OF PROPOSITIONS

The kinds of propositions that we shall consider may be sorted into two classes: *simple propositions* and *compound propositions*. A **simple proposition** is simply a declarative sentence of one clause such as "John is a good student." A **compound proposition** is a proposition made up of two or more simple propositions joined by the connectives *and, or, if ... then*, or *if and only if*; or it is a proposition prefaced by the words, *It is not true that* or *It is false that. Note:* In the English language, other words are often used in place of the connectives we have specified for these compound sentences, for example "but" or "although" for *and*, "since" or "provided" for *if ... then*, etc.

There are five basic types of compound sentences:

1. A **conjunction** is a compound sentence of two propositions joined by **and**.

EXAMPLE: John is a good student, and he is a mathematics major.

2. A **disjunction** is a compound sentence of two propositions joined by **or**.

EXAMPLE: John is a mathematics major, or he is a physics major.

3. An **implication** is a compound sentence of two propositions joined by **if ... then**.

EXAMPLE: If John is a mathematics major, then he is also a physics major.

† *Ibid.*, p. 1576.

4. A **biconditional** is a compound sentence of two propositions joined by **if and only if**, frequently abbreviated **iff**.

EXAMPLE: John is a mathematics major iff he is also a physics major.

5. A **negation** is a compound sentence formed by a single proposition prefaced by the words **It is not true that** or **It is false that**.

EXAMPLE: It is not true that John is a mathematics major.

For purposes of convenience, we shall, in this book, classify propositions by whether they are general or specific in nature. If propositions make use of such modifiers as *all, some, any,* we shall call them **general propositions.** Among the *simple propositions*, those that are *general* may take any of these four forms:

1. All x's are y's.
2. No x's are y's (or all x's are not-y's).
3. Some x's are y's.
4. Some x's are not-y's.

Those *simple propositions* that we shall call *specific* propositions may take one of these two forms:

1. This x is a y.
2. This x is not a y.

It is important to understand that we shall deal only with simple propositions that can be put into one of these forms. Most declarative sentences, however, can be altered, without change in meaning, to fit one of the forms that we are considering. Suppose, for example, we have the proposition:

Many of the trees on campus will be removed to make room for the new library.

This could be expressed in the form: "Some x's are y's," by writing it as:

Some of the trees on campus are trees that will be removed from the library site.

The correspondence of this form of the statement with "Some x's are y's" would be:

Some x's: Some of the trees on campus
    y's: trees that will be removed from the library site.

As another example, let us take the proposition:

John gets good grades.

We can express this in the form,

> John is a student who gets good grades,

where the correspondence between this form of the statement with "This x is a y" would be:

> x: John
>
> y: a student who gets good grades

3.4 EULER DIAGRAMS OF SIMPLE GENERAL PROPOSITIONS

Each of the *simple general propositions* can be interpreted in more than one way. Though we tend, in our daily nonmathematical usage, to limit the meaning of each of these statements to one interpretation, actually, as the name *general* implies, each of these propositions encompasses more than one possible formal meaning. Euler diagrams are very useful in showing us what the possible interpretations are, and in helping us to understand that these are reasonable interpretations. We use the points of a region X as the representatives of the x's and the points of a region Y as the representatives of the y's. The appropriate diagrams are

1. All x's are y's (Fig. 3.1).

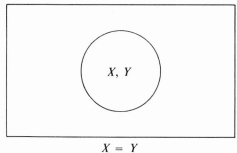

$$X = Y$$

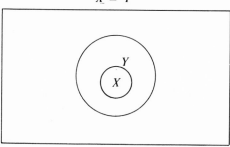

$$X \subset Y$$

Figure 3.1

2. No x's are y's, that is: All x's are not-y's (Fig. 3.2).

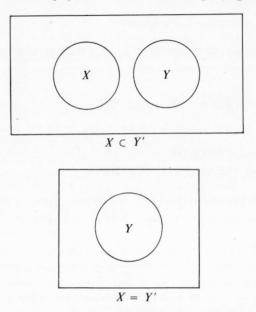

Figure 3.2

3. Some x's are y's (Fig. 3.3).

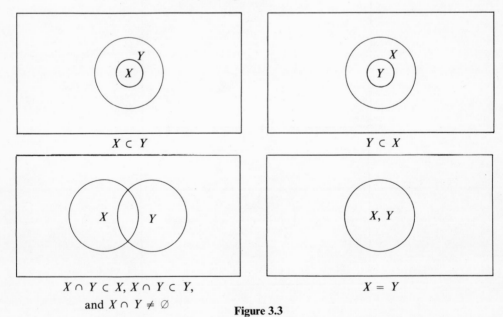

Figure 3.3

4. Some x's are not-y's (Fig. 3.4).

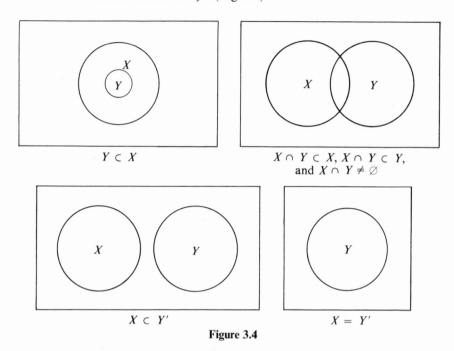

Figure 3.4

As we study these diagrams, we observe that no one interpretation belongs exclusively to one simple general proposition; there is an "overlapping" in possible interpretations between some pairs of these statements. We notice, too, that there are just six different Euler diagrams needed to express all possible interpretations of the four simple general propositions. These diagrams are shown in Figure 3.5.

It is interesting to note that as in Section 2.1 these six Euler diagrams also illustrate all possible relations between two sets X and Y.

3.5 NEGATION OF A SIMPLE PROPOSITION

We may alternately define a *negation of a given proposition* as a second proposition that is always false if the given proposition is true and always true if the given proposition is false. The original definition of a negation suggests that we can form a negation of a statement such as "John is a good student," by saying, "It is not true that John is a good student." A negation of any proposition may be formed correctly by prefixing the phrase, "It is not true that ... " or "It is false that ...." We must be more careful, however, about forming a negation by using the adverb "not" with the verb. Very often this does not give us a negation of the proposition at all. Consider the sentence: "Some x's are y's." We cannot form a

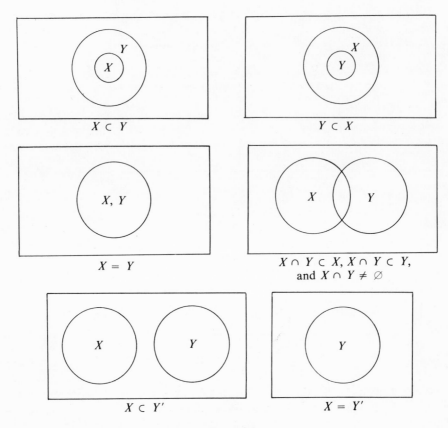

Figure 3.5

negation of this proposition by writing "Some x's are not-y's." While these two sentences do not necessarily mean exactly the same thing, there is an overlapping in meaning. Since this is the case, we cannot say that the proposition, "Some x's are not-y's," is always false if "Some x's are y's" is true. The Euler diagrams for each simple proposition can help us to decide what its negation should be; indeed, we define the negation of each of the simple general propositions in terms of Euler diagram interpretations: A **negation of a given simple general proposition** is another simple general proposition, specifically, the one that can be interpreted by just those Euler diagrams that are not interpretations of the given proposition.

As an example of this, let us consider the proposition, "Some x's are y's," once more. Its interpretations by Euler diagrams were shown in Figure 3.3. A negation of this proposition, then, would have as its possible interpretations by Euler diagrams the only remaining Euler diagrams. These were shown in Figure 3.2. The simple general proposition that has these as its only possible interpretations is "No x's are y's." Therefore, the negations of "Some x's are y's," is "No x's are y's."

The matter of the negation of any simple specific proposition can be disposed of very easily. Recall that there are just two sorts of *simple specific propositions*. They may take these forms:

1. This x is a y.
2. This x is not a y.

We define a **negation of a simple specific proposition** to be the simple specific proposition in the other form for the same x and y. The negation of "This x is a y," then, is "This x is not a y," and vice versa.

3.6 SYMBOLIC REPRESENTATION OF COMPOUND PROPOSITIONS

At this point we introduce symbolic notation to simplify our expression of compound propositions. The symbols for the four connectives are

and: $\wedge$.

or: $\vee$. *Note:* *Or*, as we use it, is *inclusive* in meaning. $p \vee q$ means either p or q or both p and q.

if... then: $\rightarrow$.

if and only if: $\leftrightarrow$.

Each simple proposition contained in a compound proposition can be expressed by a single lowercase letter, and the negation of any proposition is indicated by prefixing the symbol $\sim$ to the symbolic expression of the proposition.

EXAMPLE 1: The sun is shining and the sky is cloudless.
Let "p" represent "the sun is shining," and let "q" represent "the sky is cloudless." In our symbolic notation, then, the given sentence can be written $p \wedge q$.

EXAMPLE 2: The sun is shining or the sky is cloudless.
This sentence can be written symbolically $p \vee q$.

EXAMPLE 3: If the sun is shining, then the sky is cloudless.
Symbolically, this would be $p \rightarrow q$.

EXAMPLE 4: The sun is shining if and only if the sky is cloudless.
Symbolically, this would be $p \leftrightarrow q$.

EXAMPLE 5: The sun is not shining.
This sentence can be expressed as $\sim p$.

EXERCISES 3.6

1. Write the negation of each of the following simple propositions without using "It is not true that ..." or "It is false that ...." Draw all the Euler diagram interpretations possible for each proposition; then draw all Euler diagram interpretations possible for its negation.

   (a) No x's are y's.
   (b) Some x's are not-y's.
   (c) All x's are y's.

2. Write the negation of each of the following statements without using "It is not true that ..." or "It is false that ...."

   (a) John is a good student.
   (b) John gets good grades.
   (c) Some professors at Paragon College are authors.
   (d) Some professors have not had anything published.
   (e) Many trees on campus will be removed to make room for the new library.
   (f) Many trees on campus will not be removed to make room for the new library.
   (g) All students having an average of 90 per cent or better will get A's in the course.

3. Choose a letter to represent each simple proposition contained in the statements below, and write each statement in symbolic form. Let a letter represent an affirmative statement wherever feasible, as we have done with "q" in the example.

   Example: Charles is a wilful boy and he refuses to obey his parents, but he does very well in school.
   p: Charles is a wilful boy.
   q: Charles obeys his parents.
   r: Charles does well in school.

   Symbolic form: $p \wedge (\sim q) \wedge r$

   (a) If the student can pass a test on the United States Constitution, he is not required to take a course in it.
   (b) Each student is required to take a course in United States Constitution and then pass a test on it.
   (c) If a student takes a course in United States Constitution or passes a test on it, he has fulfilled one requirement for graduation.
   (d) If John graduated last June, he took a course in United States Constitution and passed a test on it.
   (e) One must take a course in United States Constitution or pass a test on it, and complete a major, if one is to graduate.
   (f) If he does not pass the test on the United States Constitution or take a course in it, even though he has completed a major, he will not graduate.

3.7   TRUTH TABLES

We found that drawing Euler diagrams could help us find negations of simple propositions. There is another method of analysis that we can employ in finding ways to write negations of compound propositions. We make tables (**truth tables**) of all possible combinations of the truth values of the propositions under consideration. The construction of a truth table provides a method of analyzing the truth or falsity of the whole statement under all possible combinations of truth or falsity of the simple propositions involved. Suppose, for example, that we have a compound proposition that can be symbolized by $p \wedge q$. Very little investigation is needed to disclose that there are four possible combinations of **truth values** of the simple propositions p and q:

1. p and q both true,
2. p true and q false,
3. p false and q true,

and   4. p and q both false.

In tabular form these combinations may be arranged as in Figure 3.6.

p	q
T	T
T	F
F	T
F	F

T means "True"

F means "False"

Figure 3.6

The simplest truth table would be one for the negation of a simple proposition. This is shown in Figure 3.7.

p	$\sim p$
T	F
F	T

Figure 3.7

This table simply tells us that, when a proposition is true, its negation is false, and when a proposition is false, its negation is true.

In order to write the truth tables for the other four basic compound sentences, $p \wedge q$, $p \vee q$, $p \rightarrow q$, and $p \leftrightarrow q$, we must decide under what conditions of truth or falsity of the simple propositions involved the compound proposition is true, and under what conditions the compound proposition is false. We shall agree, somewhat arbitrarily, that the following rules shall hold: (These rules may be considered *assumptions of our Algebra of Propositions*.)

1. A conjunction is true only when both propositions involved are true, that is, $p \wedge q$ is true only when both p and q are true.

2. A disjunction is false only when both propositions involved are false, that is, $p \vee q$ is false only when both p and q are false.

3. An implication is false only when a true proposition is followed by a false one, that is, $p \to q$ is false only when p is true and q is false.

4. A biconditional is true only when both propositions are true or when both are false, that is, $p \leftrightarrow q$ is true only when both p and q are true or when both p and q are false.

The truth table for the conjunction is shown in Figure 3.8.

p	q	$p \wedge q$
T	T	T
T	F	F
F	T	F
F	F	F

Figure 3.8

EXERCISES 3.7

1. Make a truth table for the disjunction $p \vee q$.
2. Make a truth table for the implication $p \to q$.
3. Make a truth table for the biconditional $p \leftrightarrow q$.

3.8 MORE TRUTH TABLES

If the proposition under consideration is more complicated in form than the basic compound propositions, we can do the truth table for it in stages. To illustrate, let us make a truth table for the proposition $(\sim p) \to (\sim q)$. The first step is to fill in the truth values for $\sim p$ as shown in Figure 3.9. Next we write in the truth values for $\sim q$ (Fig. 3.10).

p	q	$(\sim p) \to (\sim q)$
T	T	F
T	F	F
F	T	T
F	F	T

Figure 3.9

p	q	$(\sim p) \to (\sim q)$	
T	T	F	F
T	F	F	T
F	T	T	F
F	F	T	T

Figure 3.10

Now we are ready to consider the implication as a whole. We see that the only place where $\sim p$ (the first proposition) is true and $\sim q$ (the second proposition) is false is in the third row; therefore, the completed truth table would be as shown in Figure 3.11. The column that indicates the truth values for the complete statement is encircled.

p	q	$(\sim p) \to (\sim q)$		
T	T	F	T	F
T	F	F	T	T
F	T	T	F	F
F	F	T	T	T

Figure 3.11

EXERCISES 3.8

Make a truth table for:

1. $(\sim p) \wedge q$

2. $p \wedge (\sim q)$

3. $(\sim p) \wedge (\sim q)$

4. $(\sim p) \vee q$

5. $p \vee (\sim q)$

6. $(\sim p) \vee (\sim q)$

7. $(\sim p) \to q$

8. $p \to (\sim q)$

9. $(\sim p) \to (\sim q)$

10. $(\sim p) \leftrightarrow q$

11. $p \leftrightarrow (\sim q)$

12. $(\sim p) \leftrightarrow (\sim q)$

3.9 NEGATION OF A
COMPOUND PROPOSITION

Recall that a negation of any proposition is always false when the given proposition is true, and vice versa. With this in mind, let us examine the truth tables for $p \wedge q$ and $(\sim p) \vee (\sim q)$ (Fig. 3.12).

p	q	$p \wedge q$	$(\sim p) \vee (\sim q)$		
T	T	T	F	F	F
T	F	F	F	T	T
F	T	F	T	T	F
F	F	F	T	T	T

Figure 3.12

Notice that the truth table for $p \wedge q$ has a T in the first row, but the truth table for $(\sim p) \vee (\sim q)$ has an F. In the remaining three rows, the truth table for $p \wedge q$ has F's, but the truth table for $(\sim p) \vee (\sim q)$ has T's. Thus $(\sim p) \vee (\sim q)$ is always false when $p \wedge q$ is true and vice versa. Clearly, then, $(\sim p) \vee (\sim q)$ is a negation of $p \wedge q$. In Exercises 1 and 2 of the exercise set for this section, the student is to examine the truth tables for each of several compound statements. The purpose of this inspection will be to find negations of each of these compound statements and, finally, to write statements that might serve as general definitions of negations of a conjunction, a disjunction, an implication, and a biconditional. These statements will serve as important tools to aid us in writing negations of compound propositions. Therefore, the student is advised to give them a prominent place in his notes.

While the truth tables of a proposition and its negation are *opposites* if we consider *true* as *opposite* in meaning to *false*, we often see truth tables that are identical, although the propositions for which they have been made seem to be different. Observe the truth tables for $p \vee q$ and $(\sim p) \rightarrow q$ (Fig. 3.13).

p	q	$p \vee q$	$(\sim p) \rightarrow q$	
T	T	T	F	T	T
T	F	T	F	T	F
F	T	T	T	T	T
F	F	F	T	F	F

Figure 3.13

When two propositions that involve the same simple propositions have identical truth tables, we say they are **logically equivalent**. Thus $p \vee q$ is logically equivalent to $(\sim p) \rightarrow q$.

EXERCISES 3.9

1. Refer to the truth tables in the text and those constructed for Exercises 3.7 and 3.8, and use them to make a list of the one or more negations of each of the listed propositions. (Some of the statements listed may appear to have more than one negation according to the truth tables. Each statement actually has only one negation; this negation may be expressed in many different, logically equivalent, forms. Whenever you discover two statements whose truth tables indicate that they are alternate forms of the negation of a given statement, list them both.)

(a) $p \wedge q$

(b) $(\sim p) \wedge q$

(c) $p \wedge (\sim q)$

(d) $(\sim p) \wedge (\sim q)$

(e) $p \vee q$

(f) $(\sim p) \vee q$

(g) $p \vee (\sim q)$

(h) $(\sim p) \vee (\sim q)$

(i) $p \rightarrow q$

(j) $(\sim p) \rightarrow q$

(k) $p \rightarrow (\sim q)$

(l) $(\sim p) \rightarrow (\sim q)$

(m) $p \leftrightarrow q$

(n) $(\sim p) \leftrightarrow q$

(o) $p \leftrightarrow (\sim q)$

(p) $(\sim p) \leftrightarrow (\sim q)$

2. Refer to your answers in Exercise 1 and write a statement that might serve as a definition of:

(a) The negation of a conjunction.

(b) The negation of a disjunction.

(c) The negation of an implication.

(d) The negation of a biconditional.

3. Use p: Fred washes the car,

　　　q: Fred gets his allowance,

and write each of the following statements in symbolic form.

(a) If Fred washes the car, he gets his allowance.

(b) Fred washes the car, and he gets his allowance.

(c) Fred washes the car, or he does not get his allowance.

(d) Fred washes the car if and only if he gets his allowance.

(e) If Fred gets his allowance, he washes the car.

(f) If Fred does not wash the car, he does not get his allowance.

(g) If Fred does not get his allowance, he does not wash the car.

(h) Fred washes the car, or he does not get his allowance iff "Fred does not wash the car" implies "Fred does not get his allowance."

4. Make truth tables for the following symbolic statements:

(a) $(p \wedge q) \rightarrow q$

(b) $(p \vee q) \wedge (q \rightarrow p)$

(c) $(p \rightarrow q) \leftrightarrow (q \rightarrow p)$

(d) $\sim(p \vee q) \to [(\sim p) \vee (\sim q)]$

(e) $\{[(\sim p) \vee q] \wedge (\sim p)\} \leftrightarrow (p \to q)\}$

(f) $[(p \wedge q) \vee r] \leftrightarrow [(\sim p) \vee r]$

(g) $[(p \vee (\sim q)) \to r] \leftrightarrow [p \wedge (\sim r)]$

Note: Parts f and g contain three propositions. How many combinations of truth and falsity are there for three propositions, that is, how many rows are needed for the truth table?

5. There are three implications that are related to the implication $p \to q$. They are:

1. its **converse**, $q \to p$,
2. its **inverse**, $(\sim p) \to (\sim q)$

and   3. its **contrapositive**, $(\sim q) \to (\sim p)$

In Exercise 3, which of the statements is

(a) The converse of the statement in part (a)?

(b) The inverse of the statement in part (a)?

(c) The contrapositive of the statement in part (a)?

6. Which statement in Exercise 3 is logically equivalent to the statement in part (a)? to the converse of the statement in part (a)?

7. Write the negation, in words, of each of the following:

(a) Some of the boys at the Institute have no other home.

(b) None of the boys is considered a problem child.

(c) All of the boys are taking advantage of the educational program.

(d) Some of the boys attending classes do not live at the Institute.

(e) Pascal met Desargues in Paris, and he became very interested in Desargue's work in projective geometry.

(f) All of Desargue's colleagues called him crazy, and they forgot about projective geometry.

(g) A plane figure is a parabola iff it is the locus of points equidistant from a fixed point and a fixed line.

8. Write the converse, inverse, contrapositive, and the negation, in words, of each of the following implications:

(a) If Jack is nimble, he can jump over the candlestick.

(b) If this animal is a dog, then it has four legs and it has a tail.

(c) The grass will stay green if it is watered sufficiently.

(d) I cannot understand an exercise in logic if the hypotheses are not arranged in a familiar pattern.

(e) If some boys are shy, then some girls will not be asked to dance.

(f) If the test is too difficult or it is too long, then some students will fail it.

(g) If some of the students have not studied well, then some will fail the test, but all of them will take it.

(h) Since none of the theories of science are considered absolute, all of them must be subjected to continuous testing and some of them will undoubtedly be modified.

3.10   TAUTOLOGIES AND SELF-CONTRADICTIONS

In Exercise 3, part (h) (Exercises 3.9), the truth table, if done correctly, has nothing but T's in the column representing the truth values for the proposition as a whole. Let us look at it:

Since p represents "Fred washes the car," and q represents "Fred gets his allowance," the symbolic form of the proposition is $[p \lor (\sim q)] \leftrightarrow [(\sim p) \to (\sim q)]$. Its truth table is shown in Figure 3.14.

p	q	$[p \lor (\sim q)] \leftrightarrow [(\sim p) \to (\sim q)]$		
T	T	T	T	T
T	F	T	T	T
F	T	F	T	F
F	F	T	T	T

Figure 3.14

A proposition, such as this, that is always true, regardless of the truth value of the simple propositions involved, is called a **tautology**. Another simpler example of a tautology is the proposition $p \lor (\sim p)$. Its truth table is shown in Figure 3.15.

p	$\sim p$	$p \lor (\sim p)$
T	F	T
F	T	T

Figure 3.15

The negation of a tautology, of course, would have a truth table with nothing but F's in the column containing the truth values for the proposition as a whole. Such a proposition is false regardless of the truth values of the simple propositions involved; it is called a **self-contradiction**. Two examples of this are $p \land (\sim p)$, and $(p \land q) \leftrightarrow [(\sim p) \lor (\sim q)]$. Their truth tables are shown in Figure 3.16.

p	$\sim p$	$p \land (\sim p)$
T	F	F
F	T	F

p	q	$(p \land q) \leftrightarrow [(\sim p) \lor (\sim q)]$		
T	T	T	F	F
T	F	F	F	T
F	T	F	F	T
F	F	F	F	T

Figure 3.16

In the chapters on sets we found that sets and operations on sets have certain properties. We shall find, in Exercises 3.10, that propositions and "operations" on propositions have corresponding properties. By *operations* on propositions, we mean joining them by any of the connectives we have considered, or forming the negation of a proposition.

EXERCISES 3.10

1. Use a truth table and prove, or disprove, in the system of propositions:

   (a) The commutative property of conjunction. *Note:* This proof can be accomplished by showing that $p \wedge q$ is equivalent to $q \wedge p$, that is, by showing that $(p \wedge q) \leftrightarrow (q \wedge p)$ is a tautology.

   (b) The commutative property of disjunction:
   $(p \vee q) \leftrightarrow (q \vee p)$

   (c) The commutative property of implication:
   $(p \rightarrow q) \leftrightarrow (q \rightarrow p)$

   (d) The commutative property of the biconditional:
   $(p \leftrightarrow q) \leftrightarrow (q \leftrightarrow p)$

   (e) The associative property of conjunction:
   $[(p \wedge q) \wedge r] \leftrightarrow [p \wedge (q \wedge r)]$

   (f) The associative property of disjunction:
   $[(p \vee q) \vee r] \leftrightarrow [p \vee (q \vee r)]$

   (g) The associative property of implication:
   $[(p \rightarrow q) \rightarrow r] \leftrightarrow [p \rightarrow (q \rightarrow r)]$

   (h) The distributive property of conjunction over disjunction:
   $[p \wedge (q \vee r)] \leftrightarrow [(p \wedge q) \vee (p \wedge r)]$

   (i) The distributive property of disjunction over conjunction:
   $[p \vee (q \wedge r)] \leftrightarrow [(p \vee q) \wedge (p \vee r)]$

   (j) When t is used as a special symbol to represent a tautology, then t is an identity element for conjunction:
   $(p \wedge t) \leftrightarrow p$

   (k) When f is used as a symbol for a self-contradiction, f is an identity element for disjunction:
   $(p \vee f) \leftrightarrow p$

2. Prove or disprove the following theorems in our Algebra of Propositions:

   (a) An implication is logically equivalent to its converse:
   $(p \rightarrow q) \leftrightarrow (q \rightarrow p)$

   (b) An implication is logically equivalent to its inverse.

   (c) An implication is logically equivalent to its contrapositive.

3. For an Algebra of Propositions, list:

   (a) Some undefined terms.
   (b) Some defined terms.
   (c) Some assumptions.
   (d) Some theorems.

4. Prove that the following are tautologies:

(a) $\sim(p \wedge q) \leftrightarrow [(\sim p) \vee (\sim q)]$

(b) $\sim(p \vee q) \leftrightarrow [(\sim p) \wedge (\sim q)]$

Are there any properties in the Algebra of Sets that seem analogous to these propositions?

3.11 COMPARISON OF ALGEBRA OF SETS AND ALGEBRA OF PROPOSITIONS

In the preceding exercises we have found that propositions and operations on propositions have some of the same properties as sets and set operations. Each system has commutative, associative, and distributive properties. Also, we shall assume that our Algebra of Propositions has closure properties for disjunction and conjunction analogous to the closure properties of union and intersection in Algebra of Sets. As a matter of fact, the Algebra of Sets and the Algebra of Propositions have the *same structure*. These two systems, and any other having this structure, are known as **Boolean Algebras**. Since the structure of these algebras is the same, a switch in vocabulary can take us from one system to the other. Here is a table showing associated terms in these two algebras:

Algebra of Sets	*Algebra of Propositions*
Sets A, B, C, etc.	Propositions p, q, r, etc.
Union $\cup$	Disjunction $\vee$
Intersection $\cap$	Conjunction $\wedge$
Complement A'	Negation $\sim p$
Universe U	Tautology t (true proposition)
Empty set $\varnothing$	Self-contradiction f (false proposition)
Equals $=$	Is equivalent to $\leftrightarrow$
Is a subset of $\subset$	Implies $\rightarrow$

Let us now use this table to alter some statements from the Algebra of Sets. We shall substitute terms from the Algebra of Propositions for their associated terms in the Algebra of Sets. In doing this we shall find some properties of propositions that have not previously been given:

1. The intersection of any set A with the empty set is the empty set. *Symbolized:* $A \cap \varnothing = \varnothing$.

The associated property of propositions would be: The conjunction of any proposition with a self-contradiction is a self-contradiction. *Symbolized:* $p \wedge f \leftrightarrow f$.

2. The union of any set A with the empty set is the set A. *Symbolized:* $A \cup \varnothing = A$.

The associated property of propositions would be: The disjunction of any proposition with a self-contradiction is equivalent to the given proposition. *Symbolized:* $p \vee f \leftrightarrow p$.

3. The union of a set A with itself is A. *Symbolized:* $A \cup A = A$.
The associated property of propositions is: The disjunction of any proposition p with itself is equivalent to p. *Symbolized:* $p \vee p \leftrightarrow p$.

This procedure for converting statements from one algebra to the other has thus given us three new properties of Algebra of Propositions. It is important that the reader recognize that we have not, however, *proved* these properties.

EXERCISES 3.11

1. What is the identity element for disjunction in Algebra of Propositions? the identity element for conjunction?

2. Use the table of associated terms and convert the following theorems from Algebra of Sets to statements in Algebra of Propositions:

   (a) Theorem III: The complement of the empty set is the universe:
       $\emptyset' = U$.
   (b) Theorem IV: The complement of the universe is the empty set:
       $U' = \emptyset$.
   (c) Theorem V: The union of any set A with the universe U is the universe:
       $A \cup U = U \cup A = U$.
   (d) Theorem VI: The intersection of any set A with the universe U is the set A:
       $A \cap U = U \cap A = A$.
   (e) Theorem VII: The intersection of any set A with itself is the set A:
       $A \cap A = A$.

3. In Chapter 2, we proved the following theorems. Convert each one to a corresponding theorem on propositions and prove it by constructing a truth table.

   (a) If $A \subset B$, then $A \cup B = B$.
   (b) $(A \cap B) \subset A$.
   (c) $(A \cup B) \supset A$.
   (d) $A \cup (A \cap B) = A$.
   (e) $A \cap (B \cap C) = (A \cap B) \cap (A \cap C)$.
   (f) $(A \cup B) \cup (A \cup C) = A \cup (B \cup C)$.

3.12   CONSTRUCTION OF STATEMENTS
HAVING GIVEN TRUTH TABLES

We can, by now, construct truth tables for any given compound proposition involving two or three simple propositions—no matter how complicated. The converse of this problem would be to find a statement for any given table of truth values for two or three simple propositions. This can always be done; in fact, it is possible to devise a statement having a given truth table using only the connectives $\wedge, \vee,$ and $\sim$.

It is trivial to consider the problem of finding a statement that has a truth table consisting of all F's; $p \wedge (\sim p)$ is such a statement, and you have seen others. So we shall consider statements for truth tables having one or more T's. We shall begin by finding a statement that is true for each possible combination of truth and falsity for two simple propositions. Recalling that these combinations are those shown in Figure 3.17, it takes little thought to find a conjunction that would

p	q
T	T
T	F
F	T
F	F

Figure 3.17

be true for each of the four combinations and no other. The conjunction $p \wedge q$ is certainly true for the first combination, two T's, and it is true for no other combination. In Figure 3.18 we list this with those conjunctions that are true for the other combinations:

p	q	
T	T	$p \wedge q$
T	F	$p \wedge (\sim q)$
F	T	$(\sim p) \wedge q$
F	F	$(\sim p) \wedge (\sim q)$

Figure 3.18

We are now ready to use these conjunctions in constructing a proposition having a given truth table. Suppose, for example, that we are given the truth table shown in Figure 3.19. Notice that T's appear in the given truth table only

p	q	?
T	T	F
T	F	T
F	T	F
F	F	T

Figure 3.19

for the combinations where p is true and q is false, and where p and q are both false. We observe that a conjunction that is true for the combination, T-F, and no other, is $p \wedge (\sim q)$; a conjunction that is true for the combination, F-F, and no other, is $(\sim p) \wedge (\sim q)$. Therefore, one proposition having the given truth table is

$$[p \wedge (\sim q)] \vee [(\sim p) \wedge (\sim q)]$$

as may be checked by constructing its truth table.

3.13   AN APPLICATION OF ALGEBRA
OF PROPOSITIONS

An interesting application of the concept of compound propositions may be found in electrical circuit theory. Even within the scope of this chapter, we can understand some elementary theory of switching networks. A **switching network** is an arrangement of wires and switches that connect two terminals T_1 and T_2. Each of the switches involved can be open or closed. If a switch is open, no current will flow through it from T_1 to T_2; if closed, the flow of electricity is permitted. When current flows between the terminals, we say the *network* is completed.

The simplest kind of network would be a single wire connecting T_1 and T_2 with a single switch P. Such a network can be pictured as

$$T_1 \text{———} P \text{———} T_2$$

The next simplest kind of network would be a single wire containing two switches P and Q wired *in series*. This network may be pictured as:

$$T_1 \text{———} P \text{———} Q \text{———} T_2$$

When two switches are wired **in series**, current can flow between the terminals only when both switches are closed. Thus, if we let p represent the statement "Switch P is closed" and q represent the statement "Switch Q is closed," the networks described and pictured in this paragraph can be represented by propositions. In the network containing a single switch P, current will flow only if p is true; thus this network can be represented by p. In the network containing switch P and Q wired in series, current will flow only if $p \wedge q$ is true; this network, then, can be represented by the statement $p \wedge q$.

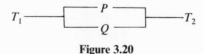

Figure 3.20

Often a network contains switches wired *in parallel*. This network may be pictured as shown in Figure 3.20. When two switches are wired **in parallel**, current flows between the terminals if either (or both) of the switches is closed. Since, in this case, current flows when either p is true, or when q is true, or when both p and q are true, the network can be represented by $p \vee q$.

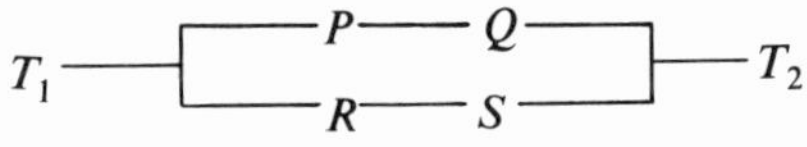

$$T_1 \quad \boxed{\begin{array}{c} P \quad Q \\ R \quad S \end{array}} \quad T_2$$

Figure 3.21

We can combine series and parallel types of wiring. One example of such combinations is shown in Figure 3.21. Can you see that the proposition $(p \wedge q) \vee (r \wedge s)$ would correctly represent this network?

$$T_1 \quad \boxed{\begin{array}{c} P \quad Q \\ P' \quad Q \\ Q' \end{array}} \quad T_2$$

Figure 3.22

Sometimes switches do not operate independently of each other; often two or more switches are **coupled**, so that *they open and close simultaneously*. In diagrams, this is indicated by giving these switches the same letter. It is also possible to **couple** switches so that *if one switch is closed, the other is opened*. This may be indicated by giving the first switch the letter P and the second the symbol P'. In the latter case of coupled switches, the proposition "P is closed" is true only if the proposition "P' is closed" is false. Therefore, if p represents "P is closed," $\sim p$ would represent "P' is closed." A network containing coupled switches is pictured in Figure 3.22. The proposition

$$(p \wedge q) \vee [(\sim p) \wedge q] \vee (\sim q)$$

correctly represents this network. Let us look at its truth table (Fig. 3.23).

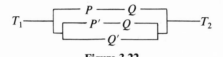

p	q	$(p$	$\wedge$	$q)$	$\vee$	$\{(\sim p)$	$\wedge$	$q]$	$\vee$	$(\sim q)\}$
T	T		T		T		F		F	F
T	F		F		T		F		T	T
F	T		F		T		T		T	F
F	F		F		T		F		T	T

Figure 3.23

The fact that only T's appear in the truth table (Fig. 3.23) for the proposition as a whole indicates that current will flow between the terminals for all possible combinations of P and Q open or closed. An F in this column of the truth table would have indicated that the network would not be completed for the corresponding combination of truth and falsity of p and q. Of course, truth tables for most switching networks would have some F's in them. It is certainly not very often desirable to have a network where current cannot be shut off.

If we use the information of Section 3.12 combined with the information in this section, we can design networks having specified properties. If, for example, a network is needed in which the current will flow only when P and Q are both closed or when P is open and Q is closed, it could be represented by a proposition in whose truth table T's appear only in the cases where p and q are both true, or where p is false and q is true. Design of such circuits will be included in the exercises.

EXERCISES 3.13

1. Use the four basic conjunctions developed in Section 3.12 and construct a proposition having each of the given truth tables (Fig. 3.24).

p	q	(a)	(b)	(c)	(d)	(e)	(f)	(g)	(h)
T	T	T	T	T	T	F	F	F	F
T	F	T	F	T	F	T	F	T	F
F	T	T	T	F	F	T	T	F	F
F	F	T	F	F	T	T	F	T	T

Figure 3.24

2. Complete the table in Figure 3.25 by writing a basic conjunction that is true for each one and only for that one of the eight possible combinations of truth and falsity of p, q, and r.

p	q	r	
T	T	T	$p \wedge q \wedge r$
T	T	F	
T	F	T	
T	F	F	$p \wedge (\sim q) \wedge (\sim r)$
F	T	T	
F	T	F	$(\sim p) \wedge q \wedge (\sim r)$
F	F	T	$(\sim p) \wedge (\sim q) \wedge r$
F	F	F	

Figure 3.25

3. Use the eight basic conjunctions obtained in Exercise 2 and construct a proposition having each of the truth tables in Figure 3.26.

p	q	r	(a)	(b)	(c)	(d)	(e)	(f)
T	T	T	T	T	T	F	F	F
T	T	F	F	F	T	F	F	T
T	F	T	F	F	F	T	T	T
T	F	F	T	F	F	T	F	F
F	T	T	T	T	T	F	F	F
F	T	F	F	F	F	F	F	F
F	F	T	T	T	F	F	F	T
F	F	F	F	F	F	T	T	T

Figure 3.26

4. Write a proposition that can be associated with each of the switching networks shown in Figure 3.27, letting p represent "P is closed," q represent "Q is closed," and r represent "R is closed."

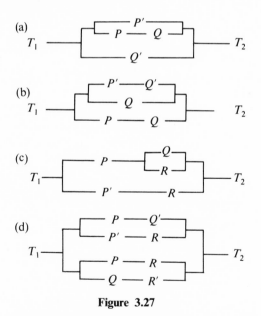

Figure 3.27

5. Find out under which combinations of open and closed switches each of the networks in Exercise 4 will be completed, by making truth tables.

6. Construct a network that could be represented by each of the following statements:

   (a) $[p \vee (p \wedge q)] \vee (\sim q)$
   (b) $\{[p \wedge (\sim q)] \vee (\sim p)\} \vee [(\sim q) \wedge (\sim p)]$
   (c) $[(p \wedge q) \vee (\sim r)] \vee [p \wedge (\sim q) \wedge r]$
   (d) $\{[(\sim p) \wedge q \wedge (\sim r)] \vee (p \wedge r)\} \vee \{(\sim p) \vee [(\sim p) \wedge r]\}$

7. Draw a network that would be completed under each of the following conditions:

   (a) When P is closed and Q is open, or when P is open and Q is closed.
   (b) When P is closed and Q is closed, or when P is closed and Q is open.
   (c) When P is closed and Q is closed.
   (d) When P is open, Q is open, and R is closed, or when P is open, Q is closed, and R is closed, or when P is open, Q is closed, and R is open.
   (e) When P is closed, Q is closed, and R is open, or when P is closed, Q is open, and R is closed, or when P is closed, Q is open, and R is open, or when P is open, Q is closed, and R is closed.

CHAPTER 4

Numbers and Numerals

4.1 THE DIFFERENCE BETWEEN NUMBERS AND NUMERALS

It is necessary, in our study of numerals and systems of numeration, to distinguish between *numerals* and *numbers*. **Numbers** are abstractions; they are ideas, concepts. **Numerals** are symbols used to represent numbers. When we say, "There are two doors into this room," we are indicating a *number* of doors, but the symbol "two" (which could have been written as "2") is not the number; it is a *numeral* that stands for the number.

4.2 THE HINDU-ARABIC SYSTEM OF NUMERATION

During the course of history, countless symbols have been used to indicate the same number. For example, "5," "V," "⊞," "five," and "cinq" are all numerals that stand for the same number. The one of these numerals that we use mos frequently is, of course, "5." It is a **digit** of our **Hindu-Arabic decimal system** numeration. In this system we can count continuously, with no theoretical lir using only the ten distinct single symbols (**digits**) 0, 1, 2, 3, 4, 5, 6, 7, 8, 9 to exp

57

the counting numbers. The power and utility of this system is a result of the fact that *the position of a digit in an array determines its value.* In the numeral 333, for example, we have the digit "3" appearing three times with a different value each time. This important aspect of our decimal system is called **positional value,** or more commonly, **place value.** Let us review how it works.

In order to facilitate our explanation, we shall insert a short discussion of exponents. An **exponent** is a number for which the numeral is always written to the right and just above the numeral representing a second number called the **base,** for example, 3^2 or 4^5. The exponents to which we shall refer in the present discussion are all counting numbers or zero. The **counting number exponent** is a number that indicates how many times the base is to be used as a factor in a product. (The zero exponent will be defined later in the discussion.) Thus 3^2 means 3×3; the exponent 2 tells us that the number 3 is to be used twice as a factor in a product. Similarly, $3^3 = 3 \times 3 \times 3$; $4^5 = 4 \times 4 \times 4 \times 4 \times 4$. The expression 3^2, consisting of the base 3 with the exponent 2, is called the *second power* of three or the *square* of three. Similarly, 3^3 is the *third power* of three or the *cube* of three, and 4^5 is called the *fifth power* of four.

Our decimal system of numeration is so called because it has a *base* ten. (The Latin root for "decimal" is "decem" meaning "ten.") This means that each *place* in a decimal numeral has a value that is a power of ten. A few powers of ten (in descending order) are

$$10^5 = 10 \times 10 \times 10 \times 10 \times 10 = \mathbf{100,000}$$

$$10^4 = 10 \times 10 \times 10 \times 10 = \mathbf{10,000}$$

$$10^3 = 10 \times 10 \times 10 = \mathbf{1000}$$

$$10^2 = 10 \times 10 = \mathbf{100}$$

$$10^1 = \mathbf{10}$$

Notice that decreasing the exponent by one results in dividing the power by ten. According to this pattern, what would 10^0 be?

Let us take the numeral 654,321 and illustrate, first, the value of the places in this numeral:

10^5	10^4	10^3	10^2	10^1	10^0
6	5	4 ,	3	2	1

This chart indicates that the value of the *place* in which the digit 6 appears in the numeral is 10^5; the value of the *place* in which the digit 5 appears is 10^4; the value of the *place* in which the digit 4 appears is 10^3; etc. Next, we indicate the value, determined by its place in the array, of each digit in the numeral; these values are: 6×10^5, 5×10^4, 4×10^3, 3×10^2, 2×10^1, and 1×10^0. The given numeral may be written in **expanded notation** as the sum of the products of each digit of the numeral by the place value of that digit: $(6 \times 10^5) + (5 \times 10^4) + (4 \times 10^3) + (3 \times 10^2) + (2 \times 10^1) + (1 \times 10^0)$, that is, $600{,}000 + 50{,}000 + 4000 + 300 + 20 + 1$. If you have decided, on the basis of the pattern of values for 10^5, 10^4, 10^3, 10^2, 10^1, that the value of 10^0 is 1, you are correct. This value for 10^0 has been assigned arbitrarily by mathematicians, but for good reason. As a matter of fact, unless $n = 0$, the **zero power** of any number n, symbolized n^0, is equal to one. The zero power of zero itself, 0^0, is an undefined, or meaningless, expression in mathematics.

EXERCISES 4.2

1. Evaluate each power:

Example: $2^3 = 2 \times 2 \times 2 = 8$.

(a) 2^4	(b) 2^5	(c) 2^6
(d) 2^0	(e) 3^4	(f) 4^3
(g) 4^0	(h) 5^3	(i) 5^4
(j) 5^5	(k) 5^6	(l) 5^0
(m) 8^2	(n) 8^3	(o) 8^0
(p) 12^2	(q) 12^3	(r) 12^4
(s) 12^5	(t) 12^0	

2. Write in expanded notation.

Example: $7963 = (7 \times 10^3) + (9 \times 10^2) + (6 \times 10^1) + (3 \times 10^0)$.

(a) 27	(b) 468
(c) 970	(d) 907
(e) 4785	(f) 45,721
(g) 40,004	(h) 40,000

3. When we add

$$482$$
$$+315,$$

we use an **algorithm** (method or procedure) in which we add the digits in each column. When we add $5 + 2 = 7$, what value does the digit 7 have in the final sum? When we add $8 + 1 = 9$, what value does the digit 9 have in the final sum? When we add $4 + 3 = 7$, what value does this digit 7 have in the final sum?

4. When we add

$$48$$
$$+24,$$

the first step is to add $8 + 4 = 12$. We write the digit 2 below the column of digits on the right, but we "carry" the digit 1 over to the column of digits on the left and add it to these digits: $1 + 4 + 2 = 7$. Explain why we "carry" the digit 1.

5. When we multiply

$$\begin{array}{r} 24 \\ \times\ 5, \\ \hline \end{array}$$

the first step is to multiply $5 \times 4 = 20$. We write the digit 0 directly below the 5 and carry the digit 2. Then we multiply $5 \times 2 = 10$ and add the digit 2 that we had carried. The final product is 120.

(a) Why do we "carry" the digit 2?

(b) Consider this method of multiplying the same numbers:

$$\begin{array}{r} 24 \\ \times\ 5 \\ \hline 100 \\ 20 \\ \hline 120 \end{array}$$

Explain how the multiplication was done.

6. Consider the following multiplication:

$$\begin{array}{r} 24 \\ \times 35 \\ \hline 120 \\ 72 \\ \hline 840 \end{array}$$

(a) Why was the product $3 \times 24 = 72$ written so that its digits are in the first and second columns (from the left) rather than in the second and third columns?

(b) Explain the following method of multiplying the same numbers:

$$\begin{array}{r} 24 \\ \times 35 \\ \hline 20 \\ 100 \\ 120 \\ 600 \\ \hline 840 \end{array}$$

4.3   THE BASE FIVE SYSTEM

Though our decimal system of numeration is used in civilized countries the world over, there is nothing sacred about it. We can have other systems with a place value structure but with different bases. As a matter of fact, the reason we have a base 10 system is probably because we have ten fingers and ten toes. Let us consider a numeration system with base 5. Just as there are ten digits in our base 10 system, there are five digits in a base 5 system. We shall use the digits 0, 1, 2, 3, and 4. When we write a base 5 numeral, we shall identify it by using "5" as a *subscript*, that is, by writing "5" to the right and slightly below the numeral. A sample base 5 numeral is 342_5 (read three four two base five). Since this system has the same structure as our decimal system of numeration, each place in the numeral has a value that is a power of the base 5. Then, 342_5 means $(3 \times 5^2) + (4 \times 5^1) + (2 \times 5^0)$. (Remember, the zero power of *any* number different from zero is 1.) In base 10 numerals, this number would be expressed as $75 + 20 + 2$, or 97.

To strengthen our understanding of the base 5 system of numeration, let us do some counting in this system. We begin 1, 2, 3, 4, ... Thus far, the process is the same as counting in the base 10. But when we get to the fifth counting number, we must write it 10_5, which means $(1 \times 5^1) + (0 \times 5^0)$, or $5 + 0$. We continue our counting. The sixth numeral would be 11_5, which means $(1 \times 5^1) + (1 \times 5^0)$, or $5 + 1$; the seventh numeral would be 12_5; etc.. We shall list the first twenty-four base 5 numerals along with the corresponding base 10 numerals. (*Note:* Numerals written without a subscript are assumed to be base 10 numerals.)

$1 \leftrightarrow 1_5$	$7 \leftrightarrow 12_5$	$13 \leftrightarrow 23_5$	$19 \leftrightarrow 34_5$
$2 \leftrightarrow 2_5$	$8 \leftrightarrow 13_5$	$14 \leftrightarrow 24_5$	$20 \leftrightarrow 40_5$
$3 \leftrightarrow 3_5$	$9 \leftrightarrow 14_5$	$15 \leftrightarrow 30_5$	$21 \leftrightarrow 41_5$
$4 \leftrightarrow 4_5$	$10 \leftrightarrow 20_5$	$16 \leftrightarrow 31_5$	$22 \leftrightarrow 42_5$
$5 \leftrightarrow 10_5$	$11 \leftrightarrow 21_5$	$17 \leftrightarrow 32_5$	$23 \leftrightarrow 43_5$
$6 \leftrightarrow 11_5$	$12 \leftrightarrow 22_5$	$18 \leftrightarrow 33_5$	$24 \leftrightarrow 44_5$

How would we write 25 as a base 5 numeral?

It is important to understand that corresponding numerals in this table represent *exactly the same numbers*. Let us demonstrate, for example, that $23 = 43_5$ by showing this *number* of things grouped in two different ways (Fig. 4.1).

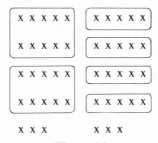

Figure 4.1

On the left, we can see that there are *two* groups of ten x's with *three* x's left over. This grouping corresponds to the decimal numeral 23. On the right we see *four* groups of five x's with three x's remaining. This grouping corresponds to the **quinary** (base 5) numeral 43_5.

In the first paragraph of this section it was explained that 342_5 means (3×5^2) $+ (4 \times 5^1) + (2 \times 5^0)$, which equals

$$(3 \times 25) + (4 \times 5) + (2 \times 1) = 75 + 20 + 2 = \underline{97}$$

in the base 10. This explanation demonstrates a way to convert base 5 numerals to base 10 numerals. Suppose, however, we wish to convert a base 10 numeral to a base 5 numeral. We offer a method by which this can be accomplished. Let us first demonstrate this method by converting 97 back to 342_5.

The first step is to find the largest power of 5 that is less than or equal to 97. Some of the powers of 5 are

$$5^0 = 1 \qquad 5^3 = 125$$
$$5^1 = 5 \qquad 5^4 = 625$$
$$5^2 = 25 \qquad 5^5 = 3125$$

We find that the largest of these powers that is less than 97 is $5^2 = 25$. Since 25 is the highest power of 5 that is less than or equal to 97, we know we are seeking digits a, b, and c such that

$$97 = (a \times 25) + (b \times 5) + (c \times 1)$$

We divide 25 into 97:

$$
\begin{array}{r}
3 \\
25\overline{)97} \\
75 \\
\hline
22 \text{ remainder}
\end{array}
$$

The quotient 3 indicates that the digit "a" in the base 5 numeral corresponding to 97 is 3. To find the digit "b," we divide the remainder 22 by the next highest power of 5. This power is 5.

$$
\begin{array}{r}
4 \\
5\overline{)22} \\
20 \\
\hline
2 \text{ remainder}
\end{array}
$$

The quotient 4 indicates that the digit "b" we are seeking is 4, and the remainder 2 tells us that the digit "c" is 2. Thus

$$97 = (a \times 25) + (b \times 5) + (c \times 1)$$
$$= (3 \times 25) + (4 \times 5) + (2 \times 1)$$
$$= 342_5$$

Let us look at one more example of this method of converting base 10 numerals into base 5 numerals. Consider the base 10 numeral 658. We see that the highest power of 5 that is less than or equal to 658 is $5^4 = 625$. Thus we are seeking digits a, b, c, d, and e such that

$$658 = (a \times 625) + (b \times 125) + (c \times 25) + (d \times 5) + (e \times 1)$$

We divide 658 by 625:

$$\begin{array}{r} 1 \\ 625\overline{)658} \\ \underline{625} \\ 33 \text{ remainder} \end{array}$$

The quotient 1 indicates that $a = 1$. Since the remainder 33 is not greater than or equal to 125 the quotient is 0 and the remainder is 33 when we divide 33 by 125; that is, $b = 0$. This illustrates a special use of zero in a place value system of numeration: its use as a **place holder**. We next divide 33 by 25:

$$\begin{array}{r} 1 \\ 25\overline{)33} \\ \underline{25} \\ 8 \text{ remainder} \end{array}$$

The quotient 1 indicates that $c = 1$. Finally we divide 8 by 5:

$$\begin{array}{r} 1 \\ 5\overline{)8} \\ \underline{5} \\ 3 \text{ remainder} \end{array}$$

The quotient 1 indicates that $d = 1$, and the remainder 3 indicates that $e = 3$. Thus

$$658 = (1 \times 625) + (0 \times 125) + (1 \times 25) + (1 \times 5) + (3 \times 1)$$

$$= 10{,}113_5$$

We suggest that numerals written in systems with bases other than ten be read orally without use of such words as "ten," "fifteen," "hundred," "thousand," etc., because these words are specifically associated with the decimal system. It is better to read a numeral having a base other than ten simply by saying its digits in order (from left to right), and stating its base. For example, $10{,}113_5$ may be read, "one zero one one three, base five."

EXERCISES 4.3

1. Express as a base 10 numeral:

Example: $123_5 = (1 \times 5^2) + (2 \times 5^1) + (3 \times 5^0)$

$$= (1 \times 25) + (2 \times 5) + (3 \times 1)$$

$$= 25 + 10 + 3$$

$$= 38$$

(a) 41_5	(b) 124_5	(c) 312_5
(d) 2140_5	(e) 3023_5	(f) 4444_5
(g) $10{,}241_5$	(h) $32{,}414_5$	(i) $430{,}213_5$

2. Use two different groupings of a set of x's to show that $36 = 121_5$.

3. Express each base 10 numeral as a base 5 numeral.

(a) 78	(b) 165	(c) 349
(d) 689	(e) 750	(f) 1000
(g) 2594	(h) 4976	(i) 5002
(j) 7899	(k) 12,459	(l) 27,404

4.4 THE DUODECIMAL SYSTEM

The Duodecimal Society of America has, for over twenty years, recommended general acceptance of a system of numeration with base 12, because of some advantages that system is alleged to have over the decimal system. Among the advantages cited for this base 12 or **duodecimal system** are (1) It would work better with commonly used units of measure, such as the inch-foot-yard units, the seconds-minutes-hours units, the minutes-seconds-degrees units, etc., and (2) fewer of the more commonly used fractions would have "decimal" fraction forms that are endless, repeating, "decimals." Though it is highly improbable that any universal changeover ever will occur, the duodecimal system is an interesting and profitable system to examine.

The base 12 system has powers of twelve (12^0, 12^1, 12^2, 12^3, etc.) as its place values, and it has twelve digits. Thus we need new digits "d" which we call **dec** for 10 and "e" which we call **elf** for 11. Then the digits for the duodecimal system are

$$0, 1, 2, 3, 4, 5, 6, 7, 8, 9, d, e$$

The numeral $76e_{12}$ means $(7 \times 12^2) + (6 \times 12^1) + (11 \times 12^0)$ in base 10, that is,

$$76e_{12} = (7 \times 144) + (6 \times 12) + (11 \times 1)$$

$$= 1008 + 72 + 11$$

$$= 1091$$

EXERCISES 4.4

1. Express as base 10 numerals:

(a) 17_{12} (b) 36_{12}

(c) 92_{12} (d) $d2_{12}$

(e) $3e4_{12}$ (f) $d9e_{12}$

(g) $7,3e6_{12}$ (h) $26,7d8_{12}$

(i) $d5,7e1_{12}$ (j) $250,d4e_{12}$

2. Express as base 12 numerals:

(a) 16   (b) 48

(c) 116   (d) 495

(e) 713   (f) 4916

(g) 9762   (h) 22,178

(i) 39,213   (j) 236,830

4.5 THE BINARY SYSTEM

In the early years of the electronic computer, input and output of data required the use of a system of numeration with base 2, called the **binary system**. The binary system requires only two digits, 0 and 1, to represent all the real numbers. The computer solves problems that can be represented by numerical models, and the simplest way to represent numbers with such an electrical device is to let an impulse of electricity stand for the digit 1 and to let the absence of an impulse stand for the digit 0.

The place values in the binary system are powers of two (2^0, 2^1, 2^2, 2^3, etc.). The numeral $110,111_2$, then, means $(1 \times 2^5) + (1 \times 2^4) + (0 \times 2^3) + (1 \times 2^2) + (1 \times 2^1) + (1 \times 2^0)$ in the base 10; that is,

$$110,111_2 = (1 \times 32) + (1 \times 16) + (0 \times 8) + (1 \times 4) + (1 \times 2) + (1 \times 1)$$

$$= 32 + 16 + 0 + 4 + 2 + 1$$

$$= 55$$

EXERCISES 4.5

1. Write the binary numerals that represent the same numbers as the decimal numerals $1, 2, 3, \ldots, 15$.

2. Express as base 10 numerals:

(a) 1001_2 (b) $111,101_2$

(c) $100,110_2$ (d) $1,011,101_2$

(e) $1,110,011_2$ (f) $11,111,111_2$

(g) $101,101,101_2$ (h) $110,111,001_2$

(i) $100,001,111_2$ (j) $1,110,110,111_2$

3. Express as base 2 numerals:

(a) 6   (b) 12
(c) 32   (d) 49
(e) 75   (f) 95
(g) 157   (h) 225
(i) 325   (j) 416

4.6 COMPUTATION WITH NUMERALS OTHER THAN DECIMAL NUMERALS

Computation with numerals of systems other than our decimal system is an interesting and enlightening activity. Let us work through some addition and multiplication problems:

Add:

$$\begin{array}{ll}
412_5 & 4(5^2) + 1(5^1) + 2(5^0) \\
311_5 = & 3(5^2) + 1(5^1) + 1(5^0) \\
\hline
& 7(5^2) + 2(5^1) + 3(5^0)
\end{array}$$

But $7(5^2) = 5(5^2) + 2(5^2) = 1(5^3) + 2(5^2)$, so that

$$412_5 + 311_5 = 1(5^3) + 2(5^2) + 2(5^1) + 3(5^0)$$

$$= \mathbf{1223_5}$$

Add:

$$\begin{array}{ll}
75d_{12} = & 7(12^2) + 5(12^1) + d(12^0) \\
641_{12} = & 6(12^2) + 4(12^1) + 1(12^0) \\
\hline
& 13(12^2) + 9(12^1) + e(12^0)
\end{array}$$

But $13(12^2) = 12(12^2) + 1(12^2) = 1(12^3) + 1(12^2)$, so that

$$75d_{12} + 641_{12} = 1(12^3) + 1(12^2) + 9(12^1) + e(12^0)$$

$$= \mathbf{119e_{12}}$$

Add:

$$\begin{array}{ll}
1011_2 = & 1(2^3) + 0(2^2) + 1(2^1) + 1(2^0) \\
101_2 = & \phantom{1(2^3) + {}} 1(2^2) + 0(2^1) + 1(2^0) \\
\hline
& 1(2^3) + 1(2^2) + 1(2^1) + 2(2^0)
\end{array}$$

But $2(2^0) = 1(2^1) + 0(2^0)$. When this change is made in the previous sum, we get $1(2^3) + 1(2^2) + 2(2^1) + 0(2^0)$. But

$$2(2^1) = 1(2^2) + 0(2^0)$$

When this change is made in the previous sum, we get $1(2^3) + 2(2^2) + 0(2^1) + 0(2^0)$. But

$$2(2^2) = 1(2^3) + 0(2^2)$$

from which we obtain another expression for the sum. After one more application of the procedure, we finally get the sum

$$1011_2 + 101_2 = 1(2^4) + 0(2^3) + 0(2^2) + 0(2^1) + 0(2^0)$$
$$= \mathbf{10{,}000_2}$$

Let us check this result by writing the original addends as base 10 numerals and adding:

$$
\begin{aligned}
1011_2 &= 8 + 0 + 2 + 1 = 11 \text{ (base 10)} \\
101_2 &= 4 + 0 + 1 = 5 \text{ (base 10)} \\
\hline
& 16 \text{ (base 10)}
\end{aligned}
$$

Since $10{,}000_2$ does, in fact, equal 16 (base 10), our work has been correct.

The expanded notation in these three examples is intended to give meaning to the process we actually use when we add. We next use the same examples and demonstrate a more mechanical, less time-consuming, method. (The reader will recognize this as being analogous to the addition algorithm he uses when adding base 10 numerals.)

$$
\begin{aligned}
&412_5 \\
+&311_5 \\
\hline
\end{aligned}
$$

We add each column and do our thinking in base 10, but we cannot write a sum until we "translate," or convert, that sum to a base 5 numeral. As we add the first and second columns (from right to left), no translation is needed because the results would be written the same in base 5 as they are in base 10. As we add $4 + 3$, however, we get 7 (base 10) which we need to translate to 12_5 in the base 5 before writing it down:

$$
\begin{aligned}
&412_5 \\
+&311_5 \\
\hline
&1223_5
\end{aligned}
$$

In the second example,

$$75d_{12}$$
$$+641_{12}$$

as we add $d + 1$, we think $10 + 1 = 11$ and write e, the base 12 numeral for 11. The sum $5 + 4$ presents no problem, because the result is written just as in base 10. As we add $7 + 6$, we think 13, and write 11, the base 12 numeral for 13.

$$75d_{12}$$
$$+641_{12}$$
$$\overline{119e_{12}}$$

The third example was

$$1011_2$$
$$+ \quad 101_2$$

When we add the column at the right, we get 2 (base 10), which we translate into 10_2 before we write anything down. Then we write the digit 0 below the column and "carry" the digit 1:

$$1$$
$$1011_2$$
$$+ \quad 101_2$$
$$\overline{0}$$

Next, we add the second column remembering to add in the "carried" digit 1. Again we get 2, which we translate to 10_2, and proceed as before:

$$11$$
$$1011_2$$
$$+ \quad 101_2$$
$$\overline{00}$$

Continuing this process, we have, finally,

$$111$$
$$1011_2$$
$$+ \quad 101_2$$
$$\overline{10000_2}$$

We now turn to multiplication with a base 5 example:

$$\begin{array}{r} 34_5 \\ \times\, 23_5 \\ \hline \end{array}$$

Our procedure will be similar to base 10 multiplication. We think $3 \times 4 = 12$ (base 10), which we write as 22_5 in the base 5. We write a 2 and carry a 2:

$$\begin{array}{r} 2 \\ 34_5 \\ \times\, 23_5 \\ \hline 2 \end{array}$$

In the next step, we multiply $3 \times 3 = 9$, but we had carried a 2. We add this 2 to the product 9 and get $9 + 2 = 11$ (base 10), which would be written 21_5:

$$\begin{array}{r} 2 \\ 34_5 \\ \times\, 23_5 \\ \hline 212 \end{array}$$

We follow the same procedure in multiplying by the digit 2:

$$\begin{array}{r} 34_5 \\ \times\, 23_5 \\ \hline 212 \\ 123 \end{array}$$

Because of place value, the digit 2 that we have just used as a multiplier, means two 5's, that is, 20_5. This accounts for the shift to the left in writing the result of this multiplication. We traditionally leave a blank space instead of writing in the zero that belongs there. Then we add the partial products:

$$\begin{array}{r} 34_5 \\ \times\, 23_5 \\ \hline 212 \\ 123 \\ \hline 1442_5 \end{array}$$

We can check our work by converting the original factors to base 10 and repeating the computation with base 10 numerals.

$$\begin{array}{rll} 34_5 = 3(5^1) + 4(5^0) = & 19 & \text{(base 10)} \\ \times\, 23_5 = 2(5^1) + 3(5^0) = & 13 & \text{(base 10)} \\ \hline & 57 & \\ & 19 & \\ \hline & 247 & \text{(base 10)} \end{array}$$

Our base five product was 1442_5; $1442_5 = 1(5^3) + 4(5^2) + 4(5^1) + 2(5^0)$, that is, in base ten numerals, $125 + 100 + 20 + 2 = 247$ (base 10). This result provides a check that our work has been correct.

4.7 A SIMPLE METHOD FOR CONVERTING DECIMAL NUMERALS

There is a simple method for converting base 10 numerals to any other base. We demonstrate:

EXAMPLE 1.   Convert 763 (base 10) to a base 5 numeral.

Step 1. Divide the base 5 into the given number, writing the quotient and remainder as shown.

$$5\overline{)763}$$
$$152 \quad r:3$$

Step 2. Divide the quotient from the first step by the base 5, writing quotient and remainder as shown.

$$5\overline{)763}$$
$$5\overline{)152} \quad r:3$$
$$30 \quad r:2$$

Continue this process until the quotient is smaller than 5.

The base 5 numeral for 763 is **11023**$_5$ found by reading from the final quotient through the remainders as shown by the arrow.

$$5\overline{)763}$$
$$5/152 \; r:3$$
$$5/ \;\; 30 \; r:2$$
$$5/ \;\;\; 6 \; r:0$$
$$1 \; r:1$$

EXAMPLE 2.   Convert 5963 (base 10) to a base 12 numeral. *Note:* When a remainder of 10 or 11 occurs, we must write it as a base twelve digit.

$$5963 \text{ (base 10)} = \mathbf{354}e_{12}$$

$$12\overline{)5963}$$
$$12/ \;\; 496 \; r:e$$
$$12/ \;\;\;\; 41 \; r:4$$
$$3 \; r:5$$

The reason that this algorithm works is not immediately obvious. To clarify the procedure, we shall use Example 1, performing the same successive divisions by five as a first step:

$$763 = (152 \times 5) + 3$$
$$152 = (30 \times 5) + 2$$
$$30 = (6 \times 5) + 0$$
$$6 = (1 \times 5) + 1$$

Next, let us substitute the expanded form of 6 (in the fourth statement) into the third statement. Then we shall substitute this expanded form of 30 into the second

statement. Finally, we shall substitute the expanded form of 152 into the first
statement and express the result in terms of powers of 5.

$$6 = (1 \times 5) + 1$$

$$30 = (6 \times 5) + 0$$

$$= \{[(1 \times 5) + 1] \times 5\} + 0$$

$$= (5 + 1) \times 5$$

$$152 = (30 \times 5) + 2$$

$$= [(5 + 1) \times 5] \times 5 + 2$$

$$763 = (152 \times 5) + 3$$

$$= \{[(5 + 1) \times 5] \times 5 + 2\} \times 5 + 3$$

$$= [(5^2 + 5) \times 5 + 2] \times 5 + 3$$

$$= [(5^3 + 5^2 + 2) \times 5] + 3$$

$$= [5^4 + 5^3 + (2 \times 5)] + 3$$

Thus

$$753 = (1 \times 5^4) + (1 \times 5^3) + (0 \times 5^2) + (2 \times 5^1) + (3 \times 5^0)$$

$$= 11,023_5$$

EXERCISES 4.7

1. Make addition and multiplication tables for the base 5 digits. As a guide, the base
five addition table has been partially completed for you.

+	0_5	1_5	2_5	3_5	4_5
0_5	0_5	1_5	2_5	3_5	4_5
1_5	1_5	2_5	3_5	4_5	10_5
2_5	2_5	3_5	4_5		
3_5					
4_5					

2. Make addition and multiplication tables for the base 12 digits.

3. Add:

(a) 3430_5	(b) 134_5	(c) 4211_5	(d) 431_5
444_5	21_5	3231_5	134_5
	403_5	1042_5	341_5
			142_5

(e) 1679_{12} (f) $d45e_{12}$ (g) 798_{12} (h) 679_{12}
   4222_{12} 798_{12} 280_{12} $4d5_{12}$
                                                       934_{12} $176e_{12}$

(i) 4971_{12} (j) 110_2 (k) 1011_2 (l) 110_2
   9047_{12} 10_2 111_2 101_2
   8053_{12} 1101_2 111_2

(m) 1101_2
    110_2
   1101_2

4. Multiply:

(a) 402_5 (b) 334_5 (c) 3041_5 (d) 4323_5
   33_5 23_5 234_5 341_5

(e) 469_{12} (f) 976_{12} (g) $7e59_{12}$ (h) 4738_{12}
   76_{12} 384_{12} $70d_{12}$ $5e9_{12}$

(i) 110_2 (j) 1101_2 (k) 10101_2 (l) 111010_2
   11_2 101_2 1101_2 10111_2

5. The base 8 or **octal system** of numeration is often used to express results given in binary numerals by computers. It is very easy to convert a base 2 numeral to a base 8 numeral and vice versa. Can you discover the shortcut by inspection of these examples?

$$10,101_2 = 25_8$$

$$111,001_2 = 71_8$$

$$11,101,110_2 = 356_8$$

$$111,111,111,111_2 = 7777_8$$

$$10,100,010,100_2 = 2424_8$$

If you have found the shortcut, convert each base 2 numeral to base 8 and each base 8 numeral to base 2.

(a) 1011_2 (b) $11,101_2$ (c) $101,101_2$
(d) $1,011,001_2$ (e) $110,001,111_2$ (f) $11,010,100,101,110_2$
(g) 76_8 (h) 543_8 (i) 7402_8
(j) $65,135_8$ (k) $342,762_8$ (l) $1,234,567_8$

6. Use the method demonstrated in Section 4.7 to make these conversions:

(a) Convert 377 (base 10) to a base 5 numeral.
(b) Convert 1643 (base 10) to a base 5 numeral.
(c) Convert 27,493 (base 10) to a base 5 numeral.
(d) Convert 4765 (base 10) to a base 12 numeral.

(e) Convert 5172 (base 10) to a base 12 numeral.

(f) Convert 6694 (base 10) to a base 12 numeral.

(g) Convert 18,674 (base 10) to a base 12 numeral.

(h) Convert 47 (base 10) to a base 2 numeral.

(i) Convert 102 (base 10) to a base 2 numeral.

(j) Convert 173 (base 10) to a base 2 numeral.

(k) Convert 395 (base 10) to a base 2 numeral.

4.8 COMPUTER ARITHMETIC

We have said that the binary system is used in work with computers. The way it is used, however, is somewhat different from the way we use it in our pencil and paper computation. In order to demonstrate how computing is done by many computers, we shall invent a considerably simplified imaginary machine, which we shall call IMAC (imaginary automatic computer).

In our hypothetical computer IMAC, information is handled in the form of sets of binary digits. These binary digits are called **bits**; sets of bits are called **words**. IMAC can handle five-bit words, for example,

$$\text{Word: } 0 \underbrace{\; 1 \; 1 \; 0 \; 1 \;}$$

Sign bit ⟶   ⟵ Magnitude bits

The bit on the left is called the **sign bit**; the remaining four bits are called **magnitude bits**. Now, the digit 0, as a sign bit, denotes the **positive sign**; thus the numeral we have written as an example denotes $^+1101_2$ or $^+13$ (base 10). The largest positive number that IMAC can handle would be expressed as 01111, which equals $^+15$ (base 10). **Negative numbers** are represented by words having 1 as the sign bit, but this is not the whole story of expressing negative numbers. To understand how negative numbers are represented, we must first define what is called the *one's complement* of a numeral, or word : The **one's complement** of a word is a new word formed by changing all 0's in the original word to 1's, and changing all 1's to 0's. Here are a few examples :

The one's complement of 00111 is 11000.

The one's complement of 01101 is 10010.

The one's complement of 11010 is 00101.

Now, in IMAC, the **negative** of any number is expressed as the one's complement of the IMAC word for that number, that is, the magnitude bits of the IMAC word for a negative number are in the one's complement form. Here are a few examples :

$$11100 = {}^-0011_2 = {}^-3 \text{ (base 10)}$$

$$10111 = {}^-1000_2 = {}^-8 \text{ (base 10)}$$

The largest negative number that IMAC can handle would be expressed as 10000. Translating this word, we get

$$10000 = {}^-1111_2 = {}^-15 \text{ (base 10)}$$

In order to compute with IMAC, we must realize that this computer has just one basic trick: *It can add binary numerals* whose sum will not have an absolute value greater than fifteen. The machine gets the same sum that we obtain by adding with pencil and paper, for example,

$$
\begin{array}{r}
\text{Add } 3 + 4: \quad 00011 \\
00100 \\
\hline
00111 = 7 \text{ (base 10)}
\end{array}
$$

$$
\begin{array}{r}
\text{Add } 9 + 3: \quad 01001 \\
00011 \\
\hline
01100 = 12 \text{ (base 12)}
\end{array}
$$

IMAC cannot *subtract*, but we can overcome this deficiency in the machine by *expressing our subtraction problems in terms of addition.* In high school algebra class, we learned that subtracting a number can be accomplished by adding its negative. For example, if we wish to subtract 12 from 7, we think of the problem as $7 + {}^-12$. Now, ${}^-12 = {}^-(01100) = 10011$ as an IMAC word. The problem becomes

$$
\begin{array}{r}
7 + {}^-12: \quad 00111 \\
10011 \\
\hline
11010 = {}^-0101 = {}^-5 \text{ (base 10)}
\end{array}
$$

Here is another example: Subtract 15 from 9. We write this as $9 + {}^-15$. We translate ${}^-15$ into computer language: ${}^-15 = {}^-(01111) = 10000$. Thus, the problem becomes

$$
\begin{array}{r}
9 + {}^-15: \quad 01001 \\
10000 \\
\hline
11001 = {}^-0110_2 = {}^-6 \text{ (base 10)}
\end{array}
$$

Subtraction, however, is not always so simple. Consider $12 - 7$, that is, $12 + {}^-7$. Following the procedure demonstrated above, we get

$$ {}^-7 = {}^-(00111) = 11000 $$

$$
\begin{array}{r}
12 + {}^-7: \quad 01100 \\
11000 \\
\hline
100100
\end{array}
$$

The result is a six-bit word, but IMAC can handle only five-bit words. At this point, the machine performs an operation called an **end-carry**. It, in effect, removes the sixth digit (on the left) and enters it below the other five digits and adds again as illustrated:

$$
\begin{array}{r}
01100 \\
11000 \\
\hline
\textcircled{1}00100 \\
\qquad\quad\longrightarrow 1 \\
\hline
00101 = {}^{+}5 \text{ (base 10)}
\end{array}
$$

An interesting question is: What will IMAC do if it is asked to do a problem beyond its scope, for example, if we attempt to add $9 + 8$. Since the largest number that can be expressed by the four magnitude bits we are allowed has an absolute value of 15, the sum 17 cannot be expressed by IMAC. Nevertheless, let us enter the problem and proceed as in the problems above:

$$
\begin{array}{r}
9 + 8: \quad 01001 \\
01000 \\
\hline
10001
\end{array}
$$

The IMAC word 10001 is read $^{-}1110 = {}^{-}13$ (base 10). Thus, the apparent result of adding two positive numbers is a negative number. This is the signal that the machine has been asked to do the impossible. If, when adding two positive numbers the result is a negative number, or, when adding two negative numbers the result is a positive number, the machine registers an **overflow**. Can we ever get a result that the machine cannot handle when adding a positive and a negative number?

EXERCISES 4.8

Perform each computation as IMAC would do it and translate each result into base 10.

1. $5 + 3$	**2.** $6 + 4$
3. $2 + 7$	**4.** $9 + 5$
5. $12 + 2$	**6.** $13 + 4$
7. $3 + 4 + 6$	**8.** $7 + 1 + 5$
9. $1 + 5 + 7$	**10.** $2 + 3 + 4 + 5$
11. $6 + {}^{-}7$	**12.** $^{-}15 + 8$
13. $^{-}3 + {}^{-}9$	**14.** $^{-}3 + 14$
15. $^{-}7 + 11$	**16.** $5 - 3$
17. $6 - 4$	**18.** $2 - 7$
19. $9 - 5$	**20.** $12 - 2$
21. $13 - 4$	**22.** $6 - {}^{-}7$
23. $^{-}15 - {}^{-}8$	**24.** $^{-}3 - 9$
25. $^{-}3 - {}^{-}14$	**26.** $^{-}7 - 11$
27. $6 - {}^{-}3$	**28.** $^{-}14 - {}^{-}11$
29. $^{-}11 - {}^{-}14$	**30.** $9 - {}^{-}4$

CHAPTER 5

The Whole Numbers

5.1 AN EARLY COUNTING METHOD

The earliest quantitative activity engaged in by man was probably keeping track of his possessions. The original ways of doing this, if we are interested in specific techniques, are a matter of conjecture. It is pretty safe to say, however, that he knew *how many* things he had because he had set up a one-to-one correspondence between his possessions and an equivalent *standard set*. Suppose one of the ancients had a flock of sheep. We can imagine that he might have dropped a stone into a pouch for each sheep as it passed through the gate to the pasture. The set of stones in the pouch, then, was *equivalent* to the set of sheep that he possessed. This means that the *number* of sheep was *equal* to the *number* of stones. If we let the set of sheep be represented as A and the set of stones as B, then $n(A) = n(B)$, using the notation of Section 1.3. It is important to understand that $n(A) = n(B)$ if and only if the set A is equivalent to the set B; that is, if and only if the distinct elements of the two sets can be placed in a one-to-one correspondence.

5.2 THE COUNTING NUMBERS

The **standard sets** that we use today to tell how many things we have are subsets of the set of **counting numbers**, also called the set of **natural numbers**. This is the

76

familiar set we use whenever we count:

$$\{1, 2, 3, 4, 5, \ldots\}$$

Indeed, we define **counting** as the act of making a one-to-one correspondence between a set of objects and a subset of the set of counting numbers. We use the counting numbers in the order 1, 2, 3, ... when we count. We agree that the first counting number is *one* and the others are used in a conventional sequence, so that the final number n that we write or say is enough to indicate the entire set of counting numbers $\{1, 2, 3, \ldots, n\}$ that has been used. For example, in counting the fingers on one hand, we say: "One, two, three, four, five," thus putting the set of fingers into a one-to-one correspondence with the set of counting numbers, $\{1, 2, 3, 4, 5\}$. In explaining **how many** fingers, it is sufficient to refer only to the last number, 5.

5.3   THE WHOLE NUMBERS

The reader will recall that the symbol $n(A)$ means the number of elements in a set A. If we exclude infinite sets from the system of sets that A may represent, it seems intuitively evident that, for any set A, $n(A)$ will be an element of the set

$$\{0, 1, 2, 3, \ldots\}$$

This set is called the set of *whole numbers*. **Whole numbers** are defined as numbers p such that, for some non-infinite set A, $n(A) = p$. We shall use two kinds of symbols to represent whole numbers: (1) lowercase letters such as a, b, c, n, x, etc. and (2) symbols suggested by the definition of whole numbers such as $n(A)$, $n(B)$, $n(C)$, $n(X)$, etc. for sets A, B, C, X, etc. Notice that the set of whole numbers contains just one element, 0, that is not contained in the set of counting numbers.

We are all very familiar with both the counting numbers and the whole numbers. This often makes it difficult to understand the need for the kind of analysis of the whole number system that we introduce in this chapter. Familiarity with these numbers is mistaken for understanding. In order to take a new look at these "old" numbers, the student must be willing to admit that he does not "already know all about them."

5.4   ADDITION OF WHOLE NUMBERS

The first operation on whole numbers that we shall consider is *addition*. Recalling that the result of adding is called the **sum** and the symbol for the operation is "$+$", we define **addition of whole numbers** as follows: If A and B are disjoint sets, that is, if $A \cap B = \emptyset$, then the sum $n(A) + n(B) = n(A \cup B)$. This definition links addition to the basic process of counting, as we shall see in our examples.

EXAMPLE 1.   If John, Mary, Eileen, and Bill sit in the first row and Joan, Bob, Olive, Anne, and Mike sit in the second row, and if only the children named sit in these rows, how many children are there altogether in these two rows?

Let $A = \{$John, Mary, Eileen, Bill$\}$, and $B = \{$Joan, Bob, Olive, Anne, Mike$\}$. (The reader will notice that sets A and B have no common elements.)

Then $n(A) + n(B) = 4 + 5 = n(A \cup B)$.†

$A \cup B = \{$John, Mary, Eileen, Bill, Joan, Bob, Olive, Anne, Mike$\}$.

$n(A \cup B) = 9$, *which we find by counting.*

Therefore, $4 + 5 = 9$.

EXAMPLE 2.   The set of consonants in the English alphabet is $\{b, c, d, f, g, h, j, k, l,$ $m, n, p, q, r, s, t, v, w, x, y, z\}$; the set of vowels is $\{a, e, i, o, u\}$. Let $C =$ the set of consonants and $V =$ the set of vowels. (Are these disjoint sets?) Then $n(C) = 21$ and $n(V) = 5$. The number of elements in the set of all letters in the alphabet is

$n(C) + n(V) = 21 + 5 = n(C \cup V)$.

$C \cup V = \{a, b, c, d, e, f, g, h, i, j, k, l, m, n, o, p, q, r, s, t, u, v, w, x, y, z\}$.

$n(C \cup V) = 26$, found by counting.

5.5 SOME PROPERTIES OF
WHOLE NUMBERS UNDER ADDITION

We have established previously that $n(A) = n(B)$ if and only if A is equivalent to B. We know that $A \cup B = B \cup A$ from Chapter 2. This implies that the set $A \cup B$ is equivalent to the set $B \cup A$ and thus that $n(A \cup B) = n(B \cup A)$. Further, since $n(A \cup B) = n(A) + n(B)$ and $n(B \cup A) = n(B) + n(A)$ by definition of whole numbers, it follows that

$$n(A) + n(B) = n(B) + n(A)$$

that is, that **addition of whole numbers is commutative.**

When a sum of three numbers is desired, the numbers must be added two at a time since addition is a binary operation. We can show that the way we choose

---

† At this point it would be well for the student to "forget" what $4 + 5$ equals. We need an element of naïve curiosity to appreciate the link between addition and counting.

to pair the numbers as we add them makes no difference in the result, that is, **addition of whole numbers is associative:**

$$[n(A) + n(B)] + n(C) = n(A) + [n(B) + n(C)]$$

We know from Section 2.2 that union of sets is associative: $(A \cup B) \cup C = A \cup (B \cup C)$. It follows that $n[(A \cup B) \cup C] = n[A \cup (B \cup C)]$. Then, by definition of addition of whole numbers, $n(A \cup B) + n(C) = n(A) + n(B \cup C)$. By the same definition, $[n(A) + n(B)] + n(C) = n(A) + [n(B) + n(C)]$.

Besides the commutative and associative properties of addition, we find that there is also an **identity element for addition of whole numbers**. Since $A \cup \varnothing = A$, $n(A \cup \varnothing) = n(A)$, or $n(A) + n(\varnothing) = n(A)$. Therefore, $n(\varnothing)$ is the identity element for addition of whole numbers.

A fourth property of whole numbers under addition is that of *closure*. We recall that a set is closed under an operation if performing the operation with any two elements of the set produces a result that is an element of the set. Translating this to **closure of the set of whole numbers under addition**, we say: The set of whole numbers is closed under addition because adding any two whole numbers always produces a whole number, that is $n(A) + n(B)$ is a whole number. This can be justified on the basis that for any given universal set, the system of sets is closed under union (see Section 2.2). By the definition of addition of whole numbers if A and B are disjoint sets,

$$n(A) + n(B) = n(A \cup B)$$

Since $A \cup B$ is a set, then $n(A \cup B)$ is the number of elements in that set, a *whole number*.

In Chapter 2 the commutative and associative properties of union, the existence of an identity element for union, and the closure property of sets under union were assumed. In this section, however, we have used these assumptions about sets and have proved the commutative and the associative properties of addition for whole numbers, the existence of an identity element for addition, and closure of whole numbers under addition. Thus, these properties of whole numbers are theorems.

EXERCISES 5.5

1. Is the set of even whole numbers closed under addition? Is the set of odd whole numbers closed under addition? Is the set $\{0, 1\}$ closed under addition? Is the set $\{0, 1, 2\}$ closed under addition?

2. Justify each of the following statements by stating one or more definitions, assumptions, or theorems.

   (a) $5 + 6 = 6 + 5$
   (b) $X \cup Y = Y \cup X$

(c) $a + b = b + a$ (*a* and *b* both whole numbers)

(d) $n(X \cup Y) = n(Y \cup X)$

(e) $(5 + 6) + 4 = 5 + (6 + 4)$

(f) $(5 + 6) + 4 = 4 + (5 + 6)$

(g) $X \cup (Y \cup Z) = (X \cup Y) \cup Z$

(h) $p + (q + r) = (p + q) + r$ (*p*, *q*, and *r* all whole numbers)

(i) $A \cup B$ is a set

(j) $7 + 6$ is a whole number

(k) $n(X) + n(Y)$ is a whole number

(l) $n\{a, b, c, d\} + n\{x, y, z\} = n(\{a, b, c, d\} \cup \{x, y, z\})$

(m) $7 + 0 = 7$

(n) $n(X) + n(\emptyset) = n(X)$

(o) $n(X) + n(\emptyset) = n(X \cup \emptyset)$

(p) $[n(U) + n(V)] + n(W) = n(U) + [n(V) + n(W)]$

(q) $[n(U) + n(V)] + n(W) = n(W) + [n(U) + n(V)]$

(r) $X \cup \emptyset = X$

(s) $X \cup \emptyset = \emptyset \cup X$

(t) $X \cup \emptyset$ is a set

3. Verify the following by computing the value of each member of the equation and state the property of whole numbers that justifies each statement (operations enclosed in parentheses must be done first).

(a) $20 + 7 = 7 + 20$

(b) $(13 + 5) + 6 = 13 + (5 + 6)$

(c) $(13 + 5) + 6 = 6 + (13 + 5)$

(d) $13 + 5 + 0 = 13 + 5$

5.6 MULTIPLICATION OF WHOLE NUMBERS

Before we define multiplication of whole numbers, we shall introduce two alternate notations for the product of two whole numbers and we shall define the term *factor*. Let *p* and *q* represent two whole numbers. Their product $p \times q$ may be written either of two other ways:

$$p \times q = p \cdot q = pq$$

In this chapter there is some danger that confusion will result if we use the sign $\times$ for both multiplication of numbers and Cartesian multiplication. Therefore, we shall use $p \cdot q$ or pq to indicate the product of two whole numbers *p* and *q*. Further, if the product of *p* and *q* is a whole number *n*, that is, if $p \cdot q = n$, then *p* is a **factor** of *n* and *q* is a **factor** of *n*.

Multiplication of whole numbers can be defined in terms of addition of whole numbers. This means that multiplication is indirectly based on counting. If *p* and *q* are whole numbers, then

$$p \cdot q = q + q + q + \ldots + q$$

where the first factor p indicates how many times the second factor q is to be used as an addend in the sum. This definition is effective for all products of two whole numbers except for the case where the first factor in the product is zero. If q is a whole number, we define the product of zero and q to be zero, that is,

$$0 \cdot q = 0$$

Using the language of sets to define **multiplication of whole numbers more** formally, *we link multiplication to counting more directly*. We give as a second definition

$$n(A) \cdot n(B) = n(A \times B)$$

where $A \times B$ is the Cartesian product of the two sets A and B. (The student may need to refer to the discussion of *Cartesian product* in Section 1.4.) Since this defin- ition establishes the product of two whole numbers as the number of elements in a set, we can *count* the elements in the set to find the product. For example, we choose a set having five elements $\{1, 2, 3, 4, 5\}$ and a set having three elements $\{1, 2, 3\}$. We arrange the Cartesian product of these sets in tabular form (Fig. 5.1). Now we may count the elements in the table to find that the product $5 \cdot 3 = 15$. Let us use this example to relate the definition of multiplication of whole numbers given in this paragraph to the definition given in the preceding paragraph. The reader will observe that there are three ordered pairs in each row and that there are five rows in the table. Therefore, the product $5 \cdot 3 = 3 + 3 + 3 + 3 + 3 = 15$.

$\times$	1	2	3	
1	(1, 1)	(1, 2)	(1, 3)	3
				+
2	(2, 1)	(2, 2)	(2, 3)	3
				+
3	(3, 1)	(3, 2)	(3, 3)	3
				+
4	(4, 1)	(4, 2)	(4, 3)	3
				+
5	(5, 1)	(5, 2)	(5, 3)	3
				—
				15

Figure 5.1

5.7   SOME PROPERTIES OF WHOLE NUMBERS UNDER MULTIPLICATION

If we arrange the elements of the Cartesian product $\{1, 2, 3\} \times \{1, 2, 3, 4, 5\}$ in tabular form, comparison with the table in the previous section will show that the two sets are *equivalent* (though not equal), that is, $n(\{1, 2, 3, 4, 5\} \times \{1, 2, 3\}) = n(\{1, 2, 3\} \times \{1, 2, 3, 4, 5\})$.

×	1	2	3	4	5	
1	(1, 1)	(1, 2)	(1, 3)	(1, 4)	(1, 5)	5 +
2	(2, 1)	(2, 2)	(2, 3)	(2, 4)	(2, 5)	5 +
3	(3, 1)	(3, 2)	(3, 3)	(3, 4)	(3, 5)	6
						15

Figure 5.2

The table in Figure 5.2 has three rows each containing five ordered pairs; therefore $n(\{1, 2, 3\} \times \{1, 2, 3, 4, 5\}) = 5 + 5 + 5 = 15$, the same number that we found using Figure 5.1 for $n(\{1, 2, 3, 4, 5\} \times \{1, 2, 3\})$.

Next, we shall consider the product $p \cdot q$, with p and q each representing any whole number. Also, we shall let $p = n(A)$ and $q = n(B)$, where

$$A = \{1, 2, 3, \ldots, p\}$$

and

$$B = \{1, 2, 3, \ldots, q\}$$

The elements of $A \times B$ may be represented in tabular form as in Figure 5.3.

×	1	2	3	$\cdots$	q	
1	(1, 1)	(1, 2)	(1, 3)	$\cdots$	(1, q)	q +
2	(2, 1)	(2, 2)	(2, 3)	$\cdots$	(2, q)	q +
3	(3, 1)	(3, 2)	(3, 3)	$\cdots$	(3, q)	q +
$\vdots$	$\vdots$	$\vdots$	$\vdots$	$\vdots$	$\vdots$	$\vdots$ +
p	(p,1)	(p, 2)	(p, 3)	$\cdots$	(p, q)	q
	p +	p +	p	$+ \cdots +$	p	

Figure 5.3

Notice that there are p rows each containing q elements in this table. Also, there are q columns each containing p elements. Therefore we can express the numbers of elements in the table as either $p \cdot q$ or $q \cdot p$.

×	1	2	3	$\cdots$	p	
1	(1, 1)	(1, 2)	(1, 3)	$\cdots$	(1, p)	p
						+
2	(2, 1)	(2, 2)	(2, 3)	$\cdots$	(2, p)	p
						+
3	(3, 1)	(3, 2)	(3, 3)	$\cdots$	(3, p)	p
						+
	$\vdots$	$\vdots$	$\vdots$	$\vdots$	$\vdots$	+
q	(q, 1)	(q, 2)	(q, 3)	$\cdots$	(q, p)	p

$$q \; + \; q \; + \; q \; + \cdots + \; q$$

Figure 5.4

The elements of $B \times A$ may be represented in tabular form as in Figure 5.4. Here we have a table with q rows each containing p elements; also, there are p columns each containing q elements. Therefore we can express the number of elements in the table as either $q \cdot p$ or $p \cdot q$.

We see from Figures 5.3 and 5.4 that $n(A \times B) = p \cdot q = q \cdot p = n(B \times A)$. By definition of multiplication of whole numbers in terms of Cartesian multiplication of sets, it follows that

$$n(A) \cdot n(B) = n(B) \cdot n(A)$$

This provides a proof that **multiplication of whole numbers is commutative**.

When a product of three whole numbers is desired, the numbers must be multiplied two at a time because multiplication is a binary operation. The way we choose to pair numbers as we multiply them makes no difference in the product, that is, **multiplication of whole numbers is associative**.

$$[n(A) \cdot n(B)] \cdot n(C) = n(A) \cdot [n(B) \cdot n(C)]$$

Verification of this property is considered in Exercises 5 through 8.

The number 1 has a special property under multiplication. If $n(A) = p$ and $n(B) = 1$, then

$$n(A) \cdot n(B) = n(A), \quad \text{that is, } p \cdot 1 = p$$

Thus **1 is the identity element for multiplication of whole numbers**. The proof of this property is based upon the fact that $n(A \times B) = n(A)$ if B contains only one element (Exercise 3).

The set of whole numbers also has the following property: If we multiply two whole numbers, we get a whole number as the product. In other words, if p and q

are whole numbers, then

$$p \cdot q \text{ is a whole number}$$

We say that the set of whole numbers is **closed under multiplication**. This property is an immediate consequence of the following fact: If $n(A)$ and $n(B)$ are whole numbers, then $n(A \times B)$ is a whole number, since $A \times B$ is a non-infinite set and the number of elements in a non-infinite set is always a whole number.

The number zero has unique properties under multiplication. If the number of elements in a non-infinite set A is p, then $p = n(A)$ is a whole number and

$$p \cdot 0 = 0, \quad \text{that is, } n(A) \cdot n(\varnothing) = n(\varnothing)$$

The following proof of this property is based upon the set definition of multiplication for whole numbers:

$$n(A) \cdot n(\varnothing) = n(A \times \varnothing)$$

Since $A \times \varnothing = \varnothing$, $n(A \times \varnothing) = n(\varnothing)$. A second property of zero may be stated as follows: If p and q are whole numbers and $p \cdot q = 0$, then p or q or both p and q equals 0.

EXERCISES 5.7

1. Verify each statement by computing both members of the equation. Then state the property of whole numbers that justifies each statement.

   (a) $16 \cdot 3 = 3 \cdot 16$
   (b) $(4 \cdot 6) \cdot 5 = 4 \cdot (6 \cdot 5)$
   (c) $(4 \cdot 6) \cdot 1 = 4 \cdot 6$
   (d) $(16 + 3) \cdot 4 = 4 \cdot (16 + 3)$

2. Justify each statement by stating a definition, an assumption, or a theorem. (Recall that all of the properties of whole numbers considered thus far can be *proved*.)

   (a) $4 \cdot 6 = 6 + 6 + 6 + 6$
   (b) $4 \cdot 6$ is a whole number
   (c) $4 \cdot 6 = 6 \cdot 4$
   (d) $n\{a, b\} \times n\{c, d, e\} = n\{(a, c), (a, d), (a, e), (b, c), (b, d), b, e)\}$
   (e) $n(X) \cdot n(Y) = n(Y) \cdot n(X)$
   (f) $(4 \cdot 6) \cdot 3 = 4 \cdot (6 \cdot 3)$
   (g) $(p \cdot q) \cdot r = r \cdot (p \cdot q)$ ($p, q,$ and r all whole numbers)
   (h) $[n(A) \cdot n(B)] \cdot n(C) = n(A) \cdot [n(B) \cdot n(C)]$
   (i) $n\{a, b, c, d\} \cdot n\{x\} = n\{a, b, c, d\}$
   (j) $n\{a, b, c, d\} \cdot n\{x\} = n\{(a, x), (b, x), (c, x), (d, x)\}$

(k) $16 \cdot 1 = 16$

(l) If $16 \cdot a = 0$, then $a = 0$

(m) $p \cdot 1 = 1 \cdot p$ (p a whole number)

(n) $p \cdot 1 = p$ (p a whole number)

(o) $p \cdot 0 = 0$ (p a whole number)

3. Use $n(A) = p$ and $n(B) = 1$ to prove that 1 is the identity element for multiplication of whole numbers.

4. Translate the statement of this property of zero into set language: If p and q are whole numbers and $p \cdot q = 0$, then p or q or both p and $q = 0$.

5. Let $A = \{1, 2, 3\}$, $B = \{a, b, c, d\}$, and $C = \{r, s\}$. Construct tables for (a) $A \times B$; (b) $(A \times B) \times C$; (c) $B \times C$; (d) $A \times (B \times C)$.

6. Use the tables constructed in Exercise 5 and explain why $(3 \cdot 4) \cdot 2 = 3 \cdot (4 \cdot 2)$.

7. As in Exercises 5 and 6, prove that $5 \cdot (2 \cdot 3) = (5 \cdot 2) \cdot 3$.

8. Let $A = \{1, 2, 3, \ldots, p\}$, $B = \{1, 2, 3, \ldots, q\}$, and $C = \{1, 2, 3, \ldots, r\}$. Prove as in Exercises 5 through 7 that $(p \cdot q) \cdot r = p \cdot (q \cdot r)$.

5.8   THE DISTRIBUTIVE PROPERTY

It can be shown that for sets the Cartesian product is distributive over union, that is, for all sets A, B, and C,

$$A \times (B \cup C) = (A \times B) \cup (A \times C)$$

We shall not attempt a general proof of this property but will verify that it is true for some specific sets. If we let $A = \{a, b\}$, $B = \{c, d, e\}$ and $C = \{f, g, h\}$, then

$$\begin{aligned} A \times (B \cup C) &= \{a, b\} \times \{c, d, e, f, g, h\} \\ &= \{(a, c), (a, d), (a, e), (a, f), (a, g), (a, h), (b, c), (b, d), \\ &\quad (b, e)\,(b, f), (b, g), (b, h)\} \end{aligned}$$

and

$$\begin{aligned} (A \times B) \cup (A \times C) &= \{(a, c), (a, d), (a, e), (b, c), (b, d), (b, e), \\ &\quad (a, f), (a, g), (a, h), (b, f), (b, g), (b, h)\} \end{aligned}$$

Inspection will show that the sets in the right members of these equations are identical. Then, since $A \times (B \cup C) = (A \times B) \cup (A \times C)$, we have $n[A \times (B \cup C)]$ $= n[(A \times B) \cup (A \times C)]$. Furthermore, by the definitions of addition and multiplication of whole numbers, $n[A \times (B \cup C)] = n(A) \cdot [n(B) + n(C)]$ and

$n[(A \times B) \cup (A \times C)] = [n(A) \cdot n(B)] + [n(A) \cdot n(C)]$. Thus

$$n(A) \cdot [n(B) + n(C)] = [n(A) \cdot n(B)] + [n(A) \cdot n(C)]$$

This is evidence that **multiplication is distributive over addition for whole numbers.**
Using p, q, and r as symbols for whole numbers, this property becomes

$$p \cdot (q + r) = p \cdot q + p \cdot r$$

EXERCISES 5.8

1. Verify each of these applications of the distributive property of multiplication
   over addition by computing the value of each member of the equation:

   (a) $6 \cdot (5 + 3) = 6 \cdot 5 + 6 \cdot 3$
   (b) $16 \cdot 4 + 16 \cdot 7 = 16 \cdot (4 + 7)$
   (c) $(16 + 6) \cdot 7 = 16 \cdot 7 + 6 \cdot 7$
   (d) $6 \cdot (5 + 3 + 9 + 4) = 6 \cdot 5 + 6 \cdot 3 + 6 \cdot 9 + 6 \cdot 4$

2. The distributive property may be used to simplify computations as in the example

   $$6 \cdot 52 = 6(50 + 2) = 6 \cdot 50 + 6 \cdot 2 = 300 + 12 = 312$$

   Use the distributive property in this way to compute each of these products.

   (a) $5 \cdot 76$ (b) $92 \cdot 8$
   (c) $12 \cdot 302$ (d) $17 \cdot 476$
   (e) $20 \cdot 562$ (f) $34 \cdot 3728$

3. Use sets $A = \{a, b\}$, $B = \{c, d, e\}$ and $C = \{f, g, h\}$ to show that $A \cup (B \times C)$
   $\neq (A \cup B) \times (A \cup C)$, thus proving that union is not distributive over Cartesian
   multiplication. Explain how this leads to the conclusion that addition of whole
   numbers is not distributive over multiplication.

5.9   SUBTRACTION AND DIVISION
OF WHOLE NUMBERS

Let us now consider the operation of **subtraction** of whole numbers *in terms
of addition.* If there exists a whole number x such that

$$q + x = p$$

then the result of subtracting q from p is x, that is,

$$p - q = x$$

We call $p - q$ a **difference** of whole numbers p and q. For example, since $4 + 3 = 7$, we define the solution of $4 + x = 7$ to be the *difference* $7 - 4$ where $7 - 4 = 3$.

Division of whole numbers is also defined in terms of another fundamental operation: *multiplication*. If there exists a whole number x such that

$$q \cdot x = p$$

then the result of dividing p by q is x, that is,

$$p \div q = x$$

We call $p \div q$ $\left(\text{also written } p/q \text{ or } \dfrac{p}{q}\right)$ a **quotient** of whole numbers p and q. For example, since $2 \cdot 4 = 8$, we define the solution of $2 \cdot x = 8$ to be the quotient of $\frac{8}{2}$ where $\frac{8}{2} = 4$.

We have found the number zero to be unique in its behavior under addition and multiplication; it also behaves uniquely under subtraction and division. Because of these maverick qualities, zero has long been confusing to the average student. Once its unusual qualities, or properties, have been made explicit and explained, this confusion can be eliminated. In subtraction zero behaves somewhat like an identity element, that is, *for any whole number a, $a - 0 = a$*.† This follows directly from the definition of subtraction: Since $0 + a = a$, then $a - 0 = a$. For whole numbers $a \neq 0$, we cannot perform the subtraction $0 - a$, because there is no whole number x such that $a + x = 0$.

The behavior of 0 is especially interesting in division. If we choose any whole number a, other than zero itself, it follows directly from our definition of division that $0/a = 0$ (see Exercise 6). Suppose we try to divide a number $a \neq 0$ by zero. We know that if $0 \cdot x = a$, then $a/0 = x$. But we have a special property of zero under multiplication that says $0 \cdot x = 0$. Thus if we try to divide a whole number $a \neq 0$ by 0, we find that a must equal 0. Because of this contradiction, division by zero is left undefined.

We have not considered the possibility of dividing zero itself by zero. Before we examine this possibility, we shall give attention to a certain characteristic of the fundamental operations. Some reflection upon your experience with whole numbers under the four operations of addition, subtraction, multiplication, and division will convince you that when any of these operations can be performed upon two whole numbers, at most one answer is obtained, that is, there are never two or more possible results. Mathematicians have decided that this characteristic of these operations must be preserved. Now we are ready to take a look at $\frac{0}{0}$. By definition of division, $\frac{0}{0} = x$ where x is a number such that $0 \cdot x = 0$. It is

---

† Zero is not a true identity element for subtraction, because the operation is not commutative; commutativity is a stipulation in the general definition of an identity element for any operation.

obvious that any whole number value of x will satisfy this equation. Therefore, the result of $\frac{0}{0}$ is not unique. For this reason, division of zero by zero is excluded in mathematics; we say it, also, is undefined. Thus, for the entire set of whole numbers, **division by zero is undefined**.

EXERCISES 5.9

1. In our definition of subtraction of whole numbers, why was it necessary to qualify the statement by saying "If there exists a whole number x such that ... "? Is the set of whole numbers closed under subtraction?

2. In our definition of division of whole numbers, why was it necessary to qualify the statement by saying "If there exists a whole number x such that ..."? Is the set of whole numbers closed under division?

3. Is subtraction of whole numbers commutative? Is division commutative? Explain your answers.

4. Choose three different whole numbers, and test them for associativity under subtraction. Is subtraction of whole numbers associative?

5. Choose three different whole numbers, and test them for associativity under division. Is division of whole numbers associative?

6. Prove that for any whole number $a \neq 0$, we have $0/a = 0$.

5.10   OPERATIONS AND
THE NUMBER LINE

The whole numbers can be represented geometrically as points on a line, customarily called the **number line.** We choose a point on a line to represent zero; this point is called the **origin** of the number line. Then we select a unit of length and mark off points to the right of the origin as in Figure 5.5. Each of these points is associated with a whole number that is called the **coordinate** of the point; the

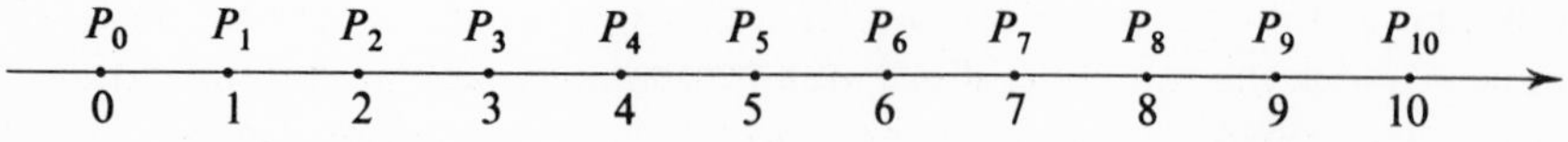

$P_0 = 0$	$P_4 = 4$	$P_7 = 7$
$P_1 = 1$	$P_5 = 5$	$P_8 = 8$
$P_2 = 2$	$P_6 = 6$	$P_9 = 9$
$P_3 = 3$		$P_{10} = 10$

Figure 5.5

point is the **graph** of the whole number. Since a line extends indefinitely, we can represent as many whole numbers as we wish. In particular, for any whole number n, we may identify the points whose coordinates are $0, 1, 2, 3, \ldots, n$ by the symbols $P_0, P_1, P_2, P_3, \ldots, P_n$ (read P sub zero, P sub one, etc.)

The number line can also be used as a geometric model to interpret the fundamental operations. We shall indicate how addition and subtraction can be illustrated on the number line.

1. **Addition of whole numbers.** In our definition of whole numbers (Section 5.4) we have shown that addition is related to the process of counting. This relationship to the counting process is obvious when we use the number line to illustrate addition. For example, suppose we desire to use the number line to find the sum $5 + 4$. We begin at the origin and *count* the points that are the endpoints of the first five units: P_1, P_2, P_3, P_4, P_5. In Figure 5.6 this counting is indicated by the looped arrow that extends from P_0 to P_5. To indicate the addition of 4, we begin at P_5 and *count* the next four points; P_6, P_7, P_8, P_9. In Figure 5.6 this counting is indicated by the looped arrow that extends four units to the right of P_5, that is the arrow that extends over P_6, P_7, P_8, P_9. The result of the addition may be represented by the coordinate 9 of the point P_9 where the second arrow terminates.

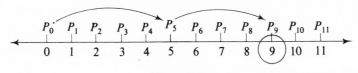

Figure 5.6

In general, if we are to add whole numbers $a + b$, we first count points $P_1, P_2, P_3, \ldots P_a$ on the number line; then we count the next b points $P_{a+1}, P_{a+2}, P_{a+3}, \ldots, P_{a+b}$. The sum may be represented by the coordinate $a + b$ of the terminal point P_{a+b} in this counting process.

2. **Subtraction of whole numbers.** In our definition of subtraction of whole numbers (Section 5.9), subtraction was defined in terms of addition. This implies that subtraction is related to the counting process, and we do indeed use counting when we use the number line to illustrate subtraction. For example, let us use the number line to show the difference $9 - 5$. We begin at the origin and *count* the points that are the end-points of the first nine units: $P_1, P_2, P_3, \ldots, P_9$. In Figure 5.7 this counting is indicated by the looped arrow that extends from P_0 to P_9. To indicate the subtraction of 5, we begin at P_9 and *count* five points lying immediately to the left of P_9: P_8, P_7, P_6, P_5, P_4. In Figure 5.7 this counting is indicated by the looped arrow extending five units to the left of P_9. The result of subtracting $9 - 5$ may be represented by the coordinate 4 of the point P_4 where the second arrow terminates.

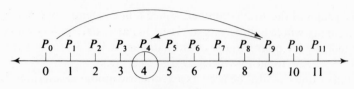

Figure 5.7

In general, if we desire to subtract whole numbers $a - b$, we first count $P_1, P_2,$ $P_3, \ldots, P_a$. Then we count b points to the left of P_a: $P_{a-1}, P_{a-2}, P_{a-3}, \ldots, P_{a-b}$. The difference may be represented by the coordinate $a - b$ of the terminal point P_{a-b} in this counting process.

EXERCISES 5.10

1. Use a number line and illustrate the following sums:

(a) $3 + 3$	(b) $4 + 3$	(c) $3 + 4$
(d) $4 + 1 + 2$	(e) $6 + 0$	(f) $2 + 4 + 0$
(g) $(4 + 3) + 5$	(h) $4 + (3 + 5)$	(i) $(5 + 3) + (4 + 2)$
(j) $5 + (3 + 4) + 2$	(k) $[(5 + 3) + 4] + 2$	(l) $5 + [3 + (4 + 2)]$

2. Use a number line and illustrate the following subtractions:

(a) $6 - 3$	(b) $6 - 5$	(c) $6 - 6$
(d) $6 - 0$	(e) $(8 - 5) - 3$	(f) $8 - (5 - 3)$

5.11   THE TRICHOTOMY PROPERTY

We read the statements

$\quad\quad a < b$ as a is less than b

$\quad\quad a > b$ as a is greater than b

$\quad\quad a \not< b$ as a is not less than b; that is,
$\quad\quad\quad\quad a$ is greater than or equal to $b, a \geq b$

$\quad\quad a \not> b$ as a is not greater than b; that is,
$\quad\quad\quad\quad a$ is less than or equal to $b, a \leq b$

An assumption we make about order of whole numbers is that, given two whole numbers a and b, a is either less than, equal to, or greater than b. This assumption is called the **trichotomy property of whole numbers**, and indicates that exactly one of the statements

$$a < b \quad\quad a = b \quad\quad a > b$$

must hold. In concise symbolic form this property may be expressed as

$$a \gtreqless b$$

We now define each of the relations involved in the trichotomy property of whole numbers. On the number line where P_1 is on the right of P_0,

$$a < b \quad \text{iff } P_a \text{ is on the left of } P_b$$
$$a = b \quad \text{iff } P_a \text{ is at } P_b$$
$$a > b \quad \text{iff } P_a \text{ is on the right of } P_b$$

For example, $6 < 9$ since P_6 is on the left of P_9 ; $7 - 5 = 2$ since $P_{(7-5)}$ is at P_2 ; $9 > 6$ since P_9 is on the right of P_6 (Fig. 5.8).

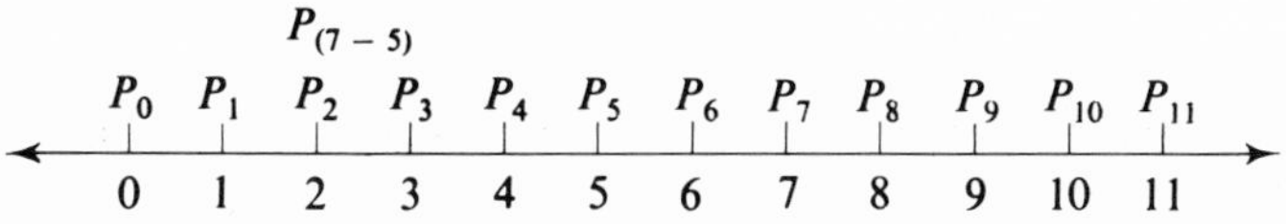

Figure 5.8

The relations in the trichotomy property may be defined more formally for whole numbers a and b, without the use of the number line:

$$a < b \quad \text{iff there exists a natural number } x \text{ such that } a + x = b$$
$$a > b \quad \text{iff there exists a natural number } x \text{ such that } b + x = a$$
$$a = b \quad \text{iff neither of these alternatives is true}$$

EXERCISES 5.11

1. Replace the asterisk with the symbol $>$, $<$ or $=$ that will make the statement true. (The symbols $a, b,$ and c represent distinct natural numbers.)

   (a) $5 * 2$
   (b) $2 * 5$
   (c) $5 - 3 * 2$
   (d) $8/4 * 8$
   (e) $8/4 * 2$
   (f) $6 * 3 \cdot 4$
   (g) $a + b * a$
   (h) $b * a \cdot b$
   (i) $b/a * b$ (assuming b/a is a natural number)
   (j) If $a - b \not> c$ and $a - b \neq c$, then $a - b * c$
   (k) If $a - b \not< c$, then $a - b * c$ or $a - b * c$
   (l) If $a - b \neq c$, then $a - b * c$ or $a - b * c$

2. Rewrite the formal definition of the three statements in the trichotomy property for whole numbers using set language to represent numbers.

5.12 RELATIONS AND THEIR PROPERTIES

Up to now the only relations we have considered are relations between numbers, that is, equality ($=$), is greater than ($>$) and is less than ($<$), and relations between sets, for example, is a subset of ($\subseteq$), is a proper subset of ($\subset$), is equal to ($=$), etc. Relations may exist between objects, ideas, figures, etc. Some of the other relations we use are

is older than

is younger than

is the same age as

is a neighbor of

is next to

Relations have properties that characterize them, just as numbers, sets, and propositions do. Equality is called an **equivalence relation**. Any relation that has the following three properties is called an **equivalence relation**. We shall use the symbol "$\circ$" to represent a relation in general.

1. **The reflexive property**. $a \circ a$. If an object (in the universe of discourse appropriate to the relation and the objects) always has the relation to itself, then the relation is said to be reflexive.

2. **The symmetric property**. If $a \circ b$, then $b \circ a$. If an object a has the relation to an object b, then the object b always has the relation to the object a. When this is true, the relation is said to be symmetric.

3. **The transitive property**. If $a \circ b$ and $b \circ c$, then $a \circ c$. If an object a has the relation to an object b, and the object b has the relation to an object c, then the object a has the relation to the object c. When this is true the relation is said to be transitive.

Equality of whole numbers certainly has all three of these properties: (1) Any whole number is equal to itself, for example, $5 = 5$. (2) If a whole number a equals another whole number b, then b is equal to a, for example, if $10 - 8 = 2$, then $2 = 10 - 8$. (3) If a whole number a is equal to a second number b, and if b is equal to a third number c, then a is equal to c; for example, if $10 - 8 = 2$ and $2 = \frac{6}{3}$, then $10 - 8 = \frac{6}{3}$. Therefore, equality of whole numbers is an equivalence relation. Some other equivalence relations are

is the same age as

is congruent to (in Euclidean geometry)

weighs the same as

is similar to (in Euclidean geometry)

is equivalent to (in Algebra of Sets)

Unlike any of these, equality of whole numbers is involved with *two names for the same thing*. If p and q represent whole numbers, $p = q$ means "p is another name for the number that q represents."

Not all relations are equivalence relations. The relation "is greater than," for example, is not reflexive; no number is greater than itself. Nor is it symmetric; $9 > 5$, but it is not true that $5 > 9$. The relation $>$ is transitive, however; since $9 > 5$ and $5 > 2$, $9 > 2$. The relation $>$ has one of the equivalence properties. Some relations have none.

Equality of whole numbers has two more properties that are important for us to consider. They are

1. **The addition property of equality.** If $a = b$, then

$$a + x = b + x$$

Adding the same number to both members of an equation preserves the equality.

2. **The multiplication property of equality.** If $a = b$, then

$$a \cdot x = b \cdot x$$

Multiplying both members of an equality by the same number preserves the equality.

EXERCISES 5.12

1. Tell whether each of the following relations is reflexive, symmetric, or transitive:

    (a) is the same color as
    (b) lives within a mile of
    (c) is taller than
    (d) is a subset of
    (e) is a proper subset of
    (f) implies
    (g) is married to
    (h) is a descendant of
    (i) is the sister of
    (j) is the son of
    (k) is as old as
    (l) is warmer than

2. (a) What restriction must be placed on x in order to state a subtraction property of equality of whole numbers? That is, if $a = b$, $a - x = b - x$ under what condition?
    (b) What restriction must be placed on x in order to state a division property of equality of whole numbers? That is, if $a = b$, $a/x = b/x$ under what condition?

3. Justify each of the following by stating a definition, assumption, or theorem:

   (a) $7 \cdot 6 = 6 \cdot 7$
   (b) $3 \cdot (6 + 2) = 3 \cdot 6 + 3 \cdot 2$
   (c) If $7 \cdot x = 0$, then $x = 0$ (x a whole number)
   (d) $7 > 3$
   (e) $7 \cdot 0 = 0$
   (f) If 7 is not equal to 3, and 7 is not less than 3, then $7 > 3$
   (g) $(3 \cdot 6) \cdot 2 = 3 \cdot (6 \cdot 2)$
   (h) $1 \cdot 9 = 9$
   (i) $9 + 0 = 9$
   (j) If $a = 9$, then $9 = a$
   (k) If m and n are whole numbers, $m + n$ is a whole number
   (l) If m and n are whole numbers, $m \cdot n$ is a whole number
   (m) $7 \cdot (6 + 4) = 7 \cdot 6 + 7 \cdot 4$
   (n) If $x = y$, then $5x = 5y$ (x and y both whole numbers)
   (o) $(7 + 6) + 3 = 3 + (7 + 6)$
   (p) If $p = q$, then $p + 7 = q + 7$ (p and q both whole numbers)
   (q) If $7 - 3 = 4$ and $4 = 3 + 1$, then $7 - 3 = 3 + 1$
   (r) $5 = 5$
   (s) $(7 + 6) + 3 = 7 + (6 + 3)$

4. Make a list of all the properties of whole numbers that we have presented in this chapter (there are twelve), and also the properties of equality (there are five). Give specific examples of each, after stating the property.

CHAPTER 6

The Integers

6.1 MEMBERSHIP OF THE SET OF INTEGERS

When we considered points on the number line as representations of the whole numbers, we used some points on that part of the line on the right of the point selected as the origin. Using the same unit length that we used to mark points at equal intervals on the right of the origin, we can mark points on the left of the origin. These points on the left are counterparts of those on the right with respect to their distances from the origin, that is, the first point on the left of the origin is the same distance from the origin as the first point on the right, the second point on the left is the same distance from the origin as the second point on the right, etc. When two points are the same distance from the origin, we say that their coordinates have the same **absolute value**. The aspect of these points on the left of the origin that makes them different from those on the right, then, is not distance from the origin. They are different because they lie in the opposite direction on the number line from the points we use to represent the counting numbers. We need a new set of numbers as coordinates of the points on the left of the origin. To emphasize that we are in a new, enlarged, number system, we use new names (**positive integers**) for the counting numbers that are coordinates of the points on the right of the origin. We name coordinates of the points on the left of the origin **negative integers**. The union of the set consisting of the number 0 (the coordinate of the origin), the set of positive integers, and the set of negative integers is the set of **integers**.

The applications of the system of integers are numerous. It is practical to use the integers (positive and negative) whenever a counting of units involves opposites. Such opposites might be: above and below, right and left, north and south, east and west, debits and credits, etc.

6.2 NUMERALS FOR
THE NEGATIVE INTEGERS

The numerals used to represent the negative integers are similar in appearance to those used for the positive integers; only a small horizontal line differentiates the numeral for a negative integer from the numeral for a positive integer; for example, $^-1$, $^-2$, $^-3$ are numerals for negative integers. (Often the line is written at a lower level as: -1, -2, -3; we shall write it in the higher position to avoid confusing it with the "minus sign," used to indicate the operation of subtraction.) The negative numerals $^-1$, $^-2$, $^-3$ are read "negative one," "negative two," "negative three." Occasionally we shall use the sign " $+$ " to indicate a positive integer, for example, $^+3$. It is understood, however, that when a numeral for an integer is written without a sign, the integer is positive.

The number line with coordinates for its **integral points** appears as in Figure 6.1.

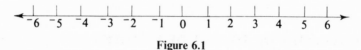

Figure 6.1

Since the **unit point** P_1 has the positive integer 1 as its coordinate and is on the right of the origin, we call the direction from left to right on this line the **positive direction**; we call the direction from right to left the **negative direction**.

Recall that when points are the same distance from the origin, their coordinates are said to have the same **absolute value** (Section 6.1). Thus $^-5$ has the same absolute value as $^+5$, $^+4$ has the same absolute value as $^-4$. In general, if n is an integer, n and ^-n have the same absolute value. Since we have considered an integer as a coordinate of a point on the number line, we can consider the absolute value of that integer as its distance from the origin without concern for its direction from the origin. To indicate the absolute value of an integer n we use the symbol $|n|$. The absolute value of $^-5$, then, is indicated as $|^-5|$, the absolute value of $^-4$ is indicated as $|^-4|$, and the absolute value of $^+4$ is indicated as $|^+4|$. Evaluating these symbols, we have

$$|^-5| = 5$$

$$|^+5| = 5$$

$$|^-4| = 4$$

$$|^+4| = 4$$

We assume a **trichotomy property of integers** that is analogous to the trichotomy property of whole numbers. For any integers a and b, a is less than, equal to, or greater than b:

$$a \lesseqqgtr b$$

The relations involved in the trichotomy property of integers are defined so that the order of the positive integers and zero is exactly the same as the order of the corresponding whole numbers. Thus for all points on the number line with integers as coordinates, if P_a with coordinate a is on the left of P_b with coordinate b, then $a < b$. If P_a is at P_b (two names for the same point), then $a = b$. If P_a is on the right of P_b, then $a > b$. We may observe in Figure 6.1, then, that $^-5 < 2$, $^-2 < 2$, $^-3 < 0$, $4 > {}^-5$, $5 > {}^-5$, etc.

The relations involved in the trichotomy property of integers may be defined without reference to the number line by saying

$a < b$ iff there exists a positive integer x such that $a + x = b$

$a > b$ iff there exists a positive integer x such that $b + x = a$

$a = b$ iff neither of these alternatives is true

We next consider the definition of addition for integers. Referring to the structure of a logical system, we recall that new words or expressions are always defined in terms of undefined words or previously defined words or expressions. In our number system the terms *number* and *one-to-one correspondence* are undefined; these are terms that we can understand readily on the basis of our common experience with them. Using these terms along with some terms from algebra of sets, we have defined *counting, counting numbers, zero, whole numbers,* and the fundamental operations of *addition, subtraction, multiplication,* and *division for whole numbers.*

We continue this "chain" of definitions in this chapter and define the fundamental operations with integers in terms of analogous operations with whole numbers. Though these operations are defined arbitrarily for integers, there are good reasons for the choices. The following discussion is an attempt to help you understand *why* operations with integers are defined as they are.

6.3 ADDITION OF INTEGERS

In defining the fundamental operations with the integers, we must recognize that, in forming this set, we have only extended the set of whole numbers. If, as we have stated previously, the counting numbers can be identified with the positive integers, then the whole numbers can be identified with the nonnegative integers. In our definitions, then, let us be guided by the stipulation that the properties of the positive integers and zero *include all properties of the whole numbers.* Indeed, we shall assume that the properties of the entire set of integers include all properties of the whole numbers; these properties were introduced in Chapter 5.

When we consider addition of integers it is very effective to represent integers on a number line by vectors. A **vector** is a line segment with a length and also an assigned direction. Each integer $a \neq 0$ may be represented by the point P_a and also by the vector $\overrightarrow{P_0P_a}$ (read P sub zero P sub a) with **initial point** P_0 and **terminal point** P_a. The direction of any such vector is from its initial point to its terminal point. Positive integers are represented by vectors that extend in the positive direction along the number line and negative integers by vectors that extend in the negative direction. The integer 0 is represented by the point P_0. We also call the point P_0 a **zero vector** $\overrightarrow{P_0P_0}$, with length 0 and no unique direction. In Figure 6.2 we have pictured $\overrightarrow{P_0P_5}$, a vector that represents the integer 5, and $\overrightarrow{P_0P_{-4}}$, a vector that represents the integer $^-4$.

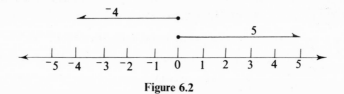

Figure 6.2

It must be noted that any other vector that is 5 units long and has a positive direction can represent the integer 5. In Figure 6.3 we show three such vectors, $\overrightarrow{P_{-5}P_0}$, $\overrightarrow{P_{-2}P_3}$, and $\overrightarrow{P_1P_6}$.

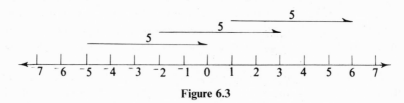

Figure 6.3

Also, any vector other than $\overrightarrow{P_0P_{-4}}$ that is 4 units long and has a negative direction may represent $^-4$. In Figure 6.4, we show three such vectors, $\overrightarrow{P_2P_{-2}}$, $\overrightarrow{P_{-2}P_{-6}}$, and $\overrightarrow{P_4P_0}$.

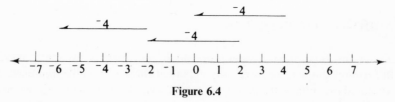

Figure 6.4

Let us illustrate the use of vectors in adding integers $4 + 5$. We first sketch a vector $\overrightarrow{P_0P_4}$ that represents 4. Beginning at the terminal point P_4 of $\overrightarrow{P_0P_4}$, we draw a second vector that represents 5. In Figure 6.5 it may be seen that the second vector is $\overrightarrow{P_4P_9}$. The sum is represented by a vector whose initial point is the initial point of $\overrightarrow{P_0P_4}$ and whose terminal point is the terminal point of $\overrightarrow{P_4P_9}$. The vector representing the sum, then, is $\overrightarrow{P_0P_9}$, and it is 9 units long. Thus $4 + 5 = 9$. *Note:* In the illustration of any operation, the vector that represents the result will be drawn as a broken line.

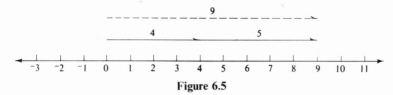

Figure 6.5

Here are two more examples of addition of integers. Consider the addition of $^-5 + {^-4}$. We first draw $\overrightarrow{P_0P_{-5}}$. Beginning at the terminal point of $\overrightarrow{P_0P_{-5}}$, we sketch a second vector that represents $^-4$, that is, a vector 4 units long extending in the negative direction. This second vector is seen to be $\overrightarrow{P_{-5}P_{-9}}$ (Fig. 6.6). The sum is represented by $\overrightarrow{P_0P_{-9}}$. Thus $^-5 + {^-4} = {^-9}$.

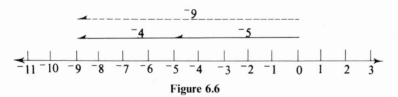

Figure 6.6

Next we illustrate the addition of $3 + {^-7}$. We draw $\overrightarrow{P_0P_3}$ to represent 3. Then we sketch a second vector that has its initial point at P_3 and represents $^-7$. This second vector is $\overrightarrow{P_3P_{-4}}$ (Fig. 6.7). The sum is represented by $\overrightarrow{P_0P_{-4}}$. Thus $3 + {^-7} = {^-4}$.

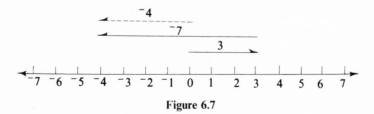

Figure 6.7

Let us generalize this procedure for addition of integers. To illustrate the addition of integers $a + b$, we first draw a vector $\overrightarrow{P_0P_a}$ that represents a. Next we draw a vector that represents b and has its initial point at the terminal point of $\overrightarrow{P_0P_a}$. The second vector is $\overrightarrow{P_aP_{a+b}}$. The vector representing the sum of $a + b$ has its initial point at the initial point of $\overrightarrow{P_0P_a}$ and its terminal point at the terminal point of $\overrightarrow{P_aP_{a+b}}$. Thus the sum is represented by the vector $\overrightarrow{P_0P_{a+b}}$.

Vectors that have the same length, but opposite directions are called **opposites**. Let us investigate what happens when we add two integers whose vector represen- tations are opposites, for example, $4 + {}^-4$. In Figure 6.8, we see that the initial point of $\overrightarrow{P_0P_4}$ (representing 4) coincides with the terminal point of $\overrightarrow{P_4P_0}$ (rep- resenting $^-4$). Thus the vector that represents $4 + {}^-4$ is the zero vector $\overrightarrow{P_0P_0}$. This indicates that $4 + {}^-4 = 0$.

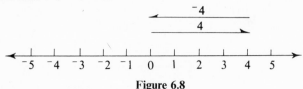

Figure 6.8

Furthermore, if we add any two integers a and ^-a whose vector representations are opposites, the sum is 0. Because of this property of integers a and ^-a, we call ^-a the **additive inverse** of a, and a the additive inverse of ^-a.

Integers a and ^-a are also called **negatives** of each other. Here it must be under- stood that the *negative of an integer* is not necessarily a *negative integer*. If a is a positive integer, its negative (additive inverse) *is* a negative integer; but if a is a *negative integer*, its negative (additive inverse) is a *positive integer*. For example, if $a = 5$, its negative (additive inverse) is $^-5$; if $a = {}^-5$, its negative (additive inverse) is $^+5$. Zero is not ordinarily considered to be either positive or negative, but it is its own additive inverse.

EXERCISES 6.3

1. Give the additive inverse of each integer:

(a) 15	(b) 1
(c) 0	(d) $^-1$
(e) $^-15$	(f) $0 + {}^-9$
(g) $5 + 4$	(h) $^-(5 + 4)$

2. Use the number line and perform the indicated operations:

(a) $^+5 + {}^+3$	(b) $^-5 + {}^-3$
(c) $5 + {}^-3$	(d) $^-5 + 3$
(e) $3 + {}^-5$	(f) $6 + {}^-6$
(g) $(4 + {}^-6) + {}^-5$	(h) $4 + ({}^-6 + {}^-5)$
(i) $(3 + {}^-2) + {}^-1$	(j) $[({}^-5 + 7) + {}^-4] + 5$
(k) $({}^-5 + 7) + ({}^-4 + 5)$	(l) $(5 + {}^-4) + (7 + {}^-5)$

3. Perform the indicated operations without the use of the number line if you can.

(a) $^+6 + {}^+2$	(b) $^-6 + 2$
(c) $6 + {}^-2$	(d) $^-6 + {}^-2$
(e) $6 + {}^-7$	(f) $^-3 + 9$
(g) $(3 + {}^-6) + {}^-4$	(h) $3 + ({}^-6 + {}^-4)$

6.4 FORMAL DEFINITION OF ADDITION FOR INTEGERS

In Sections 6.1 and 6.2, we have discussed the idea of the *absolute value* $|a|$ of an integer a. We may summarize this information by saying: If a is an integer,

$$|a| = a \quad \text{iff } a \geq 0$$
$$|a| = {}^-a \quad \text{iff } a < 0$$

Examples that illustrate these statements are

$$|5| = 5, \quad \text{since } 5 > 0$$
$$|0| = 0, \quad \text{since } 0 = 0$$
$$|{}^-5| = {}^-({}^-5), \quad \text{since } {}^-5 < 0$$

It is obvious that *the absolute value of an integer may be identified with a whole number*. This fact makes absolute value notation very convenient to use in formal definitions of operations with integers in terms of previously defined operations with whole numbers. For example, a formal definition of **addition of integers** follows:

If a and b are integers,

$$a + b = {}^+(|a| + |b|) \quad \text{iff } a > 0 \text{ and } b > 0$$
$$a + b = {}^-(|a| + |b|) \quad \text{iff } a < 0 \text{ and } b < 0$$

Iff $a > 0$ and $b < 0$,

$$a + b = {}^+(|a| - |b|) \quad \text{iff } |a| > |b|$$
$$a + b = {}^-(|b| - |a|) \quad \text{iff } |a| < |b|$$

Since the definition of addition of integers, with its several possible cases, is rather difficult to follow, we give a specific example for each case.

Let $a + b = 7 + 5$. Here $a > 0$ and $b > 0$. Therefore

$$7 + 5 = {}^+(|7| + |5|) = {}^+(7 + 5) = 12$$

Let $a + b = {}^-7 + {}^-5$. Here $a < 0$ and $b < 0$. Therefore

$$^-7 + {}^-5 = {}^-(|{}^-7| + |{}^-5|) = {}^-(7 + 5) = {}^-12$$

We shall next consider cases where $a > 0$ and $b < 0$:

Let $a + b = 7 + {}^-5$. Here $|a| > |b|$ since $|7| > |{}^-5|$. Therefore

$$7 + {}^-5 = {}^+(|7| - |{}^-5|) = {}^+(7 - 5) = 2$$

Let $a + b = 5 + {}^-7$. Here $|a| < |b|$ since $|5| < |{}^-7|$. Therefore

$$5 + {}^-7 = {}^-(|{}^-7| - |5|) = {}^-(7 - 5) = {}^-2$$

EXERCISES 6.4

1. Evaluate each expression

(a) $|{}^-6|$ (b) $|6|$

(c) $|25|$ (d) $|{}^-7|$

(e) $|{}^-7| + |6|$ (f) $|7| + |{}^-6|$

(g) $|{}^-7 + 6|$ (h) $|7 + {}^-6|$

(i) $|0|$ (j) $|0| + |25|$

(k) $|0| + |{}^-25|$ (l) $|0 + {}^-25|$

2. Add

(a) ${}^+5 + {}^+4$ (b) ${}^-5 + {}^-4$

(c) $5 + {}^-4$ (d) ${}^-4 + 5$

(e) $4 + {}^-5$ (f) ${}^-5 + 4$

(g) $(3 + 2) + {}^-4$ (h) $3 + (2 + {}^-4)$

(i) ${}^-(3 + 2) + 5$ (j) $(4 + {}^-3) + ({}^-2 + {}^-1)$

(k) $[(4 + {}^-3) + {}^-2] + {}^-1$ (l) $4 + [({}^-3 + {}^-2) + {}^-1]$

6.5 SUBTRACTION OF INTEGERS

Subtraction of integers, as well as subtraction of whole numbers, is defined in terms of addition. Suppose a and b are integers. If there exists an integer x such that

$$b + x = a$$

then the result of subtracting b from a is x, that is,

$$a - b = x$$

We now prove an important theorem concerning the difference x of subtracting integers $a - b$.

Theorem: If a and b are integers, then

$$a - b = a + {}^-b$$

PROOF

Statement	Reason
1. If $a - b = x$, then $b + x = a$	1. Definition of subtraction for integers.
2. $^-b + (b + x) = {}^-b + a$	2. Addition property of equality for integers.
3. $(^-b + b) + x = {}^-b + a$	3. Associative property of addition for integers.
4. $0 + x = {}^-b + a$	4. Additive inverse property for integers.
5. $x = {}^-b + a$	5. Identity element for addition of integers.
6. $x = a + {}^-b$	6. Commutative property of addition for integers.
7. $a - b = a + {}^-b$	7. Transitive property of equality for integers.

This theorem indicates that *subtracting an integer gives the same result as adding its additive inverse.* In other words, we can convert each subtraction problem into an addition problem by the simple device of replacing an integer with its additive inverse. Having done this, the number line procedure for addition can be used to solve the problem. For example, consider the problem $5 - 7$. By the theorem on subtraction of integers, $5 - 7 = 5 + {}^-7$. The number line solution is shown in Figure 6.9, where we see that the vector representing the difference is $\overline{P_0 P_{-2}}$. Thus $5 - 7 = {}^-2$.

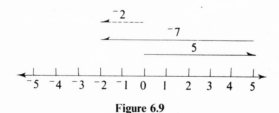

Figure 6.9

The theorem on subtraction of integers also allows us to use the formal definition of addition for integers in solving a subtraction problem. For example, consider the problem $^-6 - {}^-5$. By the theorem, $^-6 - {}^-5 = {}^-6 + 5$. We now use the definition of integers:

$$^-6 + 5 = 5 + {}^-6 \qquad \text{(commutative property of addition of integers)}$$

$$5 + {}^-6 = {}^-(|{}^-6| - |5|) = {}^-(6 - 5) = {}^-1$$

In Section 6.3 we stipulated that the set of integers must have all the properties of the set of whole numbers. Two properties of set of integers *not* shared by the set of whole numbers have been revealed in Section 6.3 (Addition of Integers) and in this section on subtraction of integers. These two properties are

1. **Additive inverse property.** Every integer a has an additive inverse ^-a, such that $a + {}^-a = {}^-a + a = 0$.

2. **Closure under subtraction.** If a and b are integers, $a - b$ is an integer.

The closure property of integers under subtraction is a consequence of the theorem on subtraction of integers $(a - b = a + {}^-b)$.

EXERCISES 6.5

1. Use the number line and perform the indicated operations.

(a) $7 - 3$ (b) $3 - 7$

(c) $^-7 - 3$ (d) $7 - {}^-3$

(e) $^-3 - {}^-7$ (f) $^-7 - {}^-3$

(g) $4 - 4$ (h) $4 - {}^-4$

(i) $5 - ({}^-3 - 4)$ (j) $(5 - {}^-3) - 4$

(k) $[5 + ({}^-3 - 4)] - {}^-2$ (l) $(5 + {}^-3) - (4 - {}^-2)$

2. Perform the indicated operations without the use of the number line.

(a) $5 - 2$ (b) $2 - 5$

(c) $^-5 - 2$ (d) $5 - {}^-2$

(e) $^-5 - {}^-2$ (f) $2 - {}^-5$

(g) $^-2 - 5$ (h) $(3 - {}^-5) - 4$

(i) $3 - ({}^-5 - 4)$ (j) $(4 - 6) - ({}^-2 - 4)$

3. Test the set of integers for closure under addition by adding several pairs of integers. Explain the final sentence in Section 6.5: The closure property of integers under subtraction is a consequence of the theorem on subtraction of integers $(a - b = a + {}^-b)$.

6.6   MULTIPLICATION AND DIVISION
WITH INTEGERS

In Section 5.6 the term *factor* was defined for the set of whole numbers. Here we define *factor* for the set of integers. If a, b, and c are integers and $a \cdot b = c$, we say a and b are **factors (divisors)** of c.

The reasonableness, or consistency, of our definitions of *multiplication of integers* is more difficult to explain than were the definitions of addition and subtraction of integers. We shall begin with the definition of multiplication of a positive integer and a negative integer. Let us take the problem $3 \cdot {}^-5$. This may be defined to mean ⁻5 + ⁻5 + ⁻5, which is fully consistent with our definition of multiplication of whole numbers. We show this product on the number line in Figure 6.10.

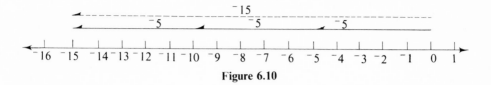

Figure 6.10

Thus we see that $3 \cdot {}^{-}5$ is ⁻15. Suppose, however, we consider the problem ⁻5·3. The sum of three negative fives is meaningful to us, but what could be meant by negative five threes? The definition of multiplication of whole numbers clearly cannot be extended to cover this product. So we use the fact that multiplication of integers is to be commutative and define ⁻5 · 3 to mean ⁻15. (If $3 \cdot {}^{-}5 = {}^{-}15$, then ⁻5·3 = ⁻15.) We can use the concept of the absolute value of an integer (Section 6.2) to make a general statement about the **product of a positive integer and a negative integer**:

$$a \cdot b = {}^{-}|a| \cdot |b|$$

iff one factor, a or b, is a positive integer and the other is a negative integer.

The next case that we shall consider in multiplication of integers is the product of two negative integers. Suppose, for example, we are multiplying ⁻5·⁻3. We certainly cannot extend the definition of multiplication of whole numbers to define this product. The method we shall use to develop a definition of the product of two negative integers is the construction of a multiplication table of integers as shown in Figure 6.11.

×	⁻4	⁻3	⁻2	⁻1	0	1	2	3	4
⁻4					0	⁻4	⁻8	⁻12	⁻16
⁻3					0	⁻3	⁻6	⁻9	⁻12
⁻2					0	⁻2	⁻4	⁻6	⁻8
⁻1					0	⁻1	⁻2	⁻3	⁻4
0	0	0	0	0	0	0	0	0	0
1	⁻4	⁻3	⁻2	⁻1	0	1	2	3	4
2	⁻8	⁻6	⁻4	⁻2	0	2	4	6	8
3	⁻12	⁻9	⁻6	⁻3	0	3	6	9	12
4	⁻16	⁻12	⁻8	⁻4	0	4	8	12	16

Figure 6.11

We have completed only that part of the table that involves products of positive integers and products that result when we multiply positive integers by negative integers. Examination of this part of the table shows a pattern: In the sixth row we see that there is a difference of 1 between the consecutive entries. In the seventh row, there is a difference of 2; in the eighth row, there is a difference of 3, etc. It is apparent that a pattern exists involving a constant difference between consecutive entries in any row or column. As we complete the patterns in the rows where there are omissions, we see that we need to define the product of two negative integers as a positive integer if we are to preserve the consistency of these patterns (Fig. 6.12).

$\times$	$^-4$	$^-3$	$^-2$	$^-1$	0	1	2	3	4
$^-4$	16	12	8	4	0	$^-4$	$^-8$	$^-12$	$^-16$
$^-3$	12	9	6	3	0	$^-3$	$^-6$	$^-9$	$^-12$
$^-2$	8	6	4	2	0	$^-2$	$^-4$	$^-6$	$^-8$
$^-1$	4	3	2	1	0	$^-1$	$^-2$	$^-3$	$^-4$
0	0	0	0	0	0	0	0	0	0
1	$^-4$	$^-3$	$^-2$	$^-1$	0	1	2	3	4
2	$^-8$	$^-6$	$^-4$	$^-2$	0	2	4	6	8
3	$^-12$	$^-9$	$^-6$	$^-3$	0	3	6	9	12
4	$^-16$	$^-12$	$^-8$	$^-4$	0	4	8	12	16

Figure 6.12

Since the product of two positive integers is also a positive integer, we can make a generalization about the **product of two integers when both are positive or both are negative**:

$$a \cdot b = |a| \cdot |b|$$

iff both factors a and b are positive integers or both are negative integers. Examination of this table will verify the useful fact that

$$^-a \cdot b = {}^-(a \cdot b) = a \cdot {}^-b$$

Division of integers, like division of whole numbers, is defined in terms of multiplication. Suppose a and b are integers. If there exists an integer x such that

$$b \cdot x = a$$

then the result of dividing a by b is x, that is,

$$a/b = x$$

For example, if $3x = {}^-12$, then ${}^-12/3 = x$. Since $3 \cdot {}^-4 = {}^-12$, $x = {}^-4$, and ${}^-12/3 = {}^-4$. Two other examples follow:

  1. If ${}^-3 \cdot x = 12$, then $12/{}^-3 = x$. Since ${}^-3 \cdot {}^-4 = 12$, $x = {}^-4$ and $12/{}^-3 = {}^-4$.

  2. If ${}^-3 \cdot x = {}^-12$, then ${}^-12/{}^-3 = x$. Since ${}^-3 \cdot 4 = {}^-12$, $x = 4$ and ${}^-12/{}^-3 = 4$.

These examples indicate that a **quotient of two integers, one positive and the other negative** (if such a quotient exists), is a negative integer; that is, if either a or b but not both are negative integers,

$$\frac{a}{b} = -\frac{|a|}{|b|}$$

Also, the **quotient of two integers when both are positive or both are negative** (if such a quotient exists), is a positive integer, that is, if a and b are both positive integers or both negative integers,

$$\frac{a}{b} = \frac{|a|}{|b|}$$

EXERCISES 6.6

  1. Use the number line and perform the indicated operations:

    (a) $5 \cdot {}^-3$ (b) $5 \cdot ({}^-4 + 2)$
    (c) $(5 \cdot {}^-4) + (5 \cdot 2)$ (d) $3(4 \cdot {}^-2)$
    (e) $(3 \cdot 4) \cdot {}^-2$

  2. Use the definitions that have been given for multiplication and division in Section 6.6 and perform the indicated operations:

    (a) ${}^-5 \cdot 3$ (b) ${}^-5 \cdot {}^-3$
    (c) $5 \cdot {}^-3$ (d) ${}^-6 \cdot {}^-1$

    (e) ${}^-6 \cdot 1$ (f) $\dfrac{16}{{}^-4}$

    (g) $\dfrac{{}^-16}{4}$ (h) $\dfrac{{}^-16}{{}^-4}$

  3. Perform the indicated operations: (The horizontal bar, indicating division is also a grouping symbol.)

    (a) $\dfrac{6 + {}^-3}{3}$ (b) $\dfrac{6 - {}^-3}{{}^-3}$

    (c) $\dfrac{6}{3} - \dfrac{{}^-3}{3}$ (d) $\dfrac{6}{3} - \dfrac{3}{{}^-3}$

(e) $\dfrac{8 \cdot {}^-4}{4 \cdot 4}$

(f) $\dfrac{\dfrac{6}{3} + \dfrac{{}^-8}{4}}{2}$

(g) $\dfrac{\left| \dfrac{9 \cdot 4}{2 \cdot 3} \right|}{{}^-3}$

(h) $\dfrac{{}^-8(7) - {}^-2({}^-5)}{6(3) - {}^-3(5)}$

(i) $\dfrac{({}^-12 - 3) \cdot (9 + {}^-4)}{\left(\dfrac{16 + {}^-12}{{}^-8 + 4} \right)}$

(j) $\dfrac{36}{12 + {}^-3} - \dfrac{{}^-4 + 16}{4}$

(k) $\left(\dfrac{36}{12} + \dfrac{36}{{}^-3} \right) - \left(\dfrac{{}^-4}{4} + \dfrac{16}{4} \right)$

4. Test the set of integers for closure under multiplication by multiplying several pairs of integers. Do you think the set of integers is closed under division? Explain your answer.

6.7   EVEN INTEGERS AND ODD INTEGERS

The set of **even integers** is a subset of the set of integers. Each element of this subset has a factor 2, that is, each element is divisible by 2. The general element, then, if we let n represent any integer, may be expressed as $2n$. The complement of the set of even integers, in the universe of integers, is called the set of **odd integers**. No element of this subset has a factor 2, that is, no element of the set of odd integers is divisible by 2. The general element of the set of odd integers, if we let n represent any integer, may be expressed as $2n + 1$. Experimenting with the expression $2n + 1$, we find that any integral value of n does, indeed, give an odd number. If we let

$$n = {}^-3, \quad \text{then } 2n + 1 = {}^-5$$

$$n = {}^-2, \quad \text{then } 2n + 1 = {}^-3$$

$$n = {}^-1, \quad \text{then } 2n + 1 = {}^-1$$

$$n = 0, \quad \text{then } 2n + 1 = 1$$

$$n = 1, \quad \text{then } 2n + 1 = 3$$

$$n = 2, \quad \text{then } 2n + 1 = 5, \text{ etc}$$

EXERCISES 6.7

1. Tell whether each integer is odd or even. If an integer is even, express it in the form $2n$ and if an integer is odd, express it in the form $2n + 1$ (n an integer).

Examples: 6 is even. $6 = 2 \cdot 3$; $^-9$ is odd. $^-9 = 2 \cdot {}^-4 + 1$.

(a) 1	(b) 2	(c) $^-8$
(d) 52	(e) $^-53$	(f) 0
(g) 17	(h) $^-1$	(i) $^-50$
(j) 101	(k) 150	(l) $^-72$

2. Test the set of even integers for closure under addition. Try to generalize your method by answering the question: Does $2m + 2n$, where m and n are integers, always equal an even integer? Test the set of even integers for closure under multiplication.

3. Test the set of odd integers for closure under addition. Test the set of odd numbers for closure under multiplication. Try to generalize your method by answering the question: Does $(2m + 1) \cdot (2n + 1)$, where m and n are integers, always equal an odd number?

4. Illustrate a one-to-one correspondence between the set of integers and the set of even integers; between the set of integers and the set of odd integers.

6.8 FACTORS AND PRIME NUMBERS

In Section 6.6 the term *factor* was defined for the set of integers: If a, b, and c are integers and $a \cdot b = c$, we say a and b are **factors** (**divisors**) of c.

Consider the positive integers. The integer 1 is a factor (divisor) of every integer n since $n = n \cdot 1$. The integer 2 has only itself and 1 as factors. Any positive integer that is greater than 1 and has only itself and 1 as positive integral factors is a **prime number**. In other words, a **prime number** is a positive integer that has *exactly two distinct positive integral factors*. Thus 3 is a prime number; 4 is not a prime number since $4 = 2 \cdot 2$. Any positive integer that is different from 1 and is not a prime number is a **composite number**. The integer 4 is a composite number. Since it may be expressed as $2 \cdot 2$ or $1 \cdot 4$, it has factors, 1, 2, and 4. Every composite number has at least one prime number as a factor; that is, has a **prime factor**. For example 12 has factors 1, 2, 3, 4, 6, and 12. Its *prime factors* are 2 and 3. We assume that factors of a positive integer are always less than or equal to the integer.

A **prime factorization** (also called **complete factorization**) of a composite number is the expression of the number as a product of prime numbers, for example, $12 = 2 \cdot 2 \cdot 3 = 2^2 \cdot 3$. We shall assume the truth of a theorem called the **fundamental theorem of arithmetic**: If n is a composite number, then the prime factorization of n is unique except for the ordering of the factors. Within the set of integers we have two units, a positive unit $^+1$ and a negative unit $^-1$. Every integer can be expressed as 0, a unit, or a product of a unit and one or more prime numbers. Here are some examples: $^-7 = {}^-1 \cdot 7$, $12 = 1 \cdot 2^2 \cdot 3$, $^-12 = {}^-1 \cdot 2^2 \cdot 3$.

Though mathematicians have known of the existence of prime numbers for centuries—Euclid proved that there are infinitely many prime numbers—relatively little is known about their properties. In fact it is often difficult to tell whether a given large number is prime, or to find the prime numbers among a given set of positive integers. About 2200 years ago, Eratosthenes devised a method of finding the prime numbers in any given set of integers 1 through n. In this array the method, called the sieve of Eratosthenes, is illustrated with integers 1 through 30.

$$
\begin{array}{cccccccccc}
\cancel{1} & 2 & 3 & \cancel{4} & 5 & \cancel{6} & 7 & \cancel{8} & \cancel{9} & \cancel{10} \\
11 & \cancel{12} & 13 & \cancel{14} & \cancel{15} & \cancel{16} & 17 & \cancel{18} & 19 & \cancel{20} \\
\cancel{21} & \cancel{22} & 23 & \cancel{24} & \cancel{25} & \cancel{26} & \cancel{27} & \cancel{28} & 29 & \cancel{30}
\end{array}
$$

We cross out all numerals representing integers that are not prime numbers, using the following procedure. First we cross out the numeral 1, because we know 1 is not a prime number. We know that 2 is a prime number; therefore, we do not cross out the numeral 2. We begin at the numeral 2 and count by twos throughout the set, crossing out 4, 6, 8, 10, etc., because they are numerals for multiples of two. We know that 3 is a prime number; therefore, its numeral is not crossed out. We begin at the numeral 3 and count by threes, crossing out any multiples of three (6, 9, 12, 15, etc.) that have not been crossed out as multiples of two. In the next round, we see that the numeral 4 has been crossed out and the first remaining numeral after 3 is 5. It is a numeral representing a prime number since it did not turn out to be a multiple of 2 or 3. Therefore 5 is not crossed out. We begin with 5 and count by fives crossing out any multiples of five (10, 15, 20, etc.) that have not been crossed out. The numeral 6 has been crossed out as a multiple of 2, so the first remaining numeral after 5 is 7. It is prime since it was not crossed out as a multiple of 2, 3 or 5. When we begin at 7 and count by sevens, we find no multiple of 7 between 7 and 30 that has not been crossed out. This means the process is finished; all remaining numbers in the array are prime numbers. Thus the prime numbers between 1 and 30 are 2, 3, 5, 7, 11, 13, 17, 19, 23, and 29.

The last prime number p whose multiples must be deleted in finding the prime numbers less than or equal to a given number N is the largest prime number p such that $p^2 \leq N$.

EXERCISES 6.8

1. Tell whether each number is prime or composite:

(a) 1	(b) 2	(c) 5
(d) 6	(e) 7	(f) 8
(g) 9	(h) 10	(i) 11
(j) 12	(k) 13	(l) 14
(m) 15	(n) 16	(o) 17
(p) 41	(q) 51	(r) 59

2. List all the positive integral factors of each integer.

(a) 16	(b) 18	(c) 23
(d) 28	(e) 30	(f) 32

(g) 39	(h) 0	(i) 51
(j) 52	(k) 57	(l) 59
(m) 101	(n) 103	(o) 130
(p) 225	(q) 256	(r) 240
(s) 400	(t) 696	(u) 1000

3. Write the prime factorization of each integer in Exercise 3.

4. Express each integer as 0, a unit, or the product of a unit and one or more prime numbers.

(a) $^-43$	(b) 21	(c) 0
(d) 54	(e) $^-63$	(f) $^-61$
(g) $^-1$	(h) 15	(i) 64
(j) $^-154$	(k) 58	(l) 72

5. Use the *sieve of Eratosthenes* to find the prime numbers in the set of integers $\{1, 2, 3, \ldots, 100\}$.

6. Are any of the even integers prime numbers? Explain your answer.

7. Make a list of the fourteen properties of the set of integers and give specific examples to illustrate each property.

8. Justify each statement by stating a definition, an assumption, or a theorem.

(a) $7 + {}^-6 = {}^+(|7| - |{}^-6|)$

(b) $7 + {}^-6 = {}^-6 + 7$

(c) $7 + {}^-6 =$ an integer

(d) $^-6 \cdot 1 = {}^-6$

(e) $^-6 + 0 = {}^-6$

(f) $({}^-5 + 7) + {}^-6 = {}^-5 + (7 + {}^-6)$

(g) $^-7 \cdot (4 + {}^-3) = {}^-7 \cdot 4 + {}^-7 \cdot {}^-3$

(h) $7 \cdot (4 + {}^-3) = (4 + {}^-3) \cdot 7$

(i) $^-7 + 7 = 0$

(j) $7 - {}^-6 = 7 + {}^+6$

(k) $7 \cdot {}^-6 = {}^-42$

(l) $7 - {}^-6 =$ an integer

(m) If $^-2 \cdot x = 0$, $x = 0$

(n) $^-9 = {}^-9$

(o) If $x = y$, then $x + {}^-3 = y + {}^-3$

(p) If $^-9 \not> {}^-7$ and $^-9 \neq {}^-7$, then $^-9 < {}^-7$

(q) If $\dfrac{{}^-56}{{}^-8} = x$, then $^-8 \cdot x = {}^-56$

(r) $\dfrac{{}^-56}{{}^-8} = 7$

(s) $^-8 \cdot 7 =$ an integer

(t) $^-8 \cdot 0 = 0$

(u) If $x = y$, then $^-8 \cdot x = {}^-8 \cdot y$

CHAPTER 7

The Rational Numbers

7.1 MEMBERSHIP OF THE SET OF RATIONAL NUMBERS

The integers, though they are very useful, are insufficient to represent solutions to many problems. For example, if a problem requires that we divide 5 by 3, we do not have an answer in the set of integers; a new type of number, called a *rational number*, is needed. A **rational number** is a number that can be represented by a numeral of the form a/b where both a and b are integers and b is not equal to zero. The numeral a/b has **numerator** a and **denominator** b. The set of rational numbers includes the set of integers as a subset, because any integer, a, can be expressed in the form a/b where $b = 1$. A **fraction** is assumed to be a rational number that is not an integer.

7.2 MULTIPLICATION AND DIVISION OF RATIONAL NUMBERS

We define the **product of two rational numbers,** a/b and c/d where a, b, c, and d are integers, $b \neq 0$ and $d \neq 0$, to be the rational number ac/bd, that is,

$$\frac{a}{b} \cdot \frac{c}{d} = \frac{ac}{bd}$$

The reader will recognize that this definition of multiplication of rational numbers is expressed in terms of multiplication of integers, since ac is the product of two integers and bd is the product of two integers.

There is an **identity element** for multiplication of rational numbers. It is the number **1**. Recall that 1 is also the identity element for multiplication of integers. A rational number, like a whole number or an integer, has many names. Among the numerals that may represent 1 are $\frac{1}{1}, \frac{2}{2}, \frac{5}{5}, \frac{16}{16}$. If we let n represent an integer other than zero, we can say that 1 may be represented by n/n, that is,

$$\frac{n}{n} = 1$$

In general, two expressions a/b and c/d for rational numbers are *defined* to name the same rational number iff $ad = bc$;

$$\frac{a}{b} = \frac{c}{d} \quad \text{iff } ad = bc$$

We use the identity element for multiplication to find different expressions for the same rational number; for any integer $n \neq 0$

$$\frac{a}{b} \cdot \frac{n}{n} = \frac{an}{bn}$$

Note that $an/bn = a/b$ since $abn = abn$.

If $a/b \neq 0$, then $b \neq 0$ *and* $a \neq 0$. Thus if $a/b \neq 0$, then b/a also represents a rational number. Notice that

$$\frac{a}{b} \cdot \frac{b}{a} = \frac{ab}{ab} = 1$$

In other words, if $a/b \neq 0$, then a/b has b/a as its *multiplicative inverse*. A rational number is defined to be the **multiplicative inverse** of a given rational number iff the product of the two numbers is the identity element 1 for multiplication of rational numbers. The multiplicative inverse b/a of a number a/b is frequently called the **reciprocal** of the number. Notice that zero has no reciprocal (multiplicative inverse), but that each nonzero element in the set of rational numbers has a reciprocal (multiplicative inverse) that is an element of the set of rational numbers. This is a property of the set of rational numbers *not* shared by the set of integers. The reciprocals of most integers are not integers.

Division of rational numbers is defined in terms of multiplication. Let a/b and c/d be rational numbers where $b \neq 0$, $c \neq 0$, and $d \neq 0$. If there exists a rational number x such that

$$\frac{c}{d} \cdot x = \frac{a}{b}$$

then the result of dividing a/b by c/d is x, that is,

$$\frac{a}{b} \div \frac{c}{d} = x$$

We prove the following theorem concerning the **quotient of two rational numbers**. We assume a/b, c/d, and x to be rational numbers where $b \neq 0$, $c \neq 0$, and $d \neq 0$. Also, we stipulate that *the rational numbers have all the properties of integers* that we have discussed in Chapter 6.

Theorem: The **quotient** of a/b divided by c/d is ad/bc. Symbolically, this may be stated

$$\frac{a}{b} \div \frac{c}{d} = \frac{ad}{bc}$$

PROOF

Statement	Reason
1. If $\dfrac{c}{d} \cdot x = \dfrac{a}{b}$, then $$\frac{a}{b} \div \frac{c}{d} = x$$	1. Definition of division for rational numbers.
2. $\dfrac{d}{c} \cdot \left(\dfrac{c}{d} \cdot x \right) = \dfrac{d}{c} \cdot \dfrac{a}{b}$	2. Multiplication property of equality of rational numbers.
3. $\left(\dfrac{d}{c} \cdot \dfrac{c}{d} \right) \cdot x = \dfrac{a}{b} \cdot \dfrac{d}{c}$	3. Associative property of multiplication and commutative property of multiplication for rational numbers.
4. $1 \cdot x = \dfrac{a}{b} \cdot \dfrac{d}{c}$	4. Multiplicative inverse property of rational numbers.
5. $1 \cdot x = \dfrac{ad}{bc}$	5. Definition of multiplication for rational numbers.
6. $x = \dfrac{ad}{bc}$	6. Identity element for multiplication of rational numbers.
7. $\dfrac{a}{b} \div \dfrac{c}{d} = \dfrac{ad}{bc}$	7. Transitive property of equality for rational numbers.

It is interesting to find that *dividing by a rational number gives the same result as multiplying by its multiplicative inverse*:

$$\frac{a}{b} \div \frac{c}{d} = \frac{ad}{bc}$$

but

$$\frac{a}{b} \cdot \frac{d}{c} = \frac{ad}{bc} \quad \text{also}$$

The following sequence of ideas results in an important and useful statement: For any integers $a \neq 0$ and $b \neq 0$,

(a) $a \div b = \dfrac{a}{1} \div \dfrac{b}{1} = \dfrac{a}{1} \cdot \dfrac{1}{b} = \dfrac{a}{b}$

(b) Since $b \cdot \dfrac{1}{b} = 1$ and 1 is positive, both b and $\dfrac{1}{b}$ are positive *or* both are negative

(c) Since $^-a \cdot b = {}^-(a \cdot b) = a \cdot {}^-b$ (Section 6.6), $^-1 \cdot \dfrac{1}{b} = {}^-\left(1 \cdot \dfrac{1}{b}\right) = 1 \cdot {}^-\left(\dfrac{1}{b}\right)$,

that is $\dfrac{^-1}{b} = {}^-\left(\dfrac{1}{b}\right) = \dfrac{1}{^-b}$

(d) $a \cdot \dfrac{^-1}{b} = {}^-\left(a \cdot \dfrac{1}{b}\right) = a \cdot \left(\dfrac{1}{^-b}\right)$, that is, $\dfrac{^-a}{b} = -\dfrac{a}{b} = \dfrac{a}{^-b}$

Though we have accepted the fact that the properties of rational numbers are consequences of definitions and of properties of integers, many of these properties can be proved. When the first exercises calling for formal deductive proofs appeared in Chapter 2, the student was urged to attempt these proofs even though he had not been adequately prepared for them. This advice is offered once more with respect to Exercise 8.

EXERCISES 7.2

1. Find the multiplicative inverse for each of the following.

(a) $\dfrac{2}{3}$ (b) $\dfrac{8}{5}$ (c) $\dfrac{3}{8}$

(d) 3   (e) 16   (f) 10

(g) 100   (h) 1   (i) $^-3$

(j) $\dfrac{^-5}{3}$ (k) $^-1$ (l) $\dfrac{3}{^-5}$

2. Express each rational number in a form a/b where $b > 0$.

(a) $-\dfrac{2}{3}$ (b) $\dfrac{^-2}{^-3}$ (c) $\dfrac{2}{^-3}$

(d) $^-3$ (e) $\dfrac{5}{^-2}$ (f) $\dfrac{^-5}{^-2}$

(g) $-\left(\dfrac{5}{2}\right)$ (h) 0

3. Justify each step in the following computations by stating a definition, an assumption or a theorem.

Example: $\dfrac{3}{4} \cdot \dfrac{2}{3} = \dfrac{3 \cdot 2}{4 \cdot 3}$ Definition of multiplication for rational numbers.

$\qquad\qquad = \dfrac{3 \cdot 2}{3 \cdot 4}$ Commutative property of multiplication of integers.

$\qquad\qquad = \dfrac{3}{3} \cdot \dfrac{2}{4}$ Definition of multiplication for rational numbers.

$\qquad\qquad = \dfrac{2}{4}$ Identity element for multiplication of rational numbers.

$\qquad\qquad = \dfrac{2 \cdot 1}{2 \cdot 2}$ *Factoring* performed.

$\qquad\qquad = \dfrac{2}{2} \cdot \dfrac{1}{2}$ Definition of multiplication for rational numbers.

$\qquad\qquad = \dfrac{1}{2}$ Identity element for multiplication of rational numbers.

(a) $\dfrac{3}{7} \cdot \dfrac{5}{6} = \dfrac{3 \cdot 5}{7 \cdot 6}$ _____

$\qquad\qquad = \dfrac{3 \cdot 5}{6 \cdot 7}$ _____

$\qquad\qquad = \dfrac{3 \cdot 5}{(3 \cdot 2) \cdot 7}$ _____

$\qquad\qquad = \dfrac{3 \cdot 5}{3 \cdot (2 \cdot 7)}$ _____

$\qquad\qquad = \dfrac{3}{3} \cdot \dfrac{5}{14}$ _____

$\qquad\qquad = \dfrac{5}{14}$ _____

(b) $\dfrac{7}{15} \div \dfrac{10}{27} = \dfrac{7}{15} \cdot \dfrac{27}{10}$ ____

$= \dfrac{7 \cdot 27}{15 \cdot 10}$ ____

$= \dfrac{27 \cdot 7}{15 \cdot 10}$ ____

$= \dfrac{(3 \cdot 9) \cdot 7}{(3 \cdot 5) \cdot 10}$ ____

$= \dfrac{3 \cdot (9 \cdot 7)}{3 \cdot (5 \cdot 10)}$ ____

$= \dfrac{3}{3} \cdot \dfrac{9 \cdot 7}{5 \cdot 10}$ ____

$= \dfrac{63}{50}$ ____

4. Compute, using the definitions and properties that we have considered.

(a) $\dfrac{2}{3} \cdot \dfrac{5}{9}$ (b) $\dfrac{2}{3} \cdot \dfrac{9}{5}$

(c) $\dfrac{^-2}{3} \cdot \dfrac{9}{8}$ (d) $^-4 \cdot \dfrac{^-5}{3}$

(e) $4 \cdot \dfrac{5}{4}$ (f) $\dfrac{3}{8} \cdot {}^-4$

(g) $\dfrac{3}{4} \div \dfrac{3}{2}$ (h) $\dfrac{2}{3} \cdot \dfrac{^-9}{5}$

(i) $^-4 \div \dfrac{4}{3}$ (j) $\dfrac{^-5}{6} \div {}^-6$

5. Identify two integers whose reciprocals are also integers. Are there any others?

6. Explain why zero does not have a multiplicative inverse.

7. Do you think the set of rational numbers is closed under division? Refer to the theorem concerning the quotient of two rational numbers in answering this question.

8. Prove that multiplication of rational numbers is

(a) Commutative, that is, $\dfrac{a}{b} \cdot \dfrac{c}{d} = \dfrac{c}{d} \cdot \dfrac{a}{b}$ (b and $d \neq 0$)

(b) Associative, that is, $\dfrac{a}{b} \cdot \left(\dfrac{c}{d} \cdot \dfrac{e}{f} \right) = \left(\dfrac{a}{b} \cdot \dfrac{c}{d} \right) \cdot \dfrac{e}{f}$ (b, d, and $f \neq 0$)

(Recall that $a, b, c, d, e,$ and f are integers; therefore, properties of integers may be used in your proofs.)

7.3   ADDITION OF RATIONAL NUMBERS

In considering *addition of rational numbers*, we shall first define **addition for rational numbers expressed with like denominators**:

$$\frac{a}{b} + \frac{c}{b} = \frac{a+c}{b} \qquad (a, b, \text{ and } c \text{ integers}, b \neq 0)$$

The problem of addition of rational numbers expressed with *unlike denominators* may be resolved by a theorem. We state and prove this theorem.

Theorem: The sum of two rational numbers a/b and c/d ($a, b, c,$ and d integers, b and $d \neq 0$) is $\dfrac{ad+bc}{bd}$. Symbolically, this may be stated

$$\frac{a}{b} + \frac{c}{d} = \frac{ad+bc}{bd}$$

PROOF

Statements	*Reasons*
1. $\dfrac{a}{b} + \dfrac{c}{d} = \left(\dfrac{a}{b} \cdot \dfrac{d}{d}\right) + \left(\dfrac{b}{b} \cdot \dfrac{c}{d}\right)$	1. Identity element for multiplication of rational numbers.
2. $\left(\dfrac{a}{b} \cdot \dfrac{d}{d}\right) + \left(\dfrac{b}{b} \cdot \dfrac{c}{d}\right) =$ $\dfrac{ad}{bd} + \dfrac{bc}{bd}$	2. Definition of multiplication for rational numbers.
3. $\dfrac{ad}{bd} + \dfrac{bc}{bd} = \dfrac{ad+bc}{bd}$	3. Definition of addition for rational numbers.
4. $\dfrac{a}{b} + \dfrac{c}{d} = \dfrac{ad+bc}{bd}$	4. Extended transitive property of equality of rational numbers.

7.4   RATIONAL NUMBERS IN SIMPLEST FORM

Before we discuss *rational numbers in simplest form*, we need to clarify the meaning of "rational number" as we use it in this expression. It is important that the reader understand that "rational number" here is an abbreviation for *the* **numeral** *that represents a rational number.* When we speak of the numerator (or denominator) of a rational number, we are again using "rational number" as an abbreviation. *Numbers* do not have simplest forms, numerators, or denominators; their numerals do.

A rational number a/b is expressed in **simplest form** (also called **reduced form**) when the numerator a is an integer, the denominator b is a positive integer, and integers a and b are relatively prime. Two (or more) integers are **relatively prime** iff they have no common prime factors. Thus we need to express numbers in terms of their prime factors in order to determine whether they are relatively prime. Consider $\frac{9}{12}$:

$$\frac{9}{12} = \frac{3 \cdot 3}{2 \cdot 2 \cdot 3}$$

But, by definition of multiplication for rational numbers,

$$\frac{3 \cdot 3}{2 \cdot 2 \cdot 3} = \frac{3}{2 \cdot 2} \cdot \frac{3}{3}.$$

Since $\frac{3}{3}$ is the identity element for multiplication of rational numbers,

$$\frac{3}{2 \cdot 2} \cdot \frac{3}{3} = \frac{3}{2 \cdot 2}$$

$$= \frac{3}{4}$$

Hence $\frac{9}{12}$, in simplest form, is expressed as $\frac{3}{4}$.

EXERCISES 7.4

1. Reduce the following rational numbers to simplest form.

(a) $\dfrac{4}{6}$ (b) $\dfrac{3}{3}$ (c) $\dfrac{7}{91}$

(d) $\dfrac{12}{56}$ (e) $\dfrac{^-16}{64}$ (f) $\dfrac{15}{^-225}$

(g) $-\dfrac{72}{144}$ (h) $\dfrac{^-63}{^-91}$ (i) $\dfrac{112}{48}$

2. Perform the operations indicated in each of the following; use only those processes that can be justified by definitions or theorems found in this section or in previous sections.

(a) $\dfrac{2}{3} + \dfrac{5}{3}$ (b) $2 + 3$

(c) $\dfrac{2}{3} + \dfrac{3}{4}$ (d) $3 + \dfrac{^-2}{3}$

3. Given the definition,

$$\frac{a}{b} - \frac{c}{b} = \frac{a - c}{b}$$

prove the following theorem for subtraction of rational numbers with unlike denominators: If a, b, c, and d are integers and b and $d \neq 0$, then

$$\frac{a}{b} - \frac{c}{d} = \frac{ad - bc}{bd}$$

4. Compute, using the instructions in Exercise 2.

(a) $3 - \dfrac{2}{3}$ (b) $\dfrac{^-2}{3} - \dfrac{5}{6}$

(c) $\dfrac{^-2}{3} - \dfrac{^-5}{6}$ (d) $\dfrac{2}{3} - 4$

5. Justify each step in the following computations by stating a definition, an assumption, or a theorem.

(a) $\dfrac{2}{5} + \dfrac{5}{6} = \dfrac{2}{5} \cdot \dfrac{6}{6} + \dfrac{5}{6} \cdot \dfrac{5}{5}$ ————

$ = \dfrac{2 \cdot 6}{5 \cdot 6} + \dfrac{5 \cdot 5}{6 \cdot 5}$ ————

$ = \dfrac{12}{30} + \dfrac{25}{30}$ Multiplication of integers performed

$ = \dfrac{37}{30}$ ————

(b) $\dfrac{5}{8} - \dfrac{3}{7} = \dfrac{5 \cdot 7 - 8 \cdot 3}{8 \cdot 7}$ ————

$ = \dfrac{35 - 24}{56}$ ————

$ = \dfrac{11}{56}$ ————

6. Compute, using only processes that can be justified by stating a definition, an assumption, or a theorem. Express result in simplest form.

(a) $\dfrac{2}{3} \cdot \dfrac{6}{5}$ (b) $\dfrac{3}{8} \cdot \dfrac{4}{7}$ (c) $\dfrac{3}{11} \cdot 33$

(d) $\dfrac{0}{12} \cdot \dfrac{24}{25}$ (e) $\dfrac{3}{7} + \dfrac{4}{7}$ (f) $\dfrac{2}{3} + \dfrac{3}{5}$

(g) $\dfrac{3}{7} + \dfrac{5}{6}$ (h) $\dfrac{3}{11} + 33$ (i) $\dfrac{0}{12} + \dfrac{24}{25}$

(j) $\dfrac{3}{4} - \dfrac{2}{3}$ (k) $\dfrac{4}{9} - \dfrac{9}{10}$ (l) $\dfrac{2}{3} \div \dfrac{6}{5}$

(m) $\dfrac{3}{7} \div \dfrac{6}{5}$ (n) $33 \div \dfrac{3}{11}$ (o) $\dfrac{0}{12} \div \dfrac{24}{25}$

(p) $5 \div \dfrac{21}{23}$ (q) $\dfrac{21}{23} \div 7$ (r) $\dfrac{5}{12} \div \dfrac{0}{11}$

7. Prove that addition of rational numbers is

(a) Commutative, that is, $\dfrac{a}{b} + \dfrac{c}{d} = \dfrac{c}{d} + \dfrac{a}{b}$ (b and $d \neq 0$)

(b) Associative, that is, $\dfrac{a}{b} + \left(\dfrac{c}{d} + \dfrac{e}{f}\right) = \left(\dfrac{a}{b} + \dfrac{c}{d}\right) + \dfrac{e}{f}$ ($b, d, f \neq 0$)

8. (a) Prove that multiplication of rational numbers is distributive over addition; that is,

$$\frac{a}{b}\left(\frac{c}{d} + \frac{e}{f}\right) = \frac{a}{b} \cdot \frac{c}{d} + \frac{a}{b} \cdot \frac{e}{f} \qquad (b, d, f \neq 0)$$

(b) Show by an example that addition of rational numbers is not distributive over multiplication; that is,

$$\frac{a}{b} + \left(\frac{c}{d} \cdot \frac{e}{f}\right) \neq \left(\frac{a}{b} + \frac{c}{d}\right) \cdot \left(\frac{a}{b} + \frac{e}{f}\right) \qquad (b, d, f \neq 0)$$

7.5 THE GREATEST COMMON FACTOR

We have defined the term *factor* and the expression of *a rational number a/b in simplest form.* Actually, in reducing the expression of a rational number a/b to simplest form, it is not merely *a* common factor of the numerator and denominator in which we are interested, but the *greatest common factor.* The **greatest common factor**, abbreviated **G.C.F.**, of two (or more) integers is the largest positive integer that is a factor of each of the integers.

When the numbers involved are not large, it is usually best to find the G.C.F. by factoring each integer in terms of its prime factors. For example, suppose we want the G.C.F. of 44, ⁻64, and 84. Prime factorizations of these numbers are

$$44 = 2 \cdot 2 \cdot 11 = 2^2 \cdot 11$$

$$^-64 = {}^-(2 \cdot 2 \cdot 2 \cdot 2 \cdot 2 \cdot 2) = {}^-(2^6)$$

$$84 = 2 \cdot 2 \cdot 3 \cdot 7 = 2^2 \cdot 3 \cdot 7$$

The greatest common factor of these three numbers would be divisible by 2, would be divisible by 2^2, would not be divisible by 3, would not be divisible by 7, and would not be divisible by 11. Since the G.C.F. must be a factor of each of the given numbers, the G.C.F. is 2^2. In general, the G.C.F. of two or more integers is the product of powers of the prime numbers that are factors of all the given integers, where the power selected is lowest to which the prime number occurs in the factorizations. Here is another example; the G.C.F. of

$$2 \cdot 3^3 \cdot 5^3 \cdot 11 \quad \text{and} \quad 3^4 \cdot 5^2 \cdot 7 \cdot 13$$

is $3^3 \cdot 5^2$.

If the numbers involved are very large, it may be tedious to obtain a complete factorization of each given integer. In this case another method of finding the G.C.F is useful. For this method we need to know the first few prime numbers. Prime numbers 2, 3, 5, 7, 11, 13, and 17 will be sufficient for the examples we shall use. This process for finding the G.C.F. of two or more numbers involves removing observable prime factors common to all the given integers until we arrive at quotients that are relatively prime. Suppose we are to find the G.C.F. of 84 and 212. We choose 2 as an obvious prime factor common to both numbers:

$$\begin{array}{cc} 42 & 106 \\ \overline{2/84} & \overline{2/212} \end{array}$$

We continue by dividing quotients 42 and 106 by 2:

$$\begin{array}{cc} 21 & 53 \\ \overline{2/42} & \overline{2/106} \end{array}$$

We observe that the only prime factors of 21 (7 and 3) are not factors of 53. Thus 21 and 53 are relatively prime. Notice that

$$84 = 2 \cdot 42 = 2 \cdot 2 \cdot 21 = 4 \cdot 21$$

and

$$212 = 2 \cdot 106 = 2 \cdot 2 \cdot 53 = 4 \cdot 53$$

Thus the G.C.F. of 84 and 212 is the product of the common factors we have removed, that is, $2 \cdot 2 = 4$. The rational number 84/212 can be expressed as

$$\frac{4 \cdot 21}{4 \cdot 53} = \frac{4}{4} \cdot \frac{21}{53} = \frac{21}{53}$$

Here is another example of this method. Suppose we need to find the G.C.F. of the three integers 210, 1785, and $^-24{,}255$. We first remove the common prime factor 5:

$$\frac{42}{5\big/210} \qquad \frac{357}{5\big/1785} \qquad \frac{^-4851}{5\ \overline{^-24255}}$$

Prime factors of 42 are 3 and 7. We test the quotients 357 and 4851 for divisibility by 3:

$$\frac{14}{3\big/42} \qquad \frac{119}{3\big/357} \qquad \frac{^-1617}{3\big/^-4851}$$

Next we try dividing the quotients 14, 119, and $^-1617$ by 7:

$$\frac{2}{7\big/14} \qquad \frac{17}{7\big/119} \qquad \frac{^-231}{7\big/^-1617}$$

The quotients 2, 17, and $^-231$ are relatively prime since 2 is a prime number, but is not a factor of 17 and $^-231$. We see that

$$210 = 5 \cdot 42 = 5 \cdot (3 \cdot 14) = 5 \cdot 3 \cdot (7 \cdot 2) = 105 \cdot 2$$

$$1785 = 5 \cdot 357 = 5 \cdot (3 \cdot 119) = 5 \cdot 3 \cdot (7 \cdot 17) = 105 \cdot 17$$

and

$$^-24{,}255 = 5 \cdot {}^-4851 = 5 \cdot (3 \cdot {}^-1617) = 5 \cdot 3 \cdot (7 \cdot {}^-231) = 105 \cdot {}^-231$$

Thus the G.C.F. of the integers 210, 1785, and $^-24{,}255$ is 105.

EXERCISES 7.5

1. Find the G.C.F. of each set of integers by the method of obtaining a prime factorization of each number.

(a) 16, 52, 60
(b) 14, 35, 98
(c) 18, $^-36$, 56
(d) 17, 31, 47
(e) $^-24$, 36, $^-144$
(f) 30, 140, 420
(g) 165, 429, $^-495$
(h) 84, 294, 2002

2. Find the G.C.F. of each set of integers by the method of removing observable common prime factors until the quotients obtained are relatively prime.

(a) 175, 385
(b) 132, $^-936$
(c) 168, 1134
(d) $^-272$, 1326
(e) 2002, 3399
(f) 30, 140, 420
(g) 165, 429, $^-495$
(h) 84, 294, 2002

3. Reduce each rational number to simplest form.

(a) $\dfrac{4}{6}$ (b) $\dfrac{3}{3}$ (c) $\dfrac{7}{91}$

(d) $\dfrac{^{-}12}{56}$ (e) $\dfrac{16}{^{-}64}$ (f) $\dfrac{15}{225}$

(g) $\dfrac{^{-}72}{144}$ (h) $\dfrac{^{-}63}{^{-}91}$ (i) $\dfrac{112}{48}$

(j) $\dfrac{315}{^{-}1449}$ (k) $\dfrac{770}{4510}$ (l) $\dfrac{^{-}1092}{^{-}2457}$

7.6 THE LEAST COMMON MULTIPLE

A **multiple** of an integer is a second integer that has the first integer as a factor, that is, is divisible by the first integer. For example, 24 is a multiple of 8. A **common multiple** of two or more integers is an integer that has each of the given integers as a factor. For example, 70 is a common multiple of 5 and 7. In addition of rational numbers expressed as a/b, c/d, e/f, etc., it is to our advantage to find the *least common multiple* of all of the denominators involved (called the least common denominator). The **least common multiple**, abbreviated **L.C.M.**, of two or more integers is the smallest positive integer that is a multiple of each of the given integers; for example, the L.C.M. of 16 and 6 is 48.

An effective method of finding the L.C.M. of two or more numbers, when it can not readily be found by inspection, is to write a complete factorization of each given number. Then all prime factors that appear in the given numbers with each prime number taken to the highest power to which it occurs in any one given number, are factors of the L.C.M.. Let us illustrate this method by finding the L.C.M. of 6, 8, and 9. Complete factorizations of the given integers are

$$6 = 2 \cdot 3$$
$$8 = 2^3$$
$$9 = 3^2$$

The L.C.M. is $2^3 \cdot 3^2$ which equals 72.

There is a "division" method of finding factors of the L.C.M. of two or more integers when writing prime factorizations of the numbers would be tedious. We illustrate this method by using it to find the L.C.M. of 90, 84, and 210. We begin by listing the numbers in a row and dividing each of the numbers by any prime number that divides one or more of them, for example, 7. The quotients are written in a row below the corresponding numbers; if one or more of the original numbers is not divisible by this factor, the number itself is listed on this second row:

$$7 / \quad 90 \quad 84 \quad 210$$

$$90 \quad 12 \quad 30$$

We repeat this process until the quotients are all distinct prime numbers or 1:

$$7 / \quad 90 \quad 84 \quad 210$$

$$5 / \quad 90 \quad 12 \quad 30$$

$$3 / \quad 18 \quad 12 \quad 6$$

$$2 / \quad 6 \quad 4 \quad 2$$

$$3 \quad 2 \quad 1$$

The L.C.M. is equal to the product of the prime numbers that have been used as divisors or obtained in the last row of quotients. For example the L.C.M. of 90, 84, and 210, then, is $7 \cdot 5 \cdot 3 \cdot 2 \cdot 3 \cdot 2$ which equals 1260.

Let us review the algorithm demonstrated in the preceding paragraph to discover how it works. In our first division, we found that

$$84 = 12 \cdot 7 \quad \text{and} \quad 210 = 30 \cdot 7$$

In our second division, we found that

$$90 = 18 \cdot 5 \quad \text{and} \quad 30 = 6 \cdot 5$$

Then substitution from $30 = 6 \cdot 5$, in the equation $210 = 30 \cdot 7$ gives

$$210 = 6 \cdot 5 \cdot 7$$

In our third division, we found that

$$18 = 6 \cdot 3, \quad 12 = 4 \cdot 3, \quad \text{and} \quad 6 = 2 \cdot 3$$

We substitute these values in the equations $90 = 18 \cdot 5, 84 = 12 \cdot 7$, and $210 = 6 \cdot 5 \cdot 7$, and we get

$$90 = 6 \cdot 3 \cdot 5, \quad 84 = 4 \cdot 3 \cdot 7, \quad \text{and} \quad 210 = 2 \cdot 3 \cdot 5 \cdot 7$$

In our fourth division, we found that

$$6 = 2 \cdot 3, \quad 4 = 2 \cdot 2, \quad \text{and} \quad 2 = 1 \cdot 2$$

We substitute these values in $90 = 6 \cdot 3 \cdot 5$, $84 = 4 \cdot 3 \cdot 7$, and $210 = 2 \cdot 3 \cdot 5 \cdot 7$, and we get

$$90 = 2 \cdot 3 \cdot 3 \cdot 5$$
$$84 = 2 \cdot 2 \cdot 3 \cdot 7$$

and

$$210 = 1 \cdot 2 \cdot 3 \cdot 5 \cdot 7$$

This shows the prime factorizations of the given numbers to be

$$90 = 2 \cdot 3^2 \cdot 5$$
$$84 = 2^2 \cdot 3 \cdot 7$$

and

$$210 = 2 \cdot 3 \cdot 5 \cdot 7$$

and the L.C.M. to be $2^2 \cdot 3^2 \cdot 5 \cdot 7$. The L.C.M. is equal to the product of these prime factors with each factor taken to the power that indicates the greatest number of times it occurred as factor.

Up to this point, addition of rational numbers a/b and c/d was expected to be done by using methods that could be justified by the definition or the theorem on addition of rational numbers. In this section, we have found methods for discovering the L.C.M. of the denominators b and d. This L.C.M. (actually a lowest common denominator, L.C.D.) may be used as the denominator of each of the rational numbers to be added. For example, suppose we wish to add

$$\frac{3}{4} + \frac{3}{14}$$

Factoring both denominators completely, we get

$$4 = 2^2 \quad \text{and} \quad 14 = 2 \cdot 7$$

The L.C.D., then, is $2^2 \cdot 7$, that is, 28. Then since $\frac{7}{7} = 1$ and $\frac{2}{2} = 1$, we have

$$\frac{3}{4} \cdot \frac{7}{7} = \frac{21}{28}; \quad \frac{3}{14} \cdot \frac{2}{2} = \frac{6}{28}$$

and thus

$$\frac{3}{4} + \frac{3}{14} = \frac{21}{28} + \frac{6}{28} = \frac{27}{28}$$

EXERCISES 7.6

1. Find the L.C.M. of each set of integers by the method of obtaining the prime factorizations of each number:

(a) 6, 10, 3   (b) 9, 15, 21
(c) 4, 12, 21   (d) 5, 20, 44
(e) 3, 5, 47   (f) 17, 31, 47
(g) 30, 140, 420   (h) 165, 429, 495

2. Find the L.C.M. of each set of integers by the method of "division" illustrated in the text.

(a) 12, 200, 210   (b) 24, 462, 1694
(c) 182, 189, 819   (d) 245, 297, 1155
(e) 36, 78, 598, 828   (f) 16, 272, 348, 408

3. Use the L.C.D. and find each sum. Express each sum in simplest form.

(a) $\dfrac{5}{6} + \dfrac{^-7}{18}$

(b) $\dfrac{5}{12} + \dfrac{7}{15}$

(c) $\dfrac{7}{20} + \dfrac{11}{32}$

(d) $\dfrac{^-13}{27} + \dfrac{2}{21}$

(e) $\dfrac{^-4}{9} + \dfrac{7}{15} + \dfrac{8}{21}$

(f) $\dfrac{1}{4} + \dfrac{5}{12} + \dfrac{^-4}{21}$

(g) $\dfrac{3}{5} + \dfrac{9}{20} + \dfrac{27}{44}$

(h) $\dfrac{4}{17} + \dfrac{^-26}{51} + \dfrac{5}{9}$

(i) $\dfrac{5}{6} + \dfrac{7}{12} + \dfrac{11}{30}$

(j) $\dfrac{^-13}{27} + \dfrac{5}{12} + \dfrac{^-13}{24}$

7.7 RATIONAL NUMBERS AND THE NUMBER LINE

We have discussed and illustrated how the set of integers may be represented by points on a number line. Any rational number may also be represented by a point on a number line. The rational number $\frac{1}{4}$, for example, may be represented by the point which is $\frac{1}{4}$ the distance from the origin to unit point P_1. The points $P_{1/4}$, $P_{2/4}$ (that is, $P_{1/2}$), and $P_{3/4}$ divide the line segment $\overline{P_0 P_1}$ into four equal parts.

The line segment $\overline{P_0P_{1/4}}$ can also be subdivided and this process can be repeated over and over as long as we like. There is a point (actually many points) that divides any given line segment into two parts; there is a rational number between any two given rational numbers. We describe this property of the set of rational numbers by saying that the set of rational numbers is a **dense set**.

On any number line each rational number a/b is the coordinate of a point $P_{a/b}$. As with integers, any two rational numbers a/b and c/d may be ordered by considering their graphs on a number line. When the unit point P_1 is on the right of the origin

$$\frac{a}{b} < \frac{c}{d} \quad \text{iff } P_{a/b} \text{ is on the left of } P_{c/d}$$

$$\frac{a}{b} = \frac{c}{d} \quad \text{iff } P_{a/b} = P_{c/d} \text{ (same point)}$$

$$\frac{a}{b} > \frac{c}{d} \quad \text{iff } P_{a/b} \text{ is on the right of } P_{c/d}$$

Thus the rational numbers have the **trichotomy property**. Exactly one of the relations

$$\frac{a}{b} < \frac{c}{d}, \quad \frac{a}{b} = \frac{c}{d}, \quad \frac{a}{b} > \frac{c}{d}$$

must hold for any two given rational numbers a/b and c/d.

The ordering of rational numbers may also be stated algebraically. Any rational number e/f may be expressed in the form a/b where $b > 0$ as in Exercises 7.2, Exercise 2. Thus any two given rational numbers may be expressed in the form $a/|b|$ and $c/|d|$. Then

$$\frac{a}{|b|} < \frac{c}{|d|} \quad \text{iff } a \cdot |d| < c \cdot |b|$$

$$\frac{a}{|b|} = \frac{c}{|d|} \quad \text{iff } a \cdot |d| = c \cdot |b|$$

$$\frac{a}{|b|} > \frac{c}{|d|} \quad \text{iff } a \cdot |d| > c \cdot |b|$$

We can put all three statements into the following concise symbolic form:

$$\frac{a}{b} \lesseqgtr \frac{c}{d} \quad \text{iff } ad \lesseqgtr bc \qquad (b \text{ and } d > 0)$$

EXERCISES 7.7

1. Is the set of integers a dense set?

2. Replace the asterisk by one of the symbols $<, =, >$ to obtain a true statement.

(a) $\dfrac{7}{16} * \dfrac{8}{17}$ (b) $\dfrac{3}{11} * \dfrac{36}{132}$

(c) $\dfrac{16}{31} * \dfrac{64}{125}$ (d) $\dfrac{20}{51} * \dfrac{4}{5}$

(e) $\dfrac{0}{6} * \dfrac{1}{5}$ (f) $\dfrac{0}{7} * \dfrac{^-1}{5}$

(g) $\dfrac{33}{132} * \dfrac{^-21}{97}$ (h) $\dfrac{^-15}{91} * \dfrac{34}{203}$

(i) $\dfrac{^-12}{132} * \dfrac{^-31}{343}$ (j) $\dfrac{11}{17} * \dfrac{42}{65}$

3. Order the numbers in the following set from smallest to largest.

$$\dfrac{^-3}{7}, \dfrac{16}{3}, 3, \dfrac{77}{79}, \dfrac{16}{32}, \dfrac{369}{371}, 0, \dfrac{14}{15}, \dfrac{1}{2}, \dfrac{35}{132}, \dfrac{^-1}{4}, \dfrac{65}{63}, \dfrac{^-7}{12}, \dfrac{^-22}{45}, \dfrac{45}{22}, \dfrac{7}{5}, \dfrac{^-64}{13}$$

4. List the 17 properties of rational numbers that have been mentioned in this chapter.

7.8 DECIMAL REPRESENTATION OF RATIONAL NUMBERS

The set of all rational numbers a/b that can be expressed as numerals where $b = 10^n$ for some positive integer n is a very important and useful subset of the rational numbers. The elements of this set are called **terminating decimal fractions**. In addition to having special denominators, the elements of this set also have a special notation involving a *decimal point* and an extension of the concept of *place value*. In order to explain this notation, we introduce and explain the idea of an *integral exponent*. The student may, at this point, wish to review the discussion of exponents in Section 4.2.

In Section 4.2 the concept of an exponent was introduced solely to develop the idea of place value. Therefore, only powers of certain positive integers were used and the discussion was limited to consideration of numerals with non-negative integral exponents. In order to understand *decimal fractions*, we still need to consider only powers of positive integers (actually just the positive integer 10). We shall, however, define a^n for any *rational number a* and any positive integer n:

$$a^n = a \cdot a \cdot a \ldots \cdot a$$

where a is used as a factor n times. For any *rational number $a \neq 0$*,

$$a^0 = 1$$

To complete our understanding of an **integral exponent**, we need one more definition. For any rational number $a \neq 0$ and any positive integer n,

$$a^{-n} = \frac{1}{a^n}$$

When $a = 10$, we have

$$10^{-1} = \frac{1}{10^1} = \frac{1}{10}$$

$$10^{-2} = \frac{1}{10^2} = \frac{1}{100}$$

$$10^{-3} = \frac{1}{10^3} = \frac{1}{1000}$$

etc. We have chosen the base 10 to illustrate negative exponents, because it is negative powers of 10 that are involved in decimal fractions. The **decimal fraction notation** of these powers are

$$10^{-1} = \frac{1}{10} = \mathbf{0.1}$$

$$10^{-2} = \frac{1}{100} = \mathbf{0.01}$$

$$10^{-3} = \frac{1}{1000} = \mathbf{0.001}$$

etc. The dot (period) in decimal fraction notation is the **decimal point**.

We can link this notation for negative powers of 10 to an extension of the place value concept of whole numbers very easily. If we notice, for example, that $0.563 = 563/1000$

$$= \frac{500}{1000} + \frac{60}{1000} + \frac{3}{1000}$$

$$= \frac{5}{10} + \frac{6}{100} + \frac{3}{1000}$$

$$= (5 \times 10^{-1}) + (6 \times 10^{-2}) + (3 \times 10^{-3})$$

we see that the consecutive **place values** to the right of the decimal point in the decimal numeral are 10^{-1}, 10^{-2}, and 10^{-3}, that is, $1/10$, $1/100$, and $1/1000$. In general, the consecutive place values to the right of the decimal point in a numeral are 10^{-1}, 10^{-2}, $10^{-3}, \ldots, 10^{-n}, \ldots$. Notice that the exponent of 10 for the nth place to the right of the decimal point is $-n$. Thus the numeral 476.563 means $(4 \times 10^2) + (7 \times 10^1) + (6 \times 10^0) + (5 \times 10^{-1}) + (6 \times 10^{-2}) + (3 \times 10^{-3})$. This is an example of the sort of simple, yet powerfully effective, patterns we find throughout mathematics.

We have described the set of rational numbers that can be expressed as numerals a/b where $b = 10^n$ for some positive integer n as an important subset of the rational numbers. It is a *proper subset* since not every rational number can be expressed exactly as a numeral with a positive integral power of 10 as its denominator; that is, not every rational number can be expressed as a terminating decimal fraction. However, every rational number can be expressed as a **periodic** (repeating) **decimal fraction**. (*Decimal fractions* are often called decimals; we shall adopt this practice for the sake of simplicity.) In other words, when we express a rational number a/b as a decimal by dividing b into a, the quotient will always contain an ordered set of digits, called its **cycle**, that is repeated indefinitely. For example, we express the number $\frac{1}{3}$ as a decimal:

$$
\begin{array}{r}
0.333\ldots \\
3\overline{\smash{\big)}\,1.000\ldots} \\
\underline{9} \\
10 \\
\underline{9} \\
10 \\
\underline{9} \\
1
\end{array}
$$

It is obvious that the decimal that represents $\frac{1}{3}$ has a *cycle* consisting of the digit 3 that repeats indefinitely. In writing the quotient, we customarily identify the cycle by placing a bar over it, for example, $\frac{1}{3} = 0.\overline{3}$.

As another example, let us express the number $\frac{5}{7}$ as a decimal.

$$
\begin{array}{r}
0.7142857\ldots \\
7\overline{\smash{\big)}\,5.0000000\ldots} \\
\underline{4\,9} \\
10 \\
\underline{7} \\
30 \\
\underline{28} \\
20 \\
\underline{14} \\
60 \\
\underline{56} \\
40 \\
\underline{35} \\
50 \\
\underline{49} \\
1
\end{array}
$$

In this division process, it is important that the reader recognize that once we have placed a decimal point in the divisor and begun adding zeros, a cycle has been obtained as soon as we get a remainder that is identical to another remainder. Thus $\frac{5}{7} = 0.\overline{714285}$

Next we express $\frac{3}{4}$ as a decimal.

$$
\begin{array}{r}
0.7500\ldots \\
4\,\overline{)\,3.0000\ldots} \\
2\,8 \\ \hline
20 \\
20 \\ \hline
00 \\
0 \\ \hline
00 \\
0 \\ \hline
0
\end{array}
$$

We find that $\frac{3}{4} = 0.75\overline{0}$; the cycle consists of the single digit 0. Thus

$$
\frac{3}{4} = 0.75\overline{0} = 0.75 = \frac{75}{100}
$$

In this case, as in every case where the cycle consists of the single digit 0, we have a **terminating decimal**.

Every periodic decimal represents a rational number and can be expressed as a common fraction. Let us look at some examples:

1. Express $0.\overline{34}$ in the form a/b. Let N represent $0.\overline{34}$ and consider $100N$. The number of digits in the cycle has determined the power of 10, in this case, 10^2, that is used as a multiplier.

$$
100N = 34.\overline{34}
$$

Subtract
$$
\frac{-N = 0.\overline{34}}{99N = 34}
$$

$$
\left(\frac{1}{99}\right) \cdot 99N = \frac{1}{99} \cdot 34
$$

$$
N = \frac{34}{99}
$$

Since

$$
N = 0.\overline{34}, \; 0.\overline{34} = \frac{34}{99}
$$

2. Express $2.0\overline{34}$ in the form a/b. Let N represent $2.0\overline{34}$. In this case we select $1000N$ and $10N$ to subtract so that the decimal point, for each numeral, precedes a cycle.

$$1000N = 2034.\overline{34}$$
$$10N = 20.\overline{34}$$
$$\overline{990N = 2014}$$

$$N = \frac{2014}{990}$$

3. Express $0.30\overline{273}$ in the form a/b. Let N represent $0.30\overline{273}$. We select $100{,}000N$ and $100N$ for subtraction so that the decimal point, for each numeral, precedes a cycle.

$$100{,}000N = 30273.\overline{273}$$
$$100N = 30.\overline{273}$$
$$\overline{99{,}900N = 30243}$$

$$N = \frac{30243}{99900}$$

EXERCISES 7.8

1. Express each number in periodic decimal form.

(a) $\frac{1}{5}$ (b) $\frac{3}{5}$ (c) $\frac{6}{5}$

(d) $\frac{1}{9}$ (e) $\frac{3}{9}$ (f) $\frac{1}{7}$

(g) $\frac{2}{7}$ (h) $\frac{3}{7}$ (i) $\frac{4}{7}$

(j) $\frac{1}{17}$ (k) $\frac{1}{16}$ (l) $\frac{1}{18}$

(m) $\frac{1}{25}$ (n) $\frac{1}{35}$

2. Express each number as a common fraction a/b.

(a) $0.\overline{23}$ (b) $0.\overline{462}$ (c) $2.\overline{47}$

(d) $2.4\overline{7}$ (e) $1.\overline{2}$ (f) $2.0\overline{47}$

(g) $0.00\overline{39}$ (h) $3.0\overline{272}$ (i) $30.\overline{03}$

(j) $30.0\overline{3}$ (k) $0.\overline{9}$

7.9 OPERATIONS WITH DECIMAL FRACTIONS

One advantage of the use of decimal fractions is the ease with which *addition* and *subtraction* can be performed. The process of rewriting each fraction in the problem as an equivalent fraction having the common denominator can be resolved by writing the decimals on separate rows with the decimal points in a vertical line. For example, in adding 35.3, 426.729, 0.0065, and 3.70682, we arrange the numerals like this and add

$$
\begin{array}{r}
35.3 \\
426.729 \\
0.0065 \\
+\,3.70682 \\
\hline
\end{array}
$$

Why does the above procedure amount to rewriting the numerals with a common denominator? To answer this question, we add zeros so that all decimals have the same number of places (in this case 5) to the right of the decimal point:

$$
\begin{array}{r}
35.30000 \\
426.72900 \\
0.00650 \\
+\,3.70682 \\
\hline
\end{array}
$$

We could express each of these numbers as common fractions with denominator 10^5, that is, 100,000. What would each numerator be?

In subtraction, as well as in addition, the process of "lining up" the decimal points serves to resolve the problem of finding the common denominator.

Let us now consider multiplication of decimals. The procedure is like that of multiplying integers, except for placing a decimal point in the product. The rule with which we are familiar in multiplication is

The number of decimal places in the product is the same as the number of combined decimal places in the factors.

Consider the product of 6.479 and 7.32:

$$
\begin{array}{r}
6.479 \\
7.32 \\
\hline
12958 \\
19437 \\
45353 \\
\hline
47.42628 \\
\hline
\end{array}
$$

The number of combined decimal places in the factors is 5; therefore we place the decimal point in the product so that there are 5 decimal places.

We shall now show that the traditional rule for multiplication of decimals can be logically justified. Consider that 6.479 can be expressed as 6479/1000, and 7.32 as 732/100. Then

$$
\frac{6479}{1000} = 6479 \cdot \frac{1}{1000}
$$

and

$$\frac{732}{100} = 732 \cdot \frac{1}{100}$$

Now

$$\frac{6479}{1000} \cdot \frac{732}{100} = \frac{6479 \cdot 732}{1000 \cdot 100} = \frac{4{,}742{,}628}{100{,}000}$$

$$= 47.42628$$

This verifies the rule for a particular case. In order to generalize our explanation, it is desirable to introduce a property (law) of exponents under multiplication where *powers of the same base* are involved. Consider that

$$\frac{6479}{1000} \cdot \frac{732}{100} = 6479 \cdot \frac{1}{1000} \cdot 732 \cdot \frac{1}{100}$$

$$= (6479 \cdot 732) \cdot \frac{1}{1000} \cdot \frac{1}{100}$$

$$= (6479 \cdot 732) \cdot 10^{-3} \cdot 10^{-2}$$

$$= 4{,}742{,}628 \cdot \frac{1}{100{,}000}$$

$$= 4{,}742{,}628 \cdot 10^{-5}$$

Notice that the exponent of 10 in the final product is the *sum of the exponents* in the expression

$$(6479 \cdot 732) \cdot 10^{-3} \cdot 10^{-2}$$

This is a consequence of the fact that

$$10^{-3} \cdot 10^{-2} = \frac{1}{10^3} \cdot \frac{1}{10^2}$$

$$= \frac{1}{10 \cdot 10 \cdot 10} \cdot \frac{1}{10 \cdot 10}$$

$$= \frac{1}{10 \cdot 10 \cdot 10 \cdot 10 \cdot 10}$$

$$= \frac{1}{10^5}$$

$$= 10^{-5}$$

Let us use any rational number a and multiply $a^5 \cdot a^2$. By definition, a^5 is the product of 5 a's and a^2 is the product of 2 a's. Thus

$$a^5 \cdot a^2 = (a \cdot a \cdot a \cdot a \cdot a) \cdot (a \cdot a)$$

$$= (a \cdot a \cdot a \cdot a \cdot a \cdot a \cdot a)$$

$$= a^7$$

Let us look at some other examples:

(1)
$$a^3 \cdot a^4 = (a \cdot a \cdot a) \cdot (a \cdot a \cdot a \cdot a)$$

$$= a \cdot a \cdot a \cdot a \cdot a \cdot a \cdot a$$

$$= a^7$$

(2)
$$a^4 \cdot a^{-3} = (a \cdot a \cdot a \cdot a) \cdot \frac{1}{a \cdot a \cdot a}$$

$$= \frac{a \cdot a \cdot a \cdot a}{a \cdot a \cdot a}$$

$$= \left(\frac{a \cdot a \cdot a}{a \cdot a \cdot a}\right) \cdot a$$

$$= 1 \cdot a$$

$$= a$$

Note: Any number a, written with no exponent, is understood to have the exponent "1," that is, $a = a^1$.

(3)
$$a^{-5} \cdot a^7 = \frac{1}{a \cdot a \cdot a \cdot a \cdot a} \cdot (a \cdot a \cdot a \cdot a \cdot a \cdot a \cdot a)$$

$$= \frac{a \cdot a \cdot a \cdot a \cdot a \cdot a \cdot a}{a \cdot a \cdot a \cdot a \cdot a}$$

$$= \frac{a \cdot a \cdot a \cdot a \cdot a}{a \cdot a \cdot a \cdot a \cdot a} \cdot (a \cdot a)$$

$$= 1 \cdot a^2$$

$$= a^2$$

(4)
$$a^{-3} \cdot a^{-6} = \frac{1}{a \cdot a \cdot a} \cdot \frac{1}{a \cdot a \cdot a \cdot a \cdot a \cdot a}$$

$$= \frac{1}{a \cdot a \cdot a \cdot a \cdot a \cdot a \cdot a \cdot a \cdot a}$$

$$= \frac{1}{a^9}$$

$$= a^{-9}$$

(5)
$$a^4 \cdot a^{-4} = (a \cdot a \cdot a \cdot a) \cdot \frac{1}{a \cdot a \cdot a \cdot a}$$

$$= \frac{a \cdot a \cdot a \cdot a}{a \cdot a \cdot a \cdot a}$$

$$= 1$$

In each of these examples, the reader will observe that the final result could have been achieved by simply adding the exponents:

(1) $\qquad\qquad a^3 \cdot a^4 = a^{3+4} = a^7$

(2) $\qquad\qquad a^4 \cdot a^{-3} = a^{4+^-3} = a^1 = a$

(3) $\qquad\qquad a^{-5} \cdot a^7 = a^{-5+7} = a^2$

(4) $\qquad\qquad a^{-3} \, a^{-6} = a^{-3+^-6} = a^{-9}$

(5) $\qquad\qquad a^4 \cdot a^{-4} = a^{4+^-4} = a^0 = 1$

The **law of integral exponents under multiplication** that we can induce from these examples may be stated symbolically as

$$\mathbf{a^m \cdot a^n = a^{m+n}} \qquad (m \text{ and } n \text{ integers})$$

Any terminating decimal can be expressed as the product of an integer and some negative integral power of 10. Thus the product of two such decimals $a \cdot 10^{-m}$ and $b \cdot 10^{-n}$ (where $a, b, m,$ and n are integers) may be expressed as

$$(a \cdot 10^{-m}) \cdot (b \cdot 10^{-n}) = (a \cdot b) \cdot 10^{-m+^-n}$$

$$= (a \cdot b) \cdot 10^{-(m+n)}$$

$$= \frac{a \cdot b}{10^{m+n}}$$

Observe that the number $m + n$ is the combined number of decimal places in the factors $a \cdot 10^{-m}$ and $b \cdot 10^{-n}$. Thus the traditional rule that we have used in multiplication of decimals is justified.

To develop a satisfactory understanding of division of decimals we need a general law for division of numbers that are powers of the same base. Such a law follows rather directly from the law for multiplication of numbers that are powers of the same base:

$$a^m \cdot a^n = a^{m+n}$$

Suppose we are to divide $a^m \div a^n$:

$$a^m \div a^n = a^m \cdot \frac{1}{a^n} = a^m \cdot a^{-n}$$

Once we have expressed the problem in terms of multiplication, we can apply the law of internal exponents under multiplication.

$$a^m \cdot a^{-n} = a^{m + -n} = a^{m-n}$$

Thus the **law of integral exponents under division** may be stated symbolically as

$$a^m \div a^n = a^{m-n} \qquad (m \text{ and } n \text{ integers})$$

The rule that most of us have learned for placing the decimal point in division is quite mechanical and not very meaningful. It involves shifting decimal points in the divisor and in the dividend, and placing the digits in the quotient very carefully so that the decimal point in the quotient will be directly above the decimal point in the dividend. We shall use the law of integral exponents under division and some properties of rational numbers to demonstrate a more meaningful procedure for division of decimals. Consider the problem $0.768 \div 0.32$.

$$0.768 \div 0.32 = \frac{768 \cdot 10^{-3}}{32 \cdot 10^{-2}} = \frac{768}{32} \cdot \frac{10^{-3}}{10^{-2}}$$

$$= \frac{768}{32} \cdot 10^{-1}$$

Thus we can divide 768 by 32 and multiply the quotient by 10^{-1} to achieve the division $0.768 \div 0.32$.

$$
\begin{array}{r}
24 \\
32\overline{\smash{\big)}768} \\
64 \\
\hline
128 \\
128 \\
\hline
0
\end{array}
$$

We find $0.768 \div 0.32 = 24 \cdot 10^{-1} = 2.4$.

The first example we have used is rather a special case since the quotient is a terminating decimal. We need to look at other examples where the quotient is "not exact" (that is, cannot be expressed as a terminating decimal). In cases of this sort we may obtain quotients to any number of decimal places (accuracy) desired. These quotients are then written as terminating decimals that are actually approximations of the exact quotients (periodic decimals). We use the symbol $\approx$ to indicate the relation "is approximately equal to." Let us examine some examples of division problems where the quotient is not exact. Consider $59.3 \div 0.24$.

$$59.3 \div 0.24 = \frac{593 \cdot 10^{-1}}{24 \cdot 10^{-2}} = \frac{593}{24} \cdot 10$$

$$
\begin{array}{r}
24 \\
24\overline{)593} \\
48 \\
\hline
113 \\
96 \\
\hline
17
\end{array}
$$

Thus far, the work we have done indicates that

$$593 \div 24 = 24\tfrac{17}{24}$$

Now we divide 17 by 24.

$$
\begin{array}{r}
0.7 \\
24\overline{)17.0} \\
16.8 \\
\hline
0.2
\end{array}
$$

Thus $\tfrac{17}{24} = 0.7 + \tfrac{0.2}{24}$ and

$$593 \div 24 = 24\tfrac{17}{24} = 24.7 + \tfrac{0.2}{24}$$

But $\tfrac{0.2}{24} = \tfrac{0.2}{24} \cdot \tfrac{10}{10} = \tfrac{2}{240}$. We divide 2 by 240.

$$
\begin{array}{r}
0.008 \\
240\overline{)2.000} \\
1.920 \\
\hline
0.080
\end{array}
$$

Thus $\tfrac{2}{240} = 0.008 + \tfrac{0.08}{240}$, and

$$593 \div 24 = 24.7 + \tfrac{0.2}{24} = 24.7 + 0.008 + \tfrac{0.08}{240}$$
$$= 24.708 + \tfrac{0.08}{240}$$

Then $\tfrac{59.3}{0.24} = \tfrac{593}{24} \cdot 10 = (24.708 + \tfrac{0.08}{240}) \cdot 10$
$$= 247.08 + \tfrac{0.8}{240}$$

If two-place accuracy is desired, we may write

$$59.3 \div 0.24 \approx 247.08$$

We have worked on this problem in parts to give meaning to the placement of the decimal point. In actual practice this division would not be fragmented, but would be accomplished as we shall show:

$$
\begin{array}{r}
24.7083 \\
24\overline{)593.0000} \\
48 \\
\hline
113 \\
96 \\
\hline
17.0 \\
16.8 \\
\hline
0.200 \\
0.192 \\
\hline
0.0080 \\
0.0072 \\
\hline
0.0008
\end{array}
$$

Recall that we are actually working on the problem $59.3 \div 0.24$, which is equal to $(593 \div 24) \cdot 10$. When the quotient we have obtained thus far is multiplied by 10, the decimal point is shifted one place to the right: $24.7083 \cdot 10 = 247.083$. Three decimal places are enough to give us a *rounded-off* two-place approximation. To **round off** a decimal, we must have it in a form that has at least one more decimal place than the number of places desired in the final approximation; that is, if n-place accuracy is desired, we need at least $n + 1$ digits following the decimal point. If the $(n + 1)$st digit is smaller than 5, we delete all digits following the nth digit; if the $(n + 1)$st digit is greater than or equal to 5, we delete all digits following the nth digit, and we increase the nth digit by 1. For example,

$$2.5763 \approx 2.576$$

$$0.0543562 \approx 0.0544$$

and

$$39.6\overline{707} \approx 39.671$$

To return to the quotient of the problem we have been working on ($59.3 \div 0.24$), the rounded-off result is

$$247.083 \ldots \approx 247.08$$

Here are two more examples.

(a) Divide 745.08 by 2.13. Express the quotient with three-place accuracy.

$$745.08 \div 2.13 = \frac{74508 \cdot 10^{-2}}{213 \cdot 10^{-2}} = \frac{74508}{213} \cdot 10^{0}$$

$$= \frac{74508}{213}$$

```
            349.8028
213 / 74508.0000
      639
      ----
      1060
       852
      ----
      2088
      1917
      -----
       171.0
       170.4
       -----
         0.60
            0
         ------
         0.600
         0.426
         ------
         0.1740
         0.1704
         ------
         0.0036
```

The four places we have obtained in the quotient are sufficient for three-place accuracy.

$$745.08 \div 2.13 \approx 349.803$$

(b) Divide 0.143 by 597.6. Express the quotient to four-place accuracy.

$$0.143 \div 597.6 = \frac{143 \cdot 10^{-3}}{5976 \cdot 10^{-1}} = \frac{143}{5976} \cdot 10^{-2}$$

```
              0.023
5976 / 143.000
         0
       -----
       143.0
           0
       ------
       143.00
       119.52
       ------
        23.480
        17.928
        ------
         5.552
```

Since we desire four-place accuracy, and since

$$0.143 \div 597.6 = \frac{143}{5976} \cdot 10^{-2}$$

$$\approx 0.023 \cdot 10^{-2} = 0.00023$$

we have proceeded far enough to round off.

$$0.143 \div 597.6 \approx 0.0002$$

EXERCISES 7.9

1. Express as a power of the indicated base:

(a) $4^2 \cdot 4$ (b) $3^2 \cdot 3^3$

(c) $2^4 \cdot 2^3$ (d) $2^4 \cdot 2^{-3}$

(e) $4^6 \cdot 4^{-2}$ (f) $4^{-2} \cdot 4^{-1}$

(g) $4^{-5} \cdot 4^2$ (h) $\dfrac{3^2}{3^3}$

(i) $\dfrac{4^2}{4}$ (j) $\dfrac{4^6}{4^{-2}}$

(k) $\dfrac{3^4}{3^{-7}}$ (l) $\dfrac{4^{-5}}{4^2}$

(m) $\dfrac{3^{-2}}{3^{-3}}$ (n) $(3^2 \cdot 3^{-3}) \cdot 3^4$

(o) $3^2 \cdot (3^{-3} \cdot 3^4)$

2. Express in decimal notation using at most four decimal places and noting approximate answers as such.

(a) $33.2 + 161.04 + 0.33$ (b) $6780.211 + 3.0041 + 72.3$

(c) $79.22 + 410.669 - 21.0335$ (d) $366 + 0.043 + 16.2 - 472.335$

(e) $3.67 \cdot 42.04$ (f) $243.09 \cdot 0.0021$

(g) $32.3 \cdot {}^{-}6.9213$ (h) $300.06 \cdot 1.011$

(i) $432.2 \div 1.53$ (j) $43.2 \div 15.3$

(k) $432.2 \div 0.0153$ (l) $43.2 \div 153$

CHAPTER 8

The Real Numbers

8.1 MEMBERSHIP OF THE
SET OF REAL NUMBERS

We have found that each rational number can be represented as a periodic decimal, and conversely, each periodic decimal represents a rational number. There are, however, infinite decimals that are not periodic; these decimals represent **irrational numbers**. It is beyond the scope and intent of this book to develop anything even approaching a thorough treatment of the irrational numbers, but it is necessary to consider them superficially in order to classify such numbers as $\sqrt{2}$ and π, which are frequently encountered in elementary school mathematics, and to develop systematically the set of real numbers.

We now define **the set of real numbers** as the union of the set of rational numbers with the set of irrational numbers. In view of the preceding paragraph we see that an alternate and quite satisfactory definition of the set of real numbers might be: the set of all numbers that may be represented by decimals—finite periodic, infinite periodic, or infinite nonperiodic. Notice that since any rational number may be expressed as a finite decimal or as an infinite periodic decimal, the set of real numbers by its very definition contains the set of rational numbers as a proper subset.

8.2   REAL NUMBERS
AND THE NUMBER LINE

In Chapter 7, we learned that the rational numbers can be represented by points
on the number line and that they are a dense set; that is, between any two rational
numbers there is another rational number. Similarly, between any two points
that are graphs of rational numbers there is a point that is the graph of a rational
number. Since this process may be repeated over and over, each segment
$\overline{P_{a/b}P_{c/d}}$ ($a/b \neq c/d$) contains infinitely many points that have rational numbers as
coordinates. This seems to suggest that *all* points on the number line represent
rational numbers. Actually, despite the density of the rational numbers, there are
more points on the number line that do not have rational numbers as coordinates
than points that do have rational numbers as coordinates.

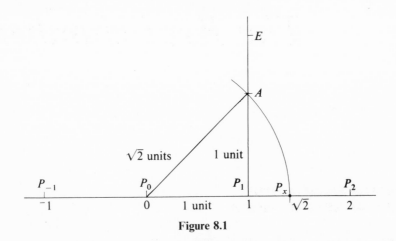

Figure 8.1

We next identify the point with the irrational number $\sqrt{2}$ as its coordinate.
As in Figure 8.1, construct a line $\overleftrightarrow{P_1E}$ that is perpendicular to the number line at P_1.
On the ray $\overrightarrow{P_1E}$ mark off a line segment $\overline{P_1A}$ one unit long. Draw the line segment
$\overline{P_0A}$ to obtain a right triangle P_0P_1A with legs $\overline{P_0P_1}$ and $\overline{P_1A}$ each one unit long.
The hypotenuse $\overline{P_0A}$, by the Pythagorean Theorem,† is $\sqrt{2}$ units long. With a
compass, we can mark the point on the number line whose coordinate is $\sqrt{2}$.
We place the metal point of the compass at the origin and open the compass until
the pencil point is on the point A. The distance between the two points of the
compass, then, is $\sqrt{2}$. Then with the origin P_0 as center we draw a circular arc
and label its intersection with $\overrightarrow{P_0P_1}$ as P_x. The length of the segment $\overline{P_0P_x}$ is the
same as the length of $\overline{P_0A}$ and thus is $\sqrt{2}$; in other words, the coordinate x is $\sqrt{2}$.

---

† The reader is reminded that the Pythagorean theorem is a statement about the lengths of the legs
and hypotenuse of a right triangle: If triangle ABC has legs whose lengths are a and b and hypotenuse
whose length is c, then $a^2 + b^2 = c^2$.

Is $\sqrt{2}$ a rational number? The answer is "No." We need to preface the proof that $\sqrt{2}$ is not a rational number, however, with the proof of another statement: If the square of an integer is even, then the integer is even, that is, if a is an integer, and a^2 is divisible by two, then a is divisible by 2.

> PROOF: Every integer is either an even number or an odd number and hence can be expressed in one of two forms: $2n$ or $2n + 1$, where n is an integer (see Section 6.7). Thus a can be written either as $2n$ or as $2n + 1$. Then $a^2 = 4n^2$ or $a^2 = 4n^2 + 4n + 1$. Under the assumption that a is an integer and a^2 is divisible by 2, we know that a^2 must equal $4n^2$ since $4n^2 + 4n + 1$ is not divisible by 2 for any integer n. Therefore $a = 2n$ and a is divisible by 2.

We are now ready to show that $\sqrt{2}$ is not a rational number. We do so by an indirect approach: We shall assume that $\sqrt{2}$ *is* a rational number and show that this assumption is false because it leads to a contradiction.

> PROOF: Let us assume that $\sqrt{2}$ is a rational number. Then we can write: $\sqrt{2} = a/b$, a rational number in simplest form;
>
> $$(\sqrt{2})^2 = (a/b)^2 = (a/b) \cdot (a/b)$$
> $$2 = a^2/b^2$$
> $$2b^2 = a^2$$
>
> and a^2 is divisible by 2. Then, from the previous proof, a is divisible by 2, and thus
>
> $$a = 2n \quad \text{where } n \text{ is an integer}$$
> $$a^2 = 4n^2$$
>
> We know already that $a^2 = 2b^2$. Therefore
>
> $$2b^2 = 4n^2$$
> $$b^2 = 2n^2$$
>
> Then b^2 is divisible by 2; by the previous theorem b is divisible by 2. It follows that a and b have a common factor, 2, which is contrary to our hypothesis that a/b is in simplest form. We must conclude, therefore, that our assumption that $\sqrt{2}$ is a rational number is false.

The next question we must consider is: Can $\sqrt{2}$ be represented as an infinite decimal expansion? That is, is $\sqrt{2}$ a real number? If $\sqrt{2}$ can be expressed as a decimal at all, it would necessarily be nonperiodic, for we have found (in Chapter 7) that all periodic decimal fractions are representations of rational numbers. We are asking, then, is $\sqrt{2}$ a member of the set of irrational numbers that we have defined as the set of all numbers that can be expressed as nonperiodic decimals? We shall say that the answer to this question is "Yes," but we will not attempt a rigorous proof of it. Rather, we shall use a method of finding successively closer decimal approximations to $\sqrt{2}$ that will lead the student to the intuitive conclusion that $\sqrt{2}$ *can* be represented as an infinite nonperiodic decimal.

By definition, $\sqrt{2}$ is a number n such that $n^2 = 2$. As a first step in expressing $\sqrt{2}$ as a decimal, we square 1, and then we square 2: $1^2 = 1$; $2^2 = 4$. Thus we find that n^2 is greater than 1^2, but smaller than 2^2, that is,

$$1^2 < (\sqrt{2})^2 < 2^2$$

and we assume that

$$1 < \sqrt{2} < 2$$

We can find a closer approximation by squaring 1.1, 1.2, 1.3, etc. Doing this, we obtain

$$(1.1)^2 = 1.21$$
$$(1.2)^2 = 1.44$$
$$(1.3)^2 = 1.69$$
$$(1.4)^2 = 1.96$$
$$(1.5)^2 = 2.25$$

Then

$$(1.4)^2 < (\sqrt{2})^2 < (1.5)^2$$

and

$$1.4 < \sqrt{2} < 1.5$$

Using more arithmetic of the same sort, we can approximate $\sqrt{2}$ more closely by squaring 1.41, 1.42, 1.43, etc.:

$$(1.41)^2 = 1.9881$$
$$(1.42)^2 = 2.0164$$

Then

$$(1.41)^2 < (\sqrt{2})^2 < (1.42)^2$$

and

$$1.41 < \sqrt{2} < 1.42$$

Continuing, we find

$$(1.411)^2 = 1.990921$$

$$(1.412)^2 = 1.993744$$

$$(1.413)^2 = 1.996569$$

$$(1.414)^2 = 1.999396$$

$$(1.415)^2 = 2.002225$$

Then

$$(1.414)^2 < (\sqrt{2})^2 < (1.415)^2$$

and

$$1.414 < \sqrt{2} < 1.415$$

If we cared to do the necessary arithmetic to continue this process, we would find, by the time we had reached seven place decimal fractions,

$$1.4142135 < \sqrt{2} < 1.4142136$$

and the process could be continued beyond this for as many decimal places as we wish. Notice that, in this way, $\sqrt{2}$ can be approximated as closely as desired. Thus we can conclude that $\sqrt{2}$ can be represented as a nonperiodic decimal and that it is an irrational number.

We can locate the point representing the square root of two on the number line by construction, as we have seen. This means there is at least one point on the number line representing a nonperiodic decimal. Actually there are infinitely many such points. For example, if a/b is any rational number other than zero, a number of the form $a/b \cdot \sqrt{2}$ can easily be located on the number line; and we can show that any number of the form $a/b \cdot \sqrt{2}$ is an irrational number. If $a/b \cdot \sqrt{2}$ were a rational number c/d, then

$$\frac{a}{b}\sqrt{2} = \frac{c}{d}$$

$$\frac{b}{a} \cdot \frac{a}{b}\sqrt{2} = \frac{b}{a} \cdot \frac{c}{d}$$

$$\sqrt{2} = \frac{bc}{ad}$$

and $\sqrt{2}$ would be a rational number. Since $\sqrt{2}$ is not a rational number, we have proved that $a/b \cdot \sqrt{2}$ represents an irrational number for any rational number a/b other than zero. Thus there are as many irrational numbers that may be expressed by $a/b \cdot \sqrt{2}$ as there are rational numbers a/b other than zero.

We know that the square of a positive integer is a positive integer. Although we shall not prove it, the square root of any positive integer n that is not the square of an integer is an irrational number. Then there are infinitely many integers n such that $\sqrt{n}$ is an irrational number. If we think about forming other irrational numbers of the form $a/b \cdot \sqrt{n}$, then something of the nature of the number of points on the number line whose coordinates are irrational numbers begins to emerge.

8.3 INFINITE SETS

Recall that in Chapter 1 we demonstrated that the elements of the set of natural numbers can be put into a one-to-one correspondence with the set of even numbers; thus the *number* of elements in these two infinite sets is the same. Further, we can show that the elements of the set of integers can be put into a one-to-one correspondence with the set of natural numbers. This may be done by arranging the integers so that 0 is the first integer, 1 the second, $^-1$ the third, 2 the fourth, $^-2$ the fifth, etc.

$$
\begin{array}{ccccccc}
1 & 2 & 3 & 4 & 5 & 6 & 7 \quad \cdots \\
\updownarrow & \updownarrow & \updownarrow & \updownarrow & \updownarrow & \updownarrow & \updownarrow \\
0 & 1 & ^-1 & 2 & ^-2 & 3 & ^-3 \quad \cdots
\end{array}
$$

An apparently paradoxical property of infinite sets is illustrated by these two examples: *An infinite set can be equivalent to one of its proper subsets.* This is certainly not true of any finite set. But it has been proved that **a set is infinite iff it is equivalent to some proper subset of itself**.

The reader may have an intuitive notion that the number of elements in any infinite set is equal to the number of elements in any other infinite set. Our experience with such infinite sets as the set of natural numbers, the set of even numbers, and the set of integers would tend to confirm this. However, the German mathematician Georg Cantor (1845–1918) discovered and proved that the set of natural numbers is not equivalent to the set of points on a line. Thus there are at least two different numbers associated with infinite sets. The number of elements in the set of natural numbers (and any equivalent infinite sets) is called **aleph null** ($\aleph_0$). Infinite sets with this number of elements are said to be **countable**. Hence the set of even numbers and the set of integers are countable. The number of elements in the set of points on a line (and any equivalent infinite sets) is expressed by the German $\mathfrak{C}$ (representing the number of a "continuum"), and $\mathfrak{C}$ is a far greater number than $\aleph_0$.

The number of elements in the set of rational numbers is $\aleph_0$, that is, the elements of the set of rational numbers can be put into a one-to-one correspondence with the elements of the set of natural numbers. One way to do this begins with an arrangement of the rational numbers in the array shown in Figure 8.2.

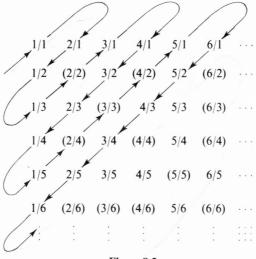

Figure 8.2

The directed line winding its way through the array indicates the sequence we shall use to arrange the positive rational numbers in setting up a one-to-one correspondence between the natural numbers and the rational numbers. The array in Figure 8.2 includes the entire set of positive rational numbers, but each number in the set appears more than once. Since we are making a one-to-one matching, we shall use a rational number the first time its name appears in the sequence and omit it thereafter. This is indicated by placing parentheses around numerals to be omitted. In our matching, we shall include zero as the first rational number and insert the negative rational numbers by placing the additive inverse of each positive number to the right of that number, as pictured in the following array:

$$
\begin{array}{ccccccccccccc}
1 & 2 & 3 & 4 & 5 & 6 & 6 & 8 & 9 & 10 & 11 & 12 & 13 & \cdots \\
\updownarrow & \updownarrow & \updownarrow & \updownarrow & \updownarrow & \updownarrow & \updownarrow & \updownarrow & \updownarrow & \updownarrow & \updownarrow & \updownarrow & \updownarrow & \\
0 & \dfrac{1}{1} & \dfrac{1}{^-1} & \dfrac{2}{1} & \dfrac{2}{^-1} & \dfrac{1}{2} & \dfrac{1}{^-2} & \dfrac{1}{3} & \dfrac{1}{^-3} & \dfrac{3}{1} & \dfrac{3}{^-1} & \dfrac{4}{1} & \dfrac{4}{^-1} & \cdots
\end{array}
$$

We have shown that the set of rational numbers is countable. This is *not* true of the set of irrational numbers. It is impossible to set up a one-to-one correspondence between the elements of the set of irrational numbers and the elements of the set of natural numbers.. No matter how we arrange the irrational numbers, we omit more of them than we include. The reader will recall that certain of the points on the number line have the rational numbers as their coordinates. Recall, further, that there are points, for example $\sqrt{2}$, on the number line that do *not* have rational numbers as their coordinates. Actually, all these points on the number line that do not correspond to rational numbers may be placed into a one-to-one correspondence with the irrational numbers. Since the set of *real numbers* has been defined as the union of the set of rational numbers with the set of irrational numbers (Section 8.1), the set of real numbers is equivalent to the set of points on the number line. Hence **the number of elements in the set of real numbers is** $\mathfrak{C}$. Also, by a property of numbers associated with infinite sets, **the number of elements in the set of irrational numbers is** $\mathfrak{C}$.

EXERCISES 8.3

1. Use the method of successive approximations that was used to find approximations of $\sqrt{2}$ and find an approximation to four decimal places of each of the following:

   (a) $\sqrt{3}$ (b) $\sqrt{5}$ (c) $\sqrt{6}$ (d) $\sqrt{7}$

2. Use an indirect method similar to the one used to prove that $\sqrt{2}$ is not a rational number, and prove that $\sqrt{3}$ is not a rational number.

3. As in Exercise 2, try to prove that $\sqrt{9}$ is not a rational number.

8.4 PROPERTIES OF THE REAL NUMBERS

The subsets of the real numbers that we have studied can be classified as in the following array:

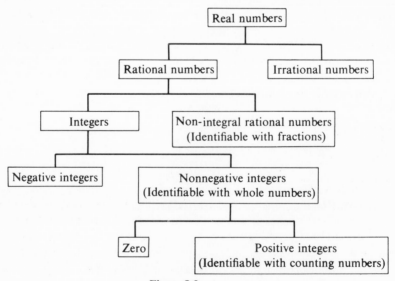

Figure 8.3

The familiar operations of addition, multiplication, subtraction, and division may be defined for real numbers so that the real numbers share the following properties:

1. **The commutative property of addition.** If a and b are real numbers, $a + b = b + a$.

2. **The commutative property of multiplication.** If a and b are real numbers, $a \cdot b = b \cdot a$.

3. **The associative property of addition.** If $a, b,$ and c are real numbers, $(a + b) + c = a + (b + c)$.

4. **The associative property of multiplication.** If $a, b,$ and c are real numbers, $(a \cdot b) \cdot c = a \cdot (b \cdot c)$.

5. **Closure under addition.** If a and b are real numbers, $a + b$ is a real number.

6. **Closure under subtraction.** If a and b are real numbers, $a - b$ is a real number.

7. **Closure under multiplication.** If a and b are real numbers, $a \cdot b$ is a real number.

8. **Closure under division.** If a and b are real numbers and $b \neq 0$, $a \div b$ is a real number.

9. **The distributive property of multiplication over addition.** If $a, b,$ and c are real numbers, $a \cdot (b + c) = (a \cdot b) + (a \cdot c)$.

10. **Identity element for addition.** If a is a real number, there exists an identity element, zero, for addition; that is, $a + 0 = 0 + a = a$.

11. **Identity element for multiplication.** If a is a real number, there exists an identity element, one, for multiplication; that is, $a \cdot 1 = 1 \cdot a = a$.

12. **First unique property of zero.** If a is a real number, then $a \cdot 0 = 0$.

13. **Second unique property of zero.** If a and b are real numbers, and $a \cdot b = 0$, then a or b, or both a and $b = 0$.

14. **The additive inverse property.** For each real number a, there exists a real number, ^-a, such that $a + {}^-a = {}^-a + a = 0$.

15. **The multiplicative inverse property.** For each real number a different from zero, there exists a real number, $1/a$, such that $a \cdot 1/a = 1/a \cdot a = 1$.

16. **The ordering property.** The real number system is ordered, so that if a and b are real numbers, $a \gtreqless b$.

17. **The density property.** The real numbers are a dense set; that is, between any two distinct real numbers, there is always another real number.

18. **The completeness property.** The real number system is complete; that is, to each point on the number line, there corresponds a single real number and conversely, to each real number there corresponds a single point on the number line.

Notice that all except the eighteenth property correspond to properties of rational numbers. Completeness is a difficult property to verify, even intuitively. An attempt to develop its meaning is delayed until Chapter 11, where it is treated in the context of geometry of the line.

The ordering of real numbers a and b is defined, as for integers and rational numbers, in terms of their graphs on a number line. Statements of equality of real numbers also have the same properties (addition, multiplication, reflexive, symmetric, transitive) as for whole numbers.

8.5   ON THE NATURE OF
THE IRRATIONAL NUMBERS

We have used $\sqrt{2}$ as an example of an irrational number. We shall now indicate something of the nature of the other numbers included in the set of irrational numbers. If a, b, and n are natural numbers and $a^n = b$, then b is called the **nth power** of a, and a is called the **nth root** of b (expressed symbolically, $a = \sqrt[n]{b}$). If a/b and c/d are rational numbers expressed in simplest form such that $a^n/b^n = c/d$, then c/d is called a **perfect nth power** of a/b, and a/b is called a **perfect nth root** of c/d. If c/d is a positive rational number expressed in simplest form, it is a perfect nth power iff both c and d are nth powers of natural numbers. For example, $\frac{9}{25}$ is a perfect second power (square), because $9 = 3^2$ and $25 = 5^2$; $\frac{1}{8}$ is a perfect third power (cube), because $1 = 1^3$ and $8 = 2^3$; and $\frac{32}{50}$, which can be written as $\frac{16}{25}$, is a perfect second power because $16 = 4^2$ and $25 = 5^2$. A number such as $\frac{3}{8}$ is not a perfect nth power because, while $8 = 2^3$, 3 is not a third power of any natural number. One subset of the set of irrational numbers is the set of *all nth roots of positive rational numbers that are not themselves perfect nth powers.*

There are infinitely many irrational numbers that can be expressed as roots of rational numbers. There are also infinitely many irrational numbers that cannot be expressed as roots of rational numbers. Among these are most of the logarithms that many of you studied in high school and most of the trigonometric ratios that we shall consider in Chapter 13.

EXERCISES 8.5

1. Identify each of the following numbers as a rational number or an irrational number.

(a) $\sqrt{6}$ (b) $3\sqrt{5}$ (c) $5.67\overline{43}$

(d) $5.67432213112\ldots$ (e) $\dfrac{3\sqrt{5}}{7\sqrt{5}}$ (f) $4 + \sqrt{9}$

(g) $4 + \sqrt{19}$ (h) $\frac{3}{7} \cdot \sqrt{5}$ (i) $\sqrt[4]{\frac{3}{7}}$

(j) $\frac{72}{11}$ (k) $1.25\overline{00}$ (l) $4.34433444333\ldots$

2. Which of the following sets do you think can be placed in one-to-one correspondence with a subset of the natural numbers? Try to show a correspondence if you think one exists.

(a) The set of all integers between $^-10$ and 10 inclusive.

(b) The set of square roots of the natural numbers.

(c) The set of odd integers.

(d) The set of all nth roots of the natural numbers. (n a natural number.)

(e) The set of all rational numbers between 1 and 2 inclusive.

(f) The set of all irrational numbers between 1 and 2 inclusive.

3. If $a^x = b$, then the **logarithm** of b, to the base a, is x. This statement is written symbolically as: If $a^x = b$, then $\log_a b = x$. Use this definition of a logarithm and decide which of the following logarithms are integers:

(a) $\log_{10} 10$ (b) $\log_{10} 1$ (c) $\log_{10} 1000$

(d) $\log_{10}(1/10)$ (e) $\log_{10}(1/10{,}000)$ (f) $\log_{10}(1/100)$

4. Logarithms of integers to an integral base are either integers or irrational numbers. Identify each given logarithm either as an integer or as an irrational number.

(a) $\log_2 8$ (b) $\log_{10} 15$ (c) $\log_3 3$

(d) $\log_3 4$ (e) $\log_{10} 20$ (f) $\log_{10} 100$

CHAPTER 9

Experimental and Formal Geometry

9.1 GEOMETRY AND THE EVOLUTION OF MATHEMATICS

The purpose of this chapter is to preface a rather careful modern development of topics in Euclidean geometry with some discussion about the nature of geometric truths and the ways in which they have been (and are) discovered.

During a class discussion on the nature of mathematics, an instructor declared that mathematics had been created by man, motivated by needs common to us all. The reaction of a student to this was, "Who needs it? No one has asked me if I need mathematics!" In a sense, however, this student had been "asked." Certainly some of his goals may be identified with the perpetual search by man to meet his physical needs and to meet the need, peculiar to the human animal, to use his brain creatively. Mathematics has constantly played a major role in meeting both of these sorts of needs.

The evolution of mathematics through the ages can probably best be illustrated by the changes in geometry. The beginnings of geometry extend backward in time to the ages when tools and artifacts man left behind were the only evidence he gave us of his existence and activities. It seems that his use of geometry in those prehistoric times was primarily to meet his physical needs, although designs on cave walls and on remains of utensils, tools, etc., indicate that the need to be creative was already present. In the period of early recorded history, the record shows that more sophistication was achieved in geometry than in any other area of mathematics. For example, the development of an effective system of numeration, which was to make advances in algebra possible, occurred relatively late in history.

The earliest geometric facts were discovered experimentally. The ancient Chinese, Babylonians, and Egyptians did not value geometry as a mental exercise, but as a tool to solve physical problems arising from such activities as architecture, road building, irrigation, astronomy, calendar construction, etc. Thus it is apparent that the first phase in the evolution of mathematics (geometry, in particular) was related principally to man's need to meet his physical needs. We call this sort of mathematics **applied mathematics**; as we have said, in ancient times it came about through experiment. The methods used to discover facts were inductive. Man learned geometric facts about physical objects by working with them.

The next phases in the evolution of mathematics may be related to man's need to use his mind creatively. There are two notable phases related to this need. The Greeks are given credit for initiating the first of these, when they introduced a new element into the search for mathematical truths. They were not so much interested in building bridges as in satisfying their intellectual curiosity. They certainly believed that the truths they were seeking were about the physical world, but they found that more could be learned by applying deductive logic to idealizations of physical objects than by experimenting with the objects themselves. Thus their mathematics was a step removed from the applied mathematics of the Egyptians. It was "purer" mathematics in the sense that it required more use of man's imagination and reasoning than of his ability to manipulate objects.

The final phase in the evolution of mathematics, which was the second phase that may be related to man's need to use his mind creatively, is very recent, and it is closely linked to geometry. The beginning of this phase did not come until the nineteenth century, when three geometers, Bolyai (a Hungarian), Lobachevsky (a Russian), and Riemann (a German), came to the conclusion that a system of mathematics need not be related to the physical universe at all. Non-Euclidean geometries were developed by each of these men, and their systems had no apparent application to the physical universe. The feeling shared by mathematicians from the time of Euclid and Pythagoras that a mathematician must derive "self-evident truths" from his environment and use them to develop more profound "truths" about that environment, was a restriction to the free exercise of man's imagination. With this restriction gone, **pure mathematics** has become truly "pure"; it is purely an activity of the mind and imagination. The only limits placed on the pure mathematician today are the limitations of his own mind. (Read the excerpt by Bertrand Russell, page 1, again.)

This change in man's concept of the nature of mathematics has been one of the most important factors in the explosive advance in mathematics during the present century. Strange to say, it does not mean that applications of mathematics to the problems of the physical world have diminished. To the contrary, applications of mathematics to the solutions of physical problems have experienced explosive advances also. Witness the wonders of modern science! Curiously, much of modern "pure" mathematics has been found to be profoundly applicable to the most complicated problems at the frontiers of every branch of physical research.

In the remaining chapters of this book we are going to be involved mainly with illustrating the second phase in the evolution of geometry—the phase introduced by Pythagoras, Euclid, and others when they began to develop geometry as a logical system. Imperfections in their approach will be discussed in this chapter, so that the logical system that we shall display will reflect modern improvements.

We shall use a geometric vocabulary in this chapter that we assume the reader has acquired in junior high school or in high school mathematics. Precise definition and understanding of terms will not be necessary until we begin our systematic development of Euclidean geometry in Chapter 10. However, we offer a list of geometric terms and notation that we shall use in this chapter. This list follows:

compass (an instrument used to draw circles and circular arcs; "compass point" refers to the metal point of the instrument; arcs and circles are drawn with the "pencil point.")

protractor (an instrument used to measure angles)

point

line
 Notation for lines: $\overleftrightarrow{AB}$, $\overleftrightarrow{CD}$, etc.

line segment
 Notation for line segments: $\overline{AB}$, $\overline{CD}$, etc.
 Notation for lengths of line segments: $m(\overline{AB})$, $m(\overline{CD})$, etc.

midpoint

endpoint

perpendicular bisector

congruent segments
 Notation: $\overline{AB} \cong \overline{CD}$

circle

arc of a circle
  Notation: $\overarc{AB}$, $\overarc{CD}$, etc.
 radius of a circle or arc

 diameter of a circle

 chord of a circle

line determined by two points

midpoint of an arc

ray
    Notation for rays: $\overrightarrow{AB}$, $\overrightarrow{CD}$, etc.

angle (union of two rays having a common endpoint)
        Notation for angles: $\angle A$, $\angle ABC$, etc.
        Notation for measures of angles: $m(\angle A)$, $m(\angle ABC)$, etc.

  vertex of an angle

  sides of an angle

  angle bisector

perpendicular lines or segments

parallel lines

triangle
        Notation: $\triangle ABC$, $\triangle DEF$, etc.

  sides of a triangle

  angles of a triangle

  vertices of a triangle

  included side

  included angle

congruent triangles

  congruence theorems for triangles

  SAS: If two sides and the included angle of one triangle are congruent respectively to two sides and the included angle of a second triangle, then the triangles are congruent.

  ASA: If two angles and the included side of one triangle are congruent respectively to two angles and the included side of a second triangle, then the triangles are congruent.

  SSS: If three sides of one triangle are congruent respectively to three sides of a second triangle, then the triangles are congruent.

acute angle

right angle

obtuse angle

altitude of a triangle

median of a triangle

centroid of a triangle

isosceles triangle

equilateral triangle

equiangular triangle

quadrilateral

parallelogram

rectangle

square

noncollinear points

Geometry may be defined as the science of space relationships. The "space" under consideration may be the physical universe in which we live and with which we have visual and tactual contact, or it may be a "space" abstracted from this physical space. In the latter case we do not study actual physical objects, but we think and reason about *idealized* objects.

9.2   EXPERIMENTAL GEOMETRY

Experimental geometry is a name that may be used to identify the study of the actual physical space mentioned earlier. Historically, it was the first kind of geometry to be considered. Problems of land measurement and surveying and construction of buildings, roads, and bridges necessitated discovery of certain properties of objects in physical space. These properties were discovered experimentally by observation and, sometimes, by manipulation of the objects. This will be recognized by the reader as an inductive process: that of coming to a conclusion based on observation of many particular cases. Such conclusions are never more than probable and, indeed, are often inaccurate. The possibility of error arises from (1) imperfection of human and mechanical instruments of observation, and (2) the fact that not all existing cases can be observed.

The value of π, the ratio of the circumference of a circle to its diameter, is important to the solution of many physical problems. It has been indicated in Chapter 8 that π is an irrational number. In order to use π in computation, a sufficiently close rational approximation must be found. Long before adequate mathematical means were available to approximate π accurately, several rational approximations were found experimentally. An ancient document called the Rhind Papyrus, for example, reveals that the Egyptians, by experimental means, approximated a value of π equivalent to 256/81 or about 3.16. This was not exact, but it was a close enough approximation for the practical purposes of the merchants, astronomers, and architects of that age.

In Exercises 9.2, the student can experience the inductive discovery of some geometric facts. Some basic compass and straightedge constructions are needed for the exercises, and the use of a protractor will also be required. The following are needed and are reviewed here to refresh your memory.

1. *Bisection of a line segment $\overline{AB}$.* Place the compass point at A. With radius greater than $\frac{1}{2}m(\overline{AB})$, draw arcs a and b with center A as in Figure 9.1. Place the compass point at B. With the same radius, draw arcs c and d with center B. Label the intersection of arcs a and c, C; label the intersection of arcs b and d, D. Draw

the line $\overleftrightarrow{CD}$ determined by points C and D. The line $\overleftrightarrow{CD}$ is called the perpendicular bisector of $\overline{AB}$. The point P where $\overleftrightarrow{CD}$ meets $\overline{AB}$ bisects $\overline{AB}$; P is called the midpoint of $\overline{AB}$.

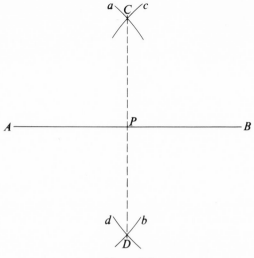

Figure 9.1

2. *Bisection of an angle.* Place the compass point at the vertex A of the angle. Then draw an arc intersecting the sides of the angle at B and C. With the compass point at B and radius greater than $\frac{1}{2}m(\overline{BC})$, draw arc a. With compass point at C, draw arc b intersecting a at D. Draw the ray $\overrightarrow{AD}$. The ray $\overrightarrow{AD}$ bisects $\angle BAC$; that is, divides the angle into two congruent angles.

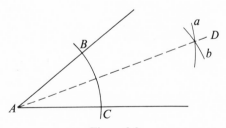

Figure 9.2

3. *Construction of a line perpendicular to a given line*
    (a) *through a point P on the*   (b) *through a point P not on*
        line.   *the line.*

In both (a) and (b), place the compass point at P and draw an arc that intersects the given line at two distinct points A and B. With center A and radius greater

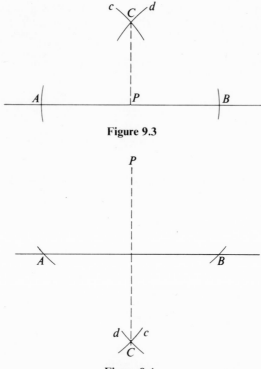

Figure 9.3

Figure 9.4

than $\frac{1}{2}m(\overline{AB})$, draw arc c. With center B and the same radius draw arc d intersecting c at C. Draw the line $\overleftrightarrow{PC}$ determined by P and C. $\overleftrightarrow{PC}$ is perpendicular to $\overleftrightarrow{AB}$.

___4. *Construction of an angle congruent to a given angle.* Given $\angle A$, draw a ray $\overrightarrow{A'X}$. Place the compass point at A, and draw an arc intersecting the sides of $\angle A$ at B and C. Use the same radius, place compass point at A', and draw an arc a intersecting the ray at P. Place the compass point at B and open compass until pencil point is at C, that is, open compass to $m(\overline{BC})$. Use this radius, place compass point at P and draw arc b intersecting arc a at Q. Draw the ray determined by A' and Q. Then, $\angle QA'P \cong \angle A$.

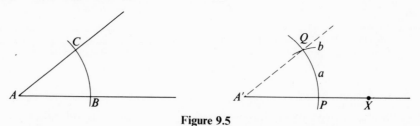

Figure 9.5

5. *Construction of a line parallel to a given line l and through a point P not on the line.* Draw $\overleftrightarrow{PQ}$ through *P*, intersecting *l* at some point *Q*. Construct $\angle QPR$ congruent to $\angle MQP$. The line $\overleftrightarrow{PR}$ is the line through the point *P* that is parallel to *l*.

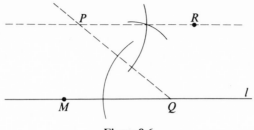

Figure 9.6

6. *Construction of a triangle congruent to a given triangle ABC.* (*Three methods.*)

    (a) SSS

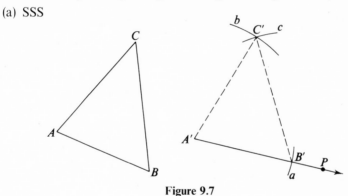

Figure 9.7

Draw a ray $\overrightarrow{A'P}$. Open the compass to $m(\overline{AB})$. With the compass point at *A'*, and radius $m(\overline{AB})$, draw arc *a* intersecting $\overrightarrow{A'P}$ at *B'*. Open compass to $m(\overline{AC})$. Place the compass point at *A'* and draw arc *b*. Open compass to $m(\overline{BC})$. Place the compass point at *B'* and draw arc *c*; label the intersection of arcs *b* and *c* as *C'*. Draw segments $\overline{A'C'}$ and $\overline{B'C'}$. Then $\triangle A'B'C' \cong \triangle ABC$.

    (b) SAS

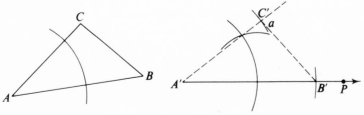

Figure 9.8

Draw a ray $\overrightarrow{A'P}$. Locate B' as in method (a). Construct an angle at A' congruent to $\angle A$. Open compass to $m(\overline{AC})$. With compass point at A', draw arc a intersecting side of $\angle A'$ at C' as shown. Draw segment $\overline{B'C'}$. Then $\triangle A'B'C' \cong \triangle ABC$.

(c) ASA

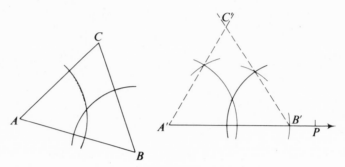

Figure 9.9

Draw a ray $\overrightarrow{A'P}$ with endpoint A'. Locate B' as in methods (a) and (b). At A' construct an angle congruent to $\angle A$. At B' construct an angle congruent to $\angle B$ as shown. Label the point where the sides of these angles intersect C'. Then $\triangle A'B'C' \cong \triangle ABC$.

EXERCISES 9.2

Note: Draw fairly large figures for each exercise; only one drawing per sheet of paper is suggested.

1. (a) Draw an acute triangle and construct the three lines that are the perpendicular bisectors of the sides of the triangle. Extend each constructed line so that it intersects each of the other two lines
   (b) Repeat part (a) for an obtuse triangle.
   (c) Do you observe any particular property of the three lines that were constructed in parts (a) and (b)?
   (d) Repeat part (a) for two or three other triangles.
   (e) What generalization seems to be confirmed by your drawings?

2. (a) Draw an acute triangle and construct the three rays that are the bisectors of the angles of the triangle. Extend each ray so that it intersects each of the other two rays.
   (b) Repeat part (a) for an obtuse triangle.
   (c) Repeat part (a) for two or three other triangles.
   (d) What generalization seems to be confirmed by your drawings?

3. (a) Draw an acute triangle and construct the three line segments that are medians of the triangle. (A median is a line segment whose endpoints are a vertex of a triangle and the midpoint of the opposite side.)

(b) Repeat part (a) for an obtuse triangle.

(c) Repeat part (a) for two or three other triangles.

(d) What generalization seems to be confirmed by your drawings?

4. (a) Draw an acute triangle and construct the three line segments that are altitudes of the triangle. Extend each altitude so that it intersects each of the other two altitudes. (An altitude of a triangle is a line segment with a vertex of the triangle as one of its endpoints, a point of the line containing the opposite side of the triangle as its other endpoint, and perpendicular to that line.)

(b) Repeat part (a) for an obtuse triangle.

(c) Repeat part (a) for two or three other triangles.

(d) What generalization seems to be confirmed by your drawings?

5. (a) Draw a triangle ABC and its medians. Call their point of interesection P. This point is called the **centroid** of the triangle.

(b) Bisect $\overline{AP}$, $\overline{BP}$, and $\overline{CP}$.

(c) Measure the three segments on each median.

(d) How do the three segments on each median compare in length?

6. (a) Draw a large right triangle.

(b) Draw the altitude from the vertex of the right angle, forming two right triangles with the altitude as a leg of each.

(c) Measure the angles of all three triangles.

(d) Compare the measures of the angles of the three triangles.

7. (a) Draw a triangle ABC and construct the midpoint D of $\overline{BC}$.

(b) Construct the line $\overleftrightarrow{DE}$ parallel to $\overleftrightarrow{AB}$ where E is determined on $\overleftrightarrow{AC}$ by this constructed line.

(c) Measure the segments $\overline{AE}$ and $\overline{EC}$.

(d) Try this same construction and measurement with four or five other triangles ABC (not congruent to each other).

(e) What generalization about the segments of $\overline{AC}$ seems to be confirmed by your drawings?

8. (a) Draw an isosceles triangle ABC. (An isosceles triangle is a triangle that has two congruent sides; let $\overline{AC} \cong \overline{BC}$.)

(b) Construct a triangle RST with $\overline{RS} \cong \overline{AB}$, $\overline{ST} \cong \overline{BC}$, and $\overline{RT} \cong \overline{AC}$.

(c) Label $\triangle RST$ so you can identify the vertices after cutting out the triangle. Then cut along the sides of the triangle and use the triangular region that comes out to represent the triangle RST.

(d) Place the paper triangle RST so that R is on A, S is on B, and T is on C. Does it fit when placed over $\triangle ABC$ this way?

(e) Place the paper triangle so that R is on B, S is on A, and T is on C. Does it still fit $\triangle ABC$?

(f) What does this seem to tell us about the angles R and S opposite the congruent sides of $\triangle RST$? About the angles A and B opposite the congruent sides of $\triangle ABC$?

9. (a) Draw a quadrilateral $ABCD$ so that $\overline{AB}, \overline{BC}, \overline{CD}$, and $\overline{DA}$ intersect only at the vertices A, B, C, and D.

(b) Find the midpoints of each of the sides of the quadrilateral.

(c) Draw the quadrilateral that has these midpoints as vertices.

(d) Repeat steps (a), (b), and (c) for several other quadrilaterals.

(e) What generalization seems to be true about a second quadrilateral formed by this process?

10. (a) Draw a circle and a diameter $\overline{AB}$.

(b) Choose a point C on the circle, different from A or B.

(c) Draw segments $\overline{AC}$ and $\overline{BC}$.

(d) Measure $\angle ACB$.

(e) Repeat parts (b), (c), and (d) for several other points "C" on the circle.

(f) How do the angles ACB compare?

11. (a) Draw a circle and a chord $\overline{AB}$ of the circle.

(b) Bisect $\overline{AB}$ and label its midpoint M.

(c) Draw two other chords $\overline{CD}$ and $\overline{EF}$, both containing the point M.

(d) Draw the chords $\overline{CE}, \overline{ED}, \overline{DF}$, and $\overline{FC}$. Label the two points of intersection of these chords with $\overline{AB}$ as P and Q.

(e) Compare the lengths of $\overline{PM}$ and $\overline{QM}$.

(f) Repeat parts (b), (c), (d), and (e) several times with chords of varying lengths.

(g) What generalization about $\overline{PM}$ and $\overline{QM}$ seems to be confirmed by your drawings and measurements?

12. (a) Draw a circle and label the center O.

(b) Locate three points A, B, and C on the circle. Draw segments $\overline{OA}, \overline{OB}, \overline{AC}$, and $\overline{BC}$.

(c) Make a heavy line over the arc AB that does not contain C.

(d) Compare the size of $\angle AOB$ (whose intercepted arc does not not contain C) and $\angle ACB$.

(e) Repeat parts (b) and (c) several times with points A, B, and C in varying locations.

(f) What seems to be the constant ratio of $m(\angle AOB)$ to $m(\angle ACB)$?

13. (a) Draw three noncollinear points A, B, and C and draw segments $\overline{AB}$ and $\overline{BC}$.

(b) Construct perpendicular bisectors of $\overline{AB}$ and $\overline{BC}$, extending them to intersect at a point P. Draw a circle with center P and radius $\overline{AP}$.

(c) What interesting observation can you make about your drawing?

(d) Try parts (a) and (b) with several other sets of three noncollinear points.

(e) What generalization seems to be confirmed by your drawings?

9.3 FORMAL GEOMETRY

In the preceding set of exercises, working experimentally, you should have discovered some important truths of what we call Euclidean geometry, that is, the ordinary geometry in which congruent figures have equal measures. We shall think of this geometry as a mathematical system concerned with sets of points in three-dimensional space. Your generalizations in the previous set of exercises are not automatically acceptable to a mathematician as properties of Euclidean geometry since these generalizations have only been observed for a few figures; they have not been "proved" to be universally true for all such figures in Euclidean geometry. You have employed an *inductive* process. To verify that your conclusions are indeed true in Euclidean geometry, a deductive process is needed; that is, these facts must be derived logically from other known or accepted facts. (A review of the discussions of a mathematical system in Chapters 1 and 2 is recommended at this point.)

When we take this mathematical (deductive) approach, we also shift from the representations of points, lines, triangles, quadrilaterals, circles, etc. as physical objects made up of pencil lead and paper, to idealizations of these objects in an "abstract space" as mentioned in Section 9.1. The deductive method of reasoning and the idealizations of physical objects are characteristic of what are called **formal geometries**.

Each *formal geometry* is a broad mathematical system including many subsystems. The Euclidean geometry that would be involved in proving the truths you may have discovered in Exercises 9.3 is just one such mathematical system. Each logical, or mathematical, system has the four components of a mathematical system that we studied in Chapter 2, that is,

1. Some undefined terms
2. Some defined terms
3. Some assumptions
4. Some theorems

9.4   UNDEFINED TERMS IN GEOMETRY

In Chapter 1 we discussed the necessity, in a logical system, of beginning with some undefined terms. Euclid, himself, though he did understand the need for the other three components, did not see the need for undefined terms. It seems clear that Euclid thought of *definitions* as *descriptions*. In his definitions he was describing idealizations of physical objects. The modern mathematician's concept of a definition involves a statement of characteristic properties that will remove all ambiguity from the meaning of the word defined; further, the word defined usually represents an idea rather than a physical object. Euclid *described all key terms* in his famous work entitled *Elements*. A few of his definitions, from Book I of the *Elements*, are listed below. These statements reveal that, as a matter of fact, he used many terms that were not defined in the sense that we now require terms to be defined.

1. A *point* is that which has no part.
2. A *line* is a breadthless length.
3. The extremities of a line are *points*.
4. A *straight line* is a line which lies evenly with the points on itself.

5. A *plane surface* is a surface which lies evenly with the straight lines on itself.

6. A *plane angle* is the inclination to one another of two lines in a plane which meet one another and do not lie in a straight line.

7. When a straight line set up on a straight line makes the adjacent angles equal to one another, each of the equal angles is *right* and the straight line standing on the other is called a *perpendicular* to that on which it stands.

8. A *boundary* is that which is an extremity of anything.

9. A *figure* is that which is contained by any boundary or boundaries.

Some of the key terms that Euclid did not define and that appear in these definitions are part, breadthless, length, evenly, surface, inclination, etc. Since Euclid did not mention that any of these terms were undefined, or **primitive** (a good word often used to identify the initial, undefined terms in a system), it appears that he was using some of his definitions as informal descriptions. Actually some of his descriptions no longer fit. For example, a line does not have any endpoints (extremities); line *segments* have endpoints. In a modern treatment of Euclidean geometry, the following are often listed as undefined terms: point, line, plane, set, on, and between.

9.5   DEFINED TERMS IN GEOMETRY

We now use the primitive terms (undefined terms) listed in the preceding section and proceed to define other key terms directly in terms of these primitive terms or in terms of these primitive terms and previously defined terms:

A **geometric figure** is a set of points.

A **line segment** (or simply a **segment**) is a figure consisting of two distinct points on a line and all points that are on the line and between the two selected points.

This definition of a *segment* uses the words *two distinct*, which is an example of the careful attempt made today to be precise in communicating mathematical ideas. The word *two* used alone would, in this definition, be ambiguous, because in certain contexts, points P and Q may really be the *same* point; we then say $P = Q$. When we wish to be sure that P and Q are *different* points, we speak of *two distinct points* or we say $P \neq Q$.

9.6   ASSUMPTIONS IN GEOMETRY

The initial assumptions of a mathematical system are required by its deductive character. We have said that, in a deductive process, we are deriving new facts from other known, or accepted, facts. These "known, or accepted, facts" are

assumptions of the system. We list Euclid's assumptions, which he divided into
a set of postulates and a set of "common notions."

Euclid's Postulates: (Assumptions about geometric figures.)

1. A straight line can be drawn from any point to any point.

2. A finite straight line can be produced continuously in a straight line.

3. A circle may be drawn with any center and any distance.

4. All right angles are equal.

5. If a straight line falling on two straight lines makes the interior angles
   on the same side less than two right angles, the straight lines, if produced
   indefinitely, meet on that side where the angles are less than two right
   angles. (This is known as **Euclid's parallel postulate**.)

Euclid's Common Notions: (Assumptions that Euclid felt are not strictly
geometric in character.)

1. Things equal to the same thing are also equal to one another.

2. If equals be added to equals, the wholes are equal.

3. If equals be subtracted from equals, the differences are equal.

4. Things that coincide with one another are equal to one another.

By today's standards, these statements are not satisfactory. They contain
key terms that Euclid did not define, for example, continuously, distance, coincide,
equal; and we have seen that the terms he did define were not defined precisely
enough. It is clear from Postulate 1 that his word "line" means "line segment"
today. His word "equal" did not mean "identical" as it does today, because he
speaks of "things that coincide"—different things—as being "equal." His idea of
equality was a fuzzy one, but we must realize that we have a great advantage over
Euclid in this matter, because we have a concept of "number" that he did not
have in his era. We can say that two distinct segments are "equal in length,"
which means that *numbers* associated with the segments are *identical*. Probably
Euclid failed to distinguish between his drawings and the lines or other figures
that they represent. The absence of this distinction would account for the fact
that his postulates were stated essentially for drawings instead of for the figures
themselves.

In his first three postulates, Euclid speaks of points, lines, and circles that
"can be drawn." Today, we have changed this emphasis on "what can be drawn"
to one of "existence." We ask what points, what lines, etc., *exist*. Whereas Euclid
said a straight line *can be drawn* from any point to any point, we now say: On any

two distinct points there *exists* a line. In addition to the question of *existence*, modern geometers are also interested in the question of *uniqueness*. If *there is* a point, or a line, or a figure, which meets certain conditions, is it the *only* point, or line, or figure, which meets the conditions? If so, we say that it is *unique*. A modern postulate for Euclidean geometry attests to the *uniqueness* as well as the *existence* of the line that is on two points: Two distinct points are on *one and only one* line.

9.7   THEOREMS IN GEOMETRY

A **theorem** is a statement that can be shown to follow logically from the postulates (assumptions) and definitions of a mathematical system. The process of "showing that it follows logically" from these assumptions is called a **proof**. Many theorems are stated in the form of an implication, and those that are not expressed as implications can be restated as implications. For example, one of the forty-eight propositions (theorems) of Euclid is:

> In an isosceles triangle, the angles at the base are congruent to one another.

This statement can be restated as:

> *If* a triangle has two congruent sides, *then* the angles opposite these sides are congruent.

The *if* clause of a theorem gives us conditions that we can assume to be true. In this particular theorem, therefore, we can assume that the statement, "A triangle has two congruent sides," is true. This is the **hypothesis** or **given statement**, which we can use as an opening statement in the proof. In order to prove this—or any other—theorem, we must show that the *then* clause, or **conclusion** as it is often called, can be obtained from the hypothesis by the deductive reasoning that we discussed in Chapter 2. Thus, in this particular theorem, we must show that the statement, "The angles opposite these sides are congruent," follows logically from the hypothesis. In a formal proof, the hypothesis of the theorem is usually the opening statement and the conclusion of the theorem is the final statement. Between these two statements are other statements that develop the reasoning compelling us to accept the conclusion. Each of these statements must be justified by a reason. This reason may be a definition, a postulate, a previously proved theorem, a given statement, or a previous statement in the proof. Usually a sketch is used to help the reader visualize the figure and relationships among the parts of the figure. We must never make assumptions based on the appearance of the figure, if those assumptions are not included in the statement of the theorem. Although statements in the proof seem to refer to the specific figure we have pictured, actually they must refer to any figure about which the hypothesis of the theorem is true.

A frequently used (though not altogether correct) proof of the theorem stated in the first paragraph of this section follows:

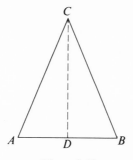

Figure 9.10

Theorem: The base angles of an isosceles triangle are congruent.

    Given: $\triangle ABC$ with $\overline{AC} \cong \overline{BC}$.

    Prove: $\angle A \cong \angle B$.

    PROOF

Statements	*Reasons*
1. Draw the bisector of $\angle C$.	1. Every angle has a bisector. (Assumed here to be a previously proved theorem.)
2. Extend the angle bisector to meet $\overline{AB}$ at D.	2. A line may be extended. (Euclid's postulate 2.)
3. In $\triangle ACD$ and $\triangle BCD$, $\overline{AC} \cong \overline{BC}$.	3. By hypothesis.
4. $\angle ACD \cong \angle BCD$.	4. Definition of angle bisector.
5. $\overline{CD} = \overline{CD}$.	5. Any figure is equal to itself.
6. $\triangle ACD \cong \triangle BCD$.	6. SAS. (Assumed here to be a previously proved theorem.)
7. $\angle A \cong \angle B$.	7. Corresponding parts of congruent triangles are congruent. (Definition of congruence.)

This proof serves to illustrate the nature of geometric proofs in general, even though the traditional high school textbook from which it was taken does not include some statements that must be included (either as postulates, theorems, or definitions) in order to overcome logical deficiencies in this and other theorems. The logical deficiencies in this proof are these:

1. In statement 2, the reason, "A line may be extended," does not justify extending the line to meet a second line. The two lines must be proved to be not parallel before we can be sure that they intersect.

2. Also, in statement 2, it is assumed that the angle bisector will meet $\overline{AB}$ at a point D *between* A and B. This assumption is not explicitly justified by any statement in the system. This is actually a case of assuming something to be true because it appears to be true about the figure we have drawn.

We next offer a valid proof of this same theorem according to modern standards of validity. This proof is credited to Pappus (A.D. 300). It employs the idea of congruence of triangles as a mapping of a triangle onto another or the same triangle (an idea also implicit in Exercise 8, Exercises 9.2).

Theorem: If a triangle is isosceles, then the base angles are congruent.

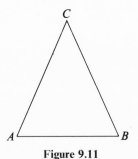

Figure 9.11

Given: $\triangle ABC$ with $\overline{AC} \cong \overline{BC}$.

Prove: $\angle BAC \cong \angle ABC$.

PROOF

Statements	*Reasons*
1. In $\triangle ABC$, $\overline{AC} \cong \overline{BC}$.	1. Given.
2. $\overline{BC} \cong \overline{AC}$.	2. Given.
3. $\angle ACB \cong \angle BCA$.	3. Identity.

Thus parts $\overline{AC}$, $\overline{BC}$, $\angle ACB$ of $\triangle ACB$ are congruent to corresponding parts $\overline{BC}$, $\overline{AC}$, $\angle BCA$ of $\triangle BCA$.

4. $\triangle ACB \cong \triangle BCA$.	4. SAS. (Assumed here to be a previously proved theorem.)
5. $\angle BAC \cong \angle ABC$.	5. Corresponding parts of congruent triangles are congruent.

Sometimes a proof of a theorem, in the form in which it is stated, is difficult to construct; in such a case it is often easier to prove a proposition that is logically equivalent (see Section 3.9) to the given theorem. Recall that the contrapositive of an implication is logically equivalent to, that is, has exactly the same truth table as, the original implication; so that, *if we can prove the contrapositive*, we thereby show that the original implication is true. Before constructing the contrapositive of the theorem above, let us recast the theorem in a more useful form. "If a triangle is isosceles, then the base angles are congruent," can be rewritten, "If a triangle has two of its sides congruent, then the angles opposite those sides are also congruent." In light of this new statement, the contrapositive would be: If the angles opposite two sides of a triangle are not congruent, then the two sides are not congruent.

Another acceptable approach in proving a theorem is to prove that *it is not true that the theorem is false*. This approach is justified by the fact that any statement must be either true or false. In this method, we begin by assuming that the theorem is false and the plan is to find a set of verifiable statements that lead to an inconsistency. Since inconsistencies are impossible in correct thinking, the one statement we have made without verification must be untrue. That statement was our beginning assumption that the theorem is false. Hence the theorem must be true. In practice, the way we usually choose to state the assumption that the theorem is false is to state that its negation is true. With the theorem we have been using as an example in this section, we would begin by assuming that the statement, "This is an isosceles triangle *and* the angles opposite the congruent sides are *not* congruent," is true. (Recall that the negation of $p \rightarrow q$ is $p \wedge (\sim q)$.) We would then show that this leads to an inconsistency. For an example of this sort of proof, the reader can refer to the proof that $\sqrt{2}$ is not a rational number, Section 8.2.

EXERCISES 9.7

1. We have listed some key words in Euclid's definitions (Section 9.5) and also in his assumptions (Section 9.7) that he left undefined. Can you find and list others?

2. State each theorem as an implication; that is, write it in "if–then" form.

   (a) Every segment is an infinite set of points.
   (b) The intersection of any two convex sets is a convex set.
   (c) The interior of any triangle is a convex set.
   (d) A segment has one and only one midpoint.
   (e) Vertical angles are congruent.
   (f) Supplements of congruent angles are congruent.
   (g) The four angles formed by perpendicular lines are right angles.
   (h) A point on the perpendicular bisector of a line segment is equidistant from the endpoints of that segment.
   (i) The measure of any exterior angle of a triangle is greater than the measure of either of the remote interior angles of that triangle.
   (j) The sum of the measures of the angles of a triangle is 180.

3. We have seen in a proof in Section 9.7 that there are logical gaps in the reasoning of Euclid that often leave proofs of theorems logically deficient. These gaps can be traced to certain tacit assumptions that are made on the basis of visual evidence or spatial intuition. We can show that the use of some of these tacit assumptions even makes it possible to prove ridiculous theorems that are quite obviously untrue. Two such theorems are presented below. Find the erroneous assumption (or assumptions) employed in each proof by constructing an accurate drawing.

(a) **Theorem:**   There exists a triangle with two right angles.

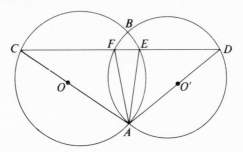

Figure 9.12

Given: Circles O and O' intersecting in points A and B with diameters $\overline{AC}$ and $\overline{AD}$ drawn from point A, and $\overline{CD}$ intersecting the circles O and O' at points E and F respectively.

Prove: $\triangle AEF$ has two right angles.

PROOF:

Statement	*Reason*
1. Points A and B are the points of intersection of circles O and O'. $\overline{AC}$ and $\overline{AD}$ are diameters of O and O' respectively. CD intersects O and O' at E and F respectively.	1. Given
2. $\angle AEC$ is a right angle and $\angle AFD$ is a right angle.	2. Any angle inscribed in a semi-circle is a right angle.

Since $\angle AEC$ and $\angle AFD$ are both angles of $\triangle AEF$, $\triangle AEF$ has two right angles.

(b) **Theorem:**   There exists an obtuse angle that is congruent to a right angle.

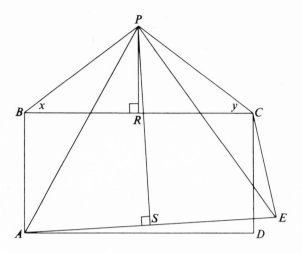

Figure 9.13

Given: Rectangle $ABCD$ with $\overline{CE}$ drawn outside the rectangle so that $\overline{CE} \cong \overline{CD}$. $\overline{PR}$ is the perpendicular bisector of $\overline{BC}$ and $\overline{PS}$ is the perpendicular bisector of $\overline{AE}$; segments $\overline{PA}$, $\overline{PB}$, $\overline{PC}$, and $\overline{PE}$ have been drawn.

Prove: Right $\angle ABC \cong$ obtuse $\angle BCE$.

PROOF

Statements	*Reasons*
1. $\overline{CE}$ is outside rectangle $ABCD$; $\overline{CE} = \overline{CD}$. $\overline{PR}$ and $\overline{PS}$ are $\perp$ bisectors of $\overline{BC}$ and $\overline{AE}$ respectively. Segments $\overline{PA}$, $\overline{PB}$, $\overline{PC}$, and $\overline{PE}$ have been drawn.	1. Given
2. $\overline{PB} \cong \overline{PC}$ and $\overline{PA} \cong \overline{PE}$	2. Each point on the perpendicular bisector of a line segment is equidistant from the endpoints of the segment.
3. $\overline{AB} \cong \overline{CD}$	3. Opposite sides of a rectangle are congruent.
4. $\overline{CD} \cong \overline{CE}$	4. Given.
5. $\overline{AB} \cong \overline{CE}$; thus $m(\overline{AB}) = m(\overline{CE})$	5. Quantities equal to the same quantity are equal to each other.

6. $\triangle ABP \cong \triangle PCE$

6. SSS.

7. $\angle ABP \cong \angle PCE$

7. Corresponding parts of congruent triangles are congruent.

8. $\angle x \cong \angle y$

8. Angles opposite congruent sides of a triangle are congruent.

9. $\angle ABC \cong \angle BCE$; thus $m(\angle ABC) = m(\angle BCE)$

9. If equals are subtracted from equals, the results are equal.

CHAPTER 10

Sets of Points in Space

10.1 POINTS

The reader will recall that the concept of a *set of elements* has been the starting point for each mathematical system that we have studied. In geometry, the fundamental set of elements is a **set of points**. The term "point" is very familiar to us; certainly each of us has an intuitive notion about its meaning. Perhaps you think of a point as an exact, fixed location. Referring to Euclid's attempt at defining the term, we find, "A point is that which has no part." Though this statement has its puzzling aspect, it does seem to suggest the characteristic of indivisible smallness. The modern mathematician shares your (and Euclid's) intuitive notions about a point, but he does not attempt to define the term. The word **point** is a primitive term in geometry—an undefined term.

We speak of *drawing* a point when we make a small "dot" on a piece of paper with a pencil, or on a chalk board with a piece of chalk. What we see when we look at this *representation of a point* is in agreement with our intuitive notion of what a point is; but we have not really created a point on the paper or chalk board. We cannot "draw" anything that is undefined—we cannot specify something that has no exact meaning. Nevertheless, in developing our understanding of concepts of geometry, we shall find our drawings, as well as our intuitive notions, an aid to understanding.

10.2 SETS OF POINTS

The set of all points in any geometry is often called **space**, and the study of geometry is simply the study of certain subsets of this universal set. A **geometric figure** is defined to be any non-empty set of points. Notice that we are defining *geometric figure* in terms of the undefined terms "point" and "set." While we are not attempting to develop a rigorous formal geometry, we shall try to define key terms in our system in terms of undefined terms and/or previously defined terms. With this as our aim, unless we introduce additional primitive terms, we shall be severely limited in the sorts of terms we can define. Therefore, we introduce three additional primitive terms: **line, plane**, and **between**.† As with the term *point* these terms are without any initial meaning, except that we shall consider the *line* and the *plane* each as a set of points, that is, as a *geometric figure*. All the primitive terms do acquire some meaning through the assumptions of a mathematical system; we shall see, for example, that the meaning of *line* in our geometry is what we have thought of intuitively as a *straight line* extending infinitely in both directions.

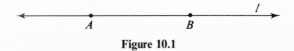

Figure 10.1

We picture a line as shown in Figure 10.1. The arrows on either end indicate that the line extends infinitely in both directions. Notice that, as we have described a line, its extent is an innate property of a line; Euclid's postulate about extending a line actually was an assumption about the picture of a line rather than about a line itself that already has extent. We name a line in one of two ways: (a) We use a lowercase letter, (for example, the line shown in Figure 10.1 can be called "*l*"), or (b) we use two points on the line, (for example, the points *A* and *B* on the same line, and we may call the line $\overleftrightarrow{AB}$).

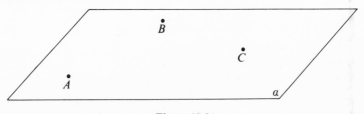

Figure 10.2

† In building a system of geometry, there is some freedom in the selection of primitive terms. For example, some authors define the terms *line* and *plane*. A line can be defined as a set of points whose coordinates satisfy a particular algebraic equation; a plane can be defined either from algebraic or geometric assumptions. These definitions would be characteristic of a more rigorous treatment than we are attempting in an elementary textbook.

The plane, as its characteristics are revealed in assumptions of the system, turns out to be what one might explain as a table top that extends infinitely in all directions. We picture the plane by drawing a part of it (Fig. 10.2). We have two conventional ways of naming the plane: (a) We use a lowercase Greek letter, (for example, the plane pictured in Figure 10.2 might be called "α"), or (b) we use three points of the plane not on a line of the plane to identify it, (for example, the same plane could be called ABC). In Exercise 6 a basis for naming the plane with three points not on the same line is introduced.

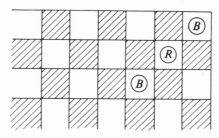

Figure 10.3

The meanings of the word "between," as the word is used outside geometry, vary with the context. Most of us would agree that Chicago is *between* Los Angeles and New York; and if we were playing checkers we would consider the red checker (R) in our drawing (Fig. 10.3) to be *between* the two black ones (B). The reader will recognize that *between* does not have quite the same meaning in these two situations. In the checker game example, it is not just a coincidence that the three checkers are in a straight line; if they were not in line, we would not consider any one checker to be between the other two. In the geographical example, however, the three cities named are not in a straight line, but the idea of "betweenness" is meaningful nevertheless. The fitting intuitive notion for us to adopt about betweenness in geometry is the same one we have about betweenness in the checker game.

As you have seen, the notation used for points consists of dots used to picture them, and capital letters used to name them, for example,

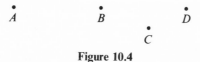

Figure 10.4

Since points A, B, and D as pictured in Figure 10.4 are in a straight line, we can say B is **between** A and D. We may use the symbols (ABD) or (DBA) to represent

the previous statement, "*B* is between *A* and *D*." We cannot use *between* to describe the relationship of *B, C*, and *D* because they are not in a straight line.

The idea of incidence as a primitive relation may be introduced here. There are
three basic **incidence relations** in plane geometry:

1. A point may be incident with a line. We shall refer to this case as a point on
a line, a line on a point, or a line containing a point.

2. A point may be incident with a plane. We shall refer to this case as a point
on (or in) a plane, or a plane containing a point.

3. A line may be incident with a plane. We shall refer to this case as a line in
(or on) a plane, or a plane containing a line.
We shall also introduce the terms *collinear* and *noncollinear*. In Figure 10.4,
points *A, B*, and *D* (as pictured) are **collinear** because they lie on the same line.
Points *A, B*, and *C* are **noncollinear** because they do not lie on the same line.

EXERCISES 10.2

1. Use the definition of *geometric figure* in Section 10.2 and state how many figures
   there would be in a geometry containing exactly (a) one point; (b) two points
   A and *B*; (c) three points *A, B*, and *C*; (d) four points *A, B, C*, and *D*.

2. What geometric figures have been introduced in Section 10.2?

3. (a) Draw three points *A, B*, and *C* such that the relation (*ABC*) holds.
   (b) Draw another picture so that, although points *A, B*, and *C* are in a different
       order, the relation (*ABC*) still holds.
   (c) Add a point *D* to each of the pictures that you have drawn so that (*BDC*).
   (d) Draw *E*, in both pictures, so that (*DCE*).
   (e) Draw *F*, in both pictures, so that (*FAB*).
   (f) What is the betweenness relationship of *A, B*, and *D*? Of *A, B*, and *E*? Of
       A, B, and *F*? Of *D, E*, and *F*? Of *B, C*, and *D*? Of *B, E*, and *F*?
   (g) Does the picture to which you refer in answering these questions [in part (f)]
       make any difference in the betweenness relationships in your answers?

4. Draw three points *A, B*, and *C* so that (*ABC*). Draw three distinct points between
   A and *B* and three more between *B* and *C*. How many additional points are
   there between *A* and *B*? Between *B* and *C*? Between *A* and *C*? Draw the figure
   consisting of *A* and *C* and all the points between them.

5. (a) How many different lines may contain a single point *A*?
   (b) How many different lines may contain two given distinct points *A* and *B*?

6. (a) How many different planes may contain a single point *A*?
   (b) How many different planes may contain two given distinct points *A* and *B*?
   (c) How many different planes may contain three given noncollinear points *A, B*
       and *C*?

10.3 SOME ASSUMPTIONS ABOUT POINTS, LINES, PLANES, AND BETWEENNESS

In the preceding set of exercises, the student could rely only on intuitive notions about the undefined terms point, line, plane, and between. It was noted in Section 10.2 that, although these terms are undefined, the assumptions of the system do tell us something about them, and it was suggested that the assumptions would confirm our intuitive notions about them. Euclid felt that his assumptions were "self-evident truths" about our physical universe. Evidence now reveals that the truth of these assumptions is highly doubtful. Therefore, the modern geometer does not concern himself with the question of "truth," but he simply states the assumptions that are to be accepted for the sake of the logical development of a system. We shall proceed to list some of the assumptions we have chosen for a geometric system of points in space with some comments on the reason for the choices.

It is possible to have a geometry with any finite number of points; we could even have a system with only one point; it would be decidedly trivial. It is our purpose, however, to examine Euclidean geometry, which involves infinitely many points. In the preceding sections, we have spoken of intuitive notions about the line and the plane that indicate that these are infinite sets of points. Therefore, we have chosen assumptions that will lead to this conclusion. We begin with:

Assumption 10.1:   There exist at least two distinct points.

Assumption 10.2:   Two distinct points are contained in one and only one line.

We pause here in listing assumptions to state an important theorem that follows from the first two assumptions.

Theorem 10.1:   Any two distinct lines intersect in at most one point.

In Theorem 10.1, we are not assuming the existence of more than one line; nor are we assuming that *if* two lines exist, their intersection cannot be empty. Care must be exercised not to "read" anything into a statement that is not explicitly there. Proof of Theorem 10.1 will be considered in Exercises 10.3.

We next introduce a sequence of assumptions that will provide a basis for our concepts of: betweenness, the infinite extension of a line, and the infinite number of points on a line.

Assumption 10.3:   For any two distinct points A and B, there exists at least one point C between them; that is, there exists at least one point C

such that (ACB). Conversely, if (ACB), then the points A, B, and C are distinct collinear points with C between A and B.

Assumption 10.4:   For any points A, B, and C, (ABC) implies (CBA).

Assumption 10.5:   For any three points A, B, and C, (ABC) implies that (BCA) is not true.

Assumption 10.6:   For any three distinct points A, B, and C on a line, exactly one of the following holds: (ABC), (BCA), or (CAB), that is, exactly one of three distinct points on a line is between the other two.

Assumption 10.7:   For any two distinct points A and B there exists at least one point D such that (ABD).

Assumption 10.8:   If (ACD) and (ABC), then (BCD).

The first eight assumptions lead to the conclusion that there are at least three points in our system. At this point, we introduce a new term: For any two distinct points A and B, the set consisting of the points A, B, and all points between A and B is defined to be a **line segment** (often called a **segment**), symbolized $\overline{AB}$ (Fig. 10.5).

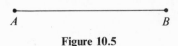

A B

Figure 10.5

The points A and B are the **endpoints** of $\overline{AB}$. The definition of the term *segment*, along with the assumptions above, makes possible the proof (Exercise 3) of the following theorem:

Theorem 10.2:   Every segment is an infinite set of points.

How does the first statement in Assumption 10.3 differ from Assumption 10.7? Assumption 10.7 is a key to the proof (Exercise 4) of the following theorem:

Theorem 10.3:   Every line is an infinite set of points, having no endpoints C_n and D_n such that for all points P on the line (C_nPD_n).

We next offer three assumptions that lead to the conclusion that every plane contains an infinite number of lines.

Assumption 10.9: For any line *l*, there exists at least one point *P* that is not on the line.

Assumption 10.10: Three distinct points not contained in the same line are contained in one and only one plane.

Assumption 10.11: If a plane contains two distinct points of a line, it contains the line.

Theorem 10.4: Every plane contains an infinite number of lines.

Proof of Theorem 10.4 will be considered in Exercise 5. Two other useful theorems also follow from Assumptions 10.1 through 10.11. They are:

Theorem 10.5: For any line *l* and any point *P* not on *l*, there is one and only one plane containing *P* and *l*.

Theorem 10.6: For any two distinct lines *l* and *m*, if $l \cap m \neq \varnothing$, then *l* and *m* are contained on one and only one plane.

We have now made enough assumptions to be able to establish, by mathematical proofs, that our geometry has an infinite number of points, that not all these points are on one line, indeed that there exists an infinite number of lines. We still cannot say, on the basis of these assumptions, that not all our points are on the same plane. Since we do wish to have a "three-dimensional" space, we make an assumption that will guarantee the existence of points not on a given plane.

Assumption 10.12: For any plane α there exists at least one point not on α.

EXERCISES 10.3

1. Refer to Assumptions 10.1 and 10.2 in this section. Explain how they lead logically to the statement in Theorem 10.1. Rephrase, if necessary, your explanation to obtain a *proof* of Theorem 10.1.

2. (a) Restate Assumption 10.4, using points *R*, *S*, and *T*. Draw and label a figure for which the statement holds.
   (b) Restate Assumption 10.5, using points *R*, *S*, and *T*. Draw and label two figures to illustrate the statement.
   (c) Restate Assumption 10.6 using points *P*, *Q*, and *R*. Draw and label three figures to illustrate the statement.

(d) Restate Assumption 10.7, using points P, Q, and R. Draw and label a figure to illustrate the statement.

(e) Restate Assumption 10.8, using points P, Q, R, and S. Draw and label a figure to illustrate the statement.

3. (a) Draw a segment $\overline{AB}$ about three inches long.

(b) Draw a point C_1 on $\overline{AB}$ so that (AC_1B); draw a point C_2 so that (AC_2C_1); draw a point C_3 so that (AC_3C_2); draw a point C_4 so that (AC_4C_3).

(c) Is there any theoretical limit to the number of distinct points one can locate on $\overline{AB}$ this way? Explain why.

(d) Reread Assumptions 10.1 through 10.3 and the definition of a *segment*. Explain how these statements lead logically to the statement in Theorem 10.2. (Your explanation should constitute a proof of Theorem 10.2.)

4. (a) Sketch a segment $\overline{AB}$.

(b) Draw a point C_1 so that (ABC_1); draw a point C_2 so that (AC_1C_2); draw a point C_3 so that (AC_2C_3); draw a point C_4 so that (AC_3C_4). Are the points A, B, C_1, C_2, C_3, C_4 distinct and all on the same line l? Explain.

(c) Draw a point D_1 so that (D_1AB); draw a point D_2 so that (D_2D_1A); draw a point D_3 so that (D_3D_2A); draw a point D_4 so that (D_4D_3A). Are the points A, B, D_1, D_2, D_3, D_4 distinct and all on the same line as C_1, C_2, C_3, C_4? Explain.

(d) Is there any theoretical limit to the number of points you can locate on l this way? Explain.

(e) Are there endpoints C_n and D_n for the line l? That is, are there points C_n and D_n such that for all other points P on l (C_nPD_n)?

(f) Refer to Theorem 10.2 and Assumptions 10.3 and 10.7. Explain how these statements lead to the statement in Theorem 10.3. (Your explanation should constitute a proof of Theorem 10.3.)

5. The constructions to be drawn and the questions asked in this exercise are designed to lead to a proof of Theorem 10.4: Every plane contains an infinite number of lines.

(a) Draw a line l and a point P not on l. Locate two distinct points A_1 and A_2 on l. Are P, A_1, and A_2 in the same plane? Why? (We shall call this plane α.)

(b) Draw the line containing A_1 and P. Is $\overleftrightarrow{A_1P}$ in α? Why?

(c) For other distinct points A_2, A_3, A_4, and A_5 on l, draw $\overleftrightarrow{A_2P}$, $\overleftrightarrow{A_3P}$, $\overleftrightarrow{A_4P}$, and $\overleftrightarrow{A_5P}$. Are they all in α?

(d) Is there any theoretical limit to the number of lines you can draw in this way? Explain why.

(e) Refer to Assumptions 10.9, 10.10, and 10.11. Explain how they lead logically to the statement of Theorem 10.4. (Your explanation should constitute a proof of Theorem 10.4.)

6. (a) When we say that a point A is *contained in* a line l, do we use "contained" to mean $A \in l$ or $A \subseteq l$?

(b) When we say that a line l is *contained in* a plane α, do we use "contained" to mean $l \in \alpha$ or $l \subseteq \alpha$?

7. Prove Theorem 10.5, that is, give a logical explanation to show that the statement

in Theorem 10.5 follows from statements in definitions, assumptions and/or theorems.

8. As in Exercise 7, prove Theorem 10.6.

9. (a) A plane α and a point P not in α are pictured in Figure 10.6. The line $\overleftrightarrow{A_1P}$, containing P and a point A_1 in α, has been drawn as well as the line $\overleftrightarrow{A_1A_2}$ where A_2 is also in α. A plane β, distinct from α, contains $\overrightarrow{A_1A_2}$ and $\overleftrightarrow{A_1P}$. Why?

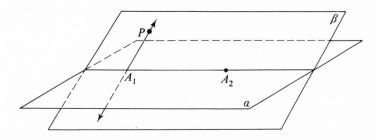

Figure 10.6

(b) How many more planes can be "determined" in a similar manner, using P, A_1, and other points A_n ($n = 2, 3, 4, \ldots$) in α?

(c) Give an explanation that will show that Theorem 10.5, or Theorem 10.6, and Assumption 10.12 lead to the conclusion that there are infinitely many planes in space.

10.4 INTERSECTIONS OF LINES AND PLANES

In studying the intersection of lines and planes, it is important to remember that lines and planes are geometric figures and, as such, are *sets of points*. We shall consider the intersection of two lines, the intersection of a line and a plane, and the intersection of two planes. We can show that non-empty intersections of each kind may exist. The existence of empty intersections cannot in all cases be proved without additional assumptions.

Lines in the same plane are called **coplanar lines**. We shall first consider the ways in which such lines may intersect. We have seen that two distinct lines intersect in at most one point (Theorem 10.1, Section 10.3), and that two lines whose intersection is non-empty are coplanar (Theorem 10.6, Section 10.3). These theorems do not imply the existence of an empty intersection of two coplanar lines; as a matter of fact, mathematicians have found that the existence of such lines cannot be proved as a theorem, regardless of the choice of assumptions. Therefore, their existence is assumed. Before giving the assumption as it is commonly stated, we name such a pair of lines: If the intersection of two distinct coplanar lines is empty, the lines are called **parallel lines**; that is, if l and m are coplanar and $l \cap m = \varnothing$, we say l **is parallel to** m, symbolized $l \parallel m$. The concept of *parallel segments* is related to the concept of parallel lines. If two segments $\overline{AB}$ and $\overline{CD}$ are subsets, respectively, of parallel lines l and m, then the segments $\overline{AB}$ and $\overline{CD}$ are called **parallel segments**. Now we give the assumption that establishes the existence of parallel lines. It is called the **parallel postulate**.

> **Assumption 10.13** (the parallel postulate):   For any line l and any point P not in l, there exists one and only one line that contains P and is parallel to l.

In general, the intersection of two distinct coplanar lines may be a set containing exactly one point, or it may be the empty set. When the intersection is a set containing exactly one point, we may refer to the lines as intersecting lines. The following useful theorem is a consequence of the definition of parallel lines and Assumption 10.13.

> **Theorem 10.7:**   If a line l in a plane α is parallel to one of two intersecting lines m and n, also in α, then l intersects the other line.

The student will prove Theorem 10.7 in Exercise 2.

Suppose two lines are not in the same plane, that is, they are **noncoplanar**. By Theorem 10.1, Section 10.3, we know that their intersection is a set that contains at most one point. Also by Theorem 10.6, we know that if the intersection of two lines is a set that contains exactly one point, then the lines are coplanar. Therefore the intersection of two noncoplanar lines, if such lines exist, must be the empty set. We must be careful, at this point, not to assume that any two lines are noncoplanar if their intersection is the empty set—the lines might be parallel. Notice that *parallel* lines are defined to be coplanar. Noncoplanar lines are called **skew lines**. The existence of skew lines can be proved from Assumptions 10.1 through 10.13, though we shall not present the proof.

In Section 10.3, Assumption 10.12 was introduced to insure the existence of a three-dimensional space. We also wish to exclude the consideration of spaces of higher dimensions; that is, we wish to restrict the space of points of our geometry to correspond to our visualization of the physical universe as a three-dimensional space. The following assumption will provide this restriction:

> **Assumption 10.14:**   If the intersection of two planes is not empty, their intersection contains at least two distinct points.

The introduction of Assumption 10.14 eliminates the possibility of a line and a plane (or two planes) being contained in different three-spaces that are themselves contained in a four-space. In a four-space a line and a plane (or two planes) might be skew.

Consider the intersection of a line and a plane. We have seen that the intersection of a line l and a plane α may be the line l itself, that is, sometimes $l \cap \alpha = l$. This follows from Assumption 10.11: If a plane contains two distinct points of a line, it contains the line. What other possibilities for the intersection of a line and a plane does this leave? Can the intersection contain three points of the line, or four, or any finite number (greater than two) of points on the line, without containing the entire line? Assumption 10.11 forces us to abandon these possibilities. This leads us to make a conclusion that we state as a theorem:

Theorem 10.8:　For any line l and any plane α, if l is not in α, then l intersects α in at most one point.

Notice that this theorem allows for the possibility that the intersection of a line and a plane may be the empty set. If the intersection is empty, we say **the line is parallel to the plane**. Actually, our system of definitions, assumptions, and theorems does make it possible to prove the existence of such a line and plane (Exercise 3). We state this as a theorem:

Theorem 10.9:　For any plane α and any point P that is not on the plane, there exist infinitely many lines that contain P and are parallel to α.

Finally, we shall consider the last of the intersections that we set out to discuss, the ways in which two planes may intersect. Experimenting with pieces of stiff paper, one can easily discover that two planes whose intersection is not the empty set seem to intersect in a line. It also appears that their intersection can be the empty set. If the intersection of two planes is the empty set, we say **the planes are parallel**.

The following theorem follows directly from Assumption 10.14 and Assumption 10.2.

Theorem 10.10:　If two distinct planes have a non-empty intersection, their intersection is a line.

The existence of parallel planes can be proved (Exercise 4).

Theorem 10.11:　For any plane α and any line l parallel to α there exists one and only one plane that contains l and is parallel to α.

EXERCISES 10.4

1. Describe all the intersection relations that may exist between a line l and a plane α in space.

2. Prove Theorem 10.7.

3. This exercise will involve the student in a proof of Theorem 10.9: For any plane α and any point P that is not on the plane, there exist infinitely many lines that contain P and are parallel to α.

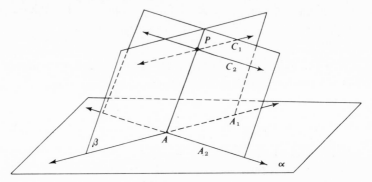

Figure 10.7

A plane α and a point P not on α are pictured in Fig. 10.7. Three noncollinear points A, A_1, and A_2 have been selected in α; lines $\overleftrightarrow{AP}$ and $\overleftrightarrow{AA_1}$ have been drawn. $\overleftrightarrow{AP}$ and $\overleftrightarrow{AA_1}$ determine a plane β. (a) Explain why $\overleftrightarrow{AP}$ and $\overleftrightarrow{AA_1}$ determine a plane β. The line $\overleftrightarrow{PC_1}$ has been drawn in β so that $\overleftrightarrow{PC_1}$ is parallel to $\overleftrightarrow{AA_1}$. (b) How do we know that such a line $\overleftrightarrow{PC_1}$ exists in β? Next $\overleftrightarrow{AA_2}$ has been drawn. In the plane of $\overleftrightarrow{PA}$ and $\overleftrightarrow{AA_2}$, $\overleftrightarrow{PC_2}$ has been drawn parallel to $\overleftrightarrow{AA_2}$. (c) In the plane α, is there any theoretical limit to the number of lines $\overleftrightarrow{AA_1}$, $\overleftrightarrow{AA_2}$, $\overleftrightarrow{AA_3}$, etc., that may be drawn? (d) Is there any theoretical limit to the number of lines that contain P and are parallel to lines in the plane α that contain A? If we can show that the lines $\overleftrightarrow{PC_1}$, $\overleftrightarrow{PC_2}$, etc., are parallel to the plane α as well as to the respective lines $\overleftrightarrow{AA_1}$, $\overleftrightarrow{AA_2}$, etc., in α, then we will know that Theorem 10.9 is true. We use an indirect method to prove this:

Suppose $\overleftrightarrow{PC_1}$ is not parallel to α. Then it intersects α in exactly one point Q. (e) Why does $\overleftrightarrow{PC_1}$ intersect α in just one point Q, if $\overleftrightarrow{PC_1}$ is not parallel to α? We know that $\overleftrightarrow{PC_1}$ and $\overleftrightarrow{AA_1}$ are coplanar. (f) How do we know that $\overleftrightarrow{PC_1}$ and $\overleftrightarrow{AA_1}$ are coplanar? We have named the plane that contains these lines β. Then $\overleftrightarrow{AA_1}$ is in both planes α and β. Consider the line $\overleftrightarrow{A_1Q}$. It lies in α. (g) Explain why $\overleftrightarrow{A_1Q}$ lies in α. But $\overleftrightarrow{A_1Q}$ is also in β. (h) Explain why $\overleftrightarrow{A_1Q}$ lies in β. Now $\overleftrightarrow{AA_1}$ and $\overleftrightarrow{A_1Q}$ either intersect in their common point A_1 or they are the same line. (i) Why? If they intersect in the point A_1, then we have two distinct lines whose intersection is not empty contained in two different planes. This contradicts Theorem 10.6. If $\overleftrightarrow{AA_1}$ and $\overleftrightarrow{A_1Q}$ are the same line, then $\overleftrightarrow{AA_1}$ and $\overleftrightarrow{PC_1}$ both contain Q and are not parallel. This contradicts an earlier statement verified in this theorem. Therefore we must abandon the hypothesis that leads to one of these two contradictions: the supposition that $\overleftrightarrow{PC}_1$ is not parallel to α. Thus PC_1 is parallel to α. By similar arguments, we can show that $\overleftrightarrow{PC}_2$, and each of the other lines $\overleftrightarrow{PC}_n$ that are parallel to lines $\overrightarrow{AA}_n$ in α, are parallel to α. This completes a proof of Theorem 10.9.

4. This exercise will involve the student in a proof of one part of Theorem 10.11: For any plane α and any line l parallel to α, there exists one and only one plane that contains l and is parallel to α.

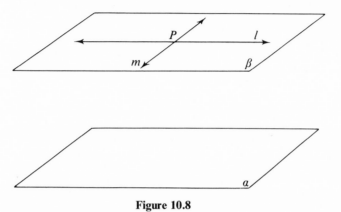

Figure 10.8

A plane α and a line l parallel to α are shown in Figure 10.8. A point P has been located on l and line m (distinct from l) containing P has been drawn parallel to α. (a) How do we know that such a line m exists? Now m and l lie in a unique plane β. (b) Explain why m and l lie in a unique plane β. We next consider the question: Is β parallel to α? Again we take an indirect approach. Suppose β intersects α in the line n. Then l and n are coplanar (both in β). This means l is either parallel to or intersects n. (c) Explain why l must either be parallel to n or intersect n. If l intersects n, then l intersects α. (d) Why does l intersect α if l intersects n? Since this statement is a contradiction of our original hypothesis that l is parallel to α, we must assume that l is parallel to n. If l is parallel to n, then m will intersect n. (e) Explain why m will intersect n if l is parallel to n. This is a contradiction of an earlier statement verified in this theorem. Therefore we must abandon the supposition that leads to one of these two contradictions: the supposition that β intersects α in a line n. This means that β is parallel to α. (f) Why?

This proves a part of Theorem 10.11, that is, we have shown that one plane exists that contains l and is parallel to α. We have not proved that β is the only such plane. Thus we have proved *existence* of such a plane; we have not proved *uniqueness*. The uniqueness proof is possible, but we shall not present it.

5. If α, β, and γ are three distinct planes, describe the possible intersections $\alpha \cap \beta \cap \gamma$.

6. Consider planes α and β that intersect in a line. If there is a line l that does not lie in either α or β, describe the possible intersections it may have with $\alpha \cup \beta$.

7. Find examples of physical situations that illustrate each of the following:
  (a) Parallel lines.
  (b) Parallel segments.
  (c) A line parallel to a plane.
  (d) Parallel planes.
  (e) Skew lines.

8. Suppose we have two distinct intersecting planes α and β. Plane α contains a line m and plane β contains a line n. Draw pictures to illustrate each of the following, and answer each question:

  (a) Line m intersects β but does not meet n. Are m and n parallel?
  (b) Line m intersects β and line n intersects α. Do m and n necessarily intersect each other?
  (c) Line m intersects line n. Does m necessarily intersect β?
  (d) Line m does not intersect β, and line n does not intersect α. Are m and n parallel?
  (e) Line m intersects β, line n intersects α, and lines m and n are skew.

10.5 SEPARATIONS ON A LINE AND ON A PLANE

We have defined one kind of subset of the set of points on a line: a *segment*. Intuitively, we know that there are others. If we locate a point P on a line, we can see that it divides the line, or **separates** the line, into three sets of points: the two sets of points on either side of P and the set containing P itself. The set of all points on one side of P, but not including P itself, is called a *half-line*. The union of the set of points on one side of P with the set containing P, that is, $\{P\}$, is called a *ray*. The point P is called the *boundary* of each half-line. These explanations of the half-line and ray are based on our spatial intuition, but we can define the terms more formally in terms of primitive terms and previously defined terms of our geometry. Such a definition of a ray might be: For any two distinct points A and B, the subset of $\overleftrightarrow{AB}$ that contains all points of $\overline{AB}$ and all points X such that (ABX) is defined to be a **ray**, symbolized $\overrightarrow{AB}$. The point A is called the **endpoint** of the ray. The ray is pictured as in Figure 10.9.

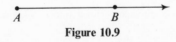

$$A \qquad\qquad B$$

Figure 10.9

A more formal definition of the half-line might be: For any two distinct points, A and B, the set containing all points of the ray $\overrightarrow{AB}$ except A is defined to be a **half-line**, symbolized $\overrightarrow{AB}$ and pictured as in Figure 10.10. The point A is called the **boundary** of $\overrightarrow{AB}$.

Suppose we have a line separated by a point P. The possible locations of two points A and B on the line fall into two cases: (1) if two distinct points A and B are

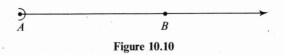

Figure 10.10

in the same half-line of this separation, that is, if P is not a point of $\overline{AB}$, then we say they are **on the same side** of P; this is illustrated in Figure 10.11. (2) If A is in one half-line and B is in the other, that is, if $P \in \overline{AB}$, we say A and B are **on opposite sides** of P; this is illustrated in Figure 10.12. The rays $\overrightarrow{PA}$ and $\overrightarrow{PB}$, in this case, are called **opposite rays**, and the half-lines $\overrightarrow{PA}$ and $\overrightarrow{PB}$ are called **opposite half-lines**. In case (1), the segment $\overline{AB}$ is a subset of $\overrightarrow{PA}$ and does not contain P; in case (2), the segment $\overline{AB}$ contains points belonging to both half-lines and contains P.

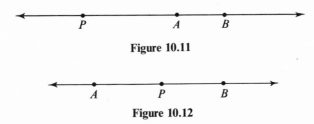

Figure 10.11

Figure 10.12

Points, rays, half-lines, and lines are each *convex sets*. The term "convex" is certainly not new to most readers. A common notion of what it means is "something that curves outward" as opposed to "concave," which is considered as "something that curves inward." Our formal definition of a "convex set" is: A set S is a **convex set** if and only if, for any two distinct points A and B in S, $\overline{AB}$ is also contained in S. This definition is in agreement with our common notion about the term "convex" as will be shown in Exercise 3.

Occasions will arise, as we proceed in the development of our geometry, when we shall need to decide whether or not a set of points is convex. Theorem 10.12, to be proved in Exercise 7, will often be useful in such cases:

Theorem 10.12:   The intersection of any two convex sets is a convex set.

A plane can be separated in many ways. One separation of a plane is suggested by the separation of a line by a point and is somewhat analogous to it. Consider a line in a plane. We can see intuitively that the line separates the plane into three sets of points: the two sets of points on opposite sides of the line and the set of points on the line itself. We call the set of points on one side of a line, not including the line itself, a *half-plane*. The line is called the *boundary* of each half-plane. It seems, on the basis of intuition, safe enough to say that the half-plane is a convex set, as is the half-line. Also, if two distinct points A and B are situated in the plane so that A is on one side of a line and B is on the other, $\overline{AB}$ would quite obviously intersect the boundary line.

The separation of a plane by a line certainly seems clear enough from the description we have given, but it does not really follow from the set of assumptions we have listed for our geometry. An additional assumption is needed and we introduce it here:

> **Assumption 10.15:** For any line l in a plane, the set of all points that are on the plane, but not in l, consists of the points of two disjoint sets such that (a) each set is a convex set, and (b) if a point A is in one set and a point B is in the other, $\overline{AB}$ intersects l in exactly one point.

The two disjoint sets mentioned in this assumption are called **half-planes,** or **sides of the plane,** and the line l is called the **boundary**, or **edge**, of each half-plane.

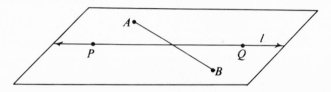

Figure 10.13

The half-plane that has the line l as edge and contains the point A can be identified by the symbol $l\overrightarrow{A}$. If the line separating the plane is identified by two of its points, for example, $\overleftrightarrow{PQ}$, (Fig. 10.13) the half-plane that has $\overleftrightarrow{PQ}$ as edge and contains A may be identified by the symbol $\overleftrightarrow{PQ}|A$. Sometimes it is useful to consider the union of a half-plane and its boundary; we shall refer to this union as a **closed half-plane**. The following theorem can now be proved, and a proof of it will be considered in Exercise 8.

> **Theorem 10.13:** If P is a point on a line l in α and Q is a point, in α, but not on l, then the half-line $\overrightarrow{PQ}$ is contained entirely in one of the two half-planes with edge l.

Another of the many ways in which a plane may be separated is by an *angle*. The explanation, or definition, of the term *angle* obtained in some past context will probably not be violated in any important sense by the definition of angle that we are about to give. Notice that this definition makes an angle another of our geometric figures: An **angle** is the union of two distinct rays having a common endpoint. The rays are called the **sides** of the angle and the common endpoint is called the **vertex**. If A is any point on one side of the angle (except the vertex), B is the vertex, and C is any point on the other side of the angle (except B), then we can identify the angle by the symbol $\angle ABC$ (or $\angle CBA$), the middle letter always denoting the vertex. When no confusion arises, this angle is often denoted simply as $\angle B$. It may be pictured as in Figure 10.14.

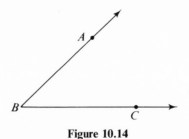

Figure 10.14

An angle whose sides are opposite rays of a line is called a **straight angle**. The name for rays not contained in the same line is **noncollinear rays**. In the following discussion of the way an angle separates a plane, we shall exclude the straight angle. (How does a straight angle separate a plane?) It seems obvious that an angle divides a plane into three sets of points. One of the sets is certainly the angle itself. In Figure 10.15 point P has been placed in one of the remaining sets, and points Q and R have been placed in the other.

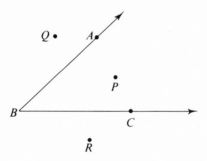

Figure 10.15

As we look at Figure 10.15, our spatial intuition indicates that, in reference to the separation of the plane by $\angle ABC$, the set of points containing P is a second set of points in the separation and the set containing Q and R is the third. We commonly refer to the set containing P as the *interior* of the angle and the set containing Q and R as the *exterior* of the angle. This discussion of the separation of a plane by an angle has been heavily dependent on the pictures we have drawn and on our intuition. We shall proceed to develop more formal definitions for the interior and exterior of an angle.

The ray $\overrightarrow{BA}$ that is one side of the angle ABC is a subset of the line $\overleftrightarrow{BA}$; similarly, $\overrightarrow{BC}$ is a subset of the line $\overleftrightarrow{BC}$. As in Figure 10.16, each of the lines $\overleftrightarrow{BA}$ and $\overleftrightarrow{BC}$ separates the plane and each is the edge of two half-planes. Two of these four

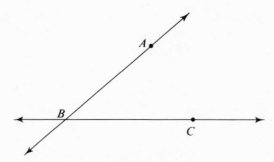

Figure 10.16

half-planes are named in our definitions of the interior and the exterior of an angle: The **interior of** $\angle$ **ABC** is the intersection of half-planes $\overleftrightarrow{BA}|C$ and $\overleftrightarrow{BC}|A$. A symbolic expression of the *interior* would be $\overleftrightarrow{BA}|C \cap \overleftrightarrow{BC}|A$. The **exterior of** $\angle$ **ABC** is the set containing those points in the plane not on the angle or in the interior of the angle. Symbolically, the *exterior* of $\angle ABC$ can be expressed as $[(\overleftrightarrow{BA}|C \cap \overleftrightarrow{BC}|A) \cup \angle ABC]'$.

Considering these definitions and Assumption 10.15, the concept of the separation of a plane by an angle is no longer merely intuitive. It is now clear that any angle whose sides are noncollinear rays separates the plane into three disjoint sets: (1) the angle itself, (2) the interior of the angle, and (3) the exterior of the angle. A straight angle does not have an exterior or an interior.

> **Theorem 10.14:** The set of all points in the plane containing any nonstraight angle consists of three disjoint sets: One is the angle itself; one is the interior of the angle and is convex; and one is the exterior of the angle and is not convex.

A proof of Theorem 10.14 will be considered in Exercise 9. A second important theorem, whose proof we shall not present, is the following:

> **Theorem 10.15:** If P is in the exterior of a nonstraight angle and Q is in the interior, the intersection of $\overline{PQ}$ and $\angle ABC$ is a set containing exactly one point.

EXERCISES 10.5

1. Can the intersection of two rays be any of the following? Draw figures to illustrate.

   (a) A set containing a point.   (b) A ray.
   (c) A line segment.   (d) The empty set.
   (e) A line.

2. Answer the questions in Exercise 1 for

(a) Half-lines instead of rays.
(b) Line segments instead of rays.
(c) A line segment and a ray.

3. Among the sets of points defined by the drawings in Figure 10.17 are some *convex sets*. Which drawings represent convex sets? Describe these sets. (Arrows in drawings indicate infinite extension.)

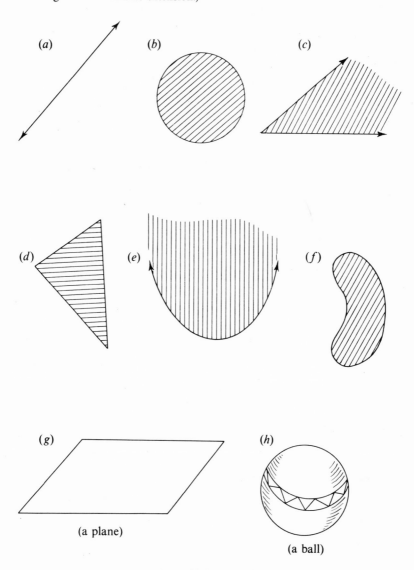

Figure 10.17

4. Refer to the drawing in Figure 10.18 and

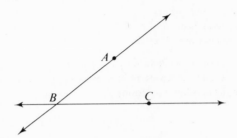

Figure 10.18

(a) Name a half-line that is in $\overleftrightarrow{BC}|A$.
(b) Name a half-line that is in $\overleftrightarrow{AB}|C$.

5. Copy the drawing for Exercise 4 and

(a) Shade $\overleftrightarrow{AB}|C$ horizontally.
(b) Shade $\overleftrightarrow{BC}|A$ vertically.
(c) Which region in the picture represents the intersection of $\overleftrightarrow{AB}|C$ and $\overleftrightarrow{BC}|A$? What do we call this intersection?

6. Copy the drawing for Exercise 4, locate a point P on $\overrightarrow{AB}$ so that (PBA), and locate a point Q on $\overleftrightarrow{BC}$ so that (QBC).

(a) Shade $\overleftrightarrow{AB}|Q$ horizontally.
(b) Shade $\overleftrightarrow{BC}|P$ vertically.
(c) Can we define the *exterior* of $\angle ABC$ as $\overleftrightarrow{AB}|Q \cup \overleftrightarrow{BC}|P$?

7. The explanation for this exercise will constitute a proof of Theorem 10.12: The intersection of any two convex sets is a convex set.

Recall that the intersection of two sets A and B is defined as the set that contains each element of A that is also an element of B and no other elements (Section 1.4). Consider any points P and Q that belong to the intersection of two convex sets A and B. Explain why $\overline{PQ}$ is entirely contained in $A \cap B$.

8. The answers to the questions in this exercise provide a basis for a proof of Theorem 10.13: If P is a point on a line l in α and Q is a point, in α, but not on l, then the half-line $\overrightarrow{PQ}$ is contained entirely in one of the two half-planes with edge l.

Consider any point R on $\overrightarrow{PQ}$ other than Q. Suppose R does not lie in $\overrightarrow{lQ}$. Then $\overline{QR}$ must intersect l at P. (a) Why? But this contradicts our hypothesis that R is on $\overrightarrow{PQ}$. (b) What must we conclude?

9. The answers to the questions in this exercise provide a basis for a proof of Theorem 10.14: The set of all points in the plane containing any nonstraight angle consists of three disjoint sets: one is the angle itself, one is the interior of the angle and is convex, and one is the exterior of the angle and is not convex.

The conclusion that the angle, its interior, and its exterior are three disjoint sets follows directly from the definitions of *angle*, *interior*, and *exterior*. (a) Why is the interior of $\angle ABC$ a convex set? The drawing in Figure 10.19 is designed to aid us in understanding the proof of the final part of this theorem: The exterior of a nonstraight angle is not convex. A nonstraight angle, $\angle ABC$, is pictured. A point P has been located on $\overleftrightarrow{AC}$ so that (PAC); a point Q has been located on $\overleftrightarrow{AC}$ so that (ACQ).

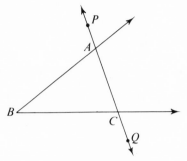

Figure 10.19

The fact that the exterior of $\angle ABC$ is not convex can be proved by showing that the exterior of $\angle ABC$ contains two points P and Q such that $\overline{PQ}$ is not entirely contained in the exterior. The point P, as we located it in the illustration, is in the exterior of $\angle ABC$. (b) Why? The point Q is also in the exterior of $\angle ABC$. (c) Why? The point A is in the segment $\overline{PQ}$. (d) Why? The point A is not in the exterior of $\angle ABC$. (e) Why? The segment $\overline{PQ}$, then, is not entirely contained in the exterior of $\angle ABC$. Therefore, the exterior of $\angle ABC$ is not convex.

10.6 SIMPLE CLOSED CURVES
IN THE PLANE

In Section 10.1, a geometric figure was defined as a nonempty set of points, and the figures we have introduced up to this point have been defined carefully. The figures known in geometry as *curves* that we introduce now, cannot be defined rigorously within the scope of this book. Therefore we shall rely on an intuitive explanation, based on visual evidence seen in drawings. A **curve** may be described as any figure whose representation can be drawn without lifting the pencil. We shall agree that, if as in Figure 10.20 we can represent a figure by a drawing that indicates a curve, except that one or two arrowheads indicate infinite extension, this figure is a curve. The curves we shall be considering will be *curves in a plane*. The drawings in Figure 10.20 are examples of curves in a plane.

A curve that does not intersect (or cross) itself is called a **simple curve**. In Figure 10.21, figures a, c, d, e, f, and h are *simple curves*. A curve that may be traced with a pencil by starting at some point in the curve and ending at the same point without lifting the pencil is called a **closed curve**. Figures c, e, f, and g are *closed curves*. A closed curve that does not intersect, or cross, itself is called a **simple closed curve**. Figures c, e, and f are *simple closed curves*.

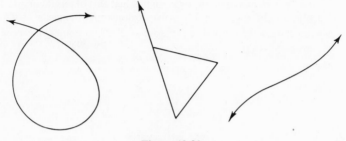

Figure 10.20

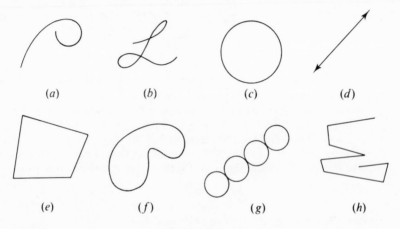

Figure 10.21

The reader will recognize from the figures that each closed curve separates the plane into three or more disjoint sets. Each simple closed curve separates the plane into exactly three disjoint sets; (1) the set of points *on the curve*, (2) the set of points *in the interior* of the curve, and (3) the set of points *in the exterior* of the curve. In the rest of this chapter, we shall be considering various polygons, beginning with triangles. All polygons are simple closed curves.

10.7   TRIANGLES

The discussion of curves in the previous section was informal, and the descriptions of curves, simple curves, closed curves, and simple closed curves were based on characteristics of drawings. In considering triangles, however, we shall return to our practice of careful definition in terms of primitive terms and/or previously defined terms.

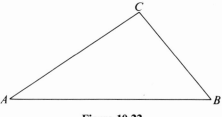

Figure 10.22

For any given noncollinear points A, B, and C a **triangle** ABC is defined as the union of segments $\overline{AB}$, $\overline{BC}$, and $\overline{AC}$:

$$\triangle ABC = \overline{AB} \cup \overline{BC} \cup \overline{AC}$$

The segments $\overline{AB}$, $\overline{BC}$, and $\overline{AC}$ are the **sides** of the triangle; the points A, B, and C are the **vertices** of triangle ABC. The singular form of "vertices" is "vertex." The triangle is denoted symbolically in terms of a small triangular symbol and its vertices, for example $\triangle ABC$.

The *interior* of a triangle is defined in terms of half-planes: The **interior of** $\triangle ABC$ is the intersection of half-planes $\overrightarrow{AB}|C$, $\overrightarrow{AC}|B$, and $\overrightarrow{BC}|A$, expressed symbolically as $\overrightarrow{AB}|C \cap \overrightarrow{AC}|B \cap \overrightarrow{BC}|A$ (Fig. 10.23).

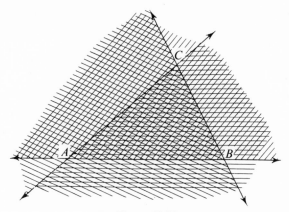

Figure 10.23

The **exterior of** $\triangle ABC$ is the set of points that are in the plane ABC but not contained in the triangle itself or in the interior of the triangle. Two theorems follow directly from these definitions. A proof of Theorem 10.16 is considered in Exercise 4; we shall not consider a proof of Theorem 10.17.

Theorem 10.16:   The interior of any triangle is a convex set.

Theorem 10.17:   Given a point P in the interior of $\triangle ABC$ and a point Q in the exterior, the intersection of $\overline{PQ}$ with $\triangle ABC$ is a set containing exactly one point.

EXERCISES 10.7

1. These figures have already been introduced in this chapter:

point	ray
line	half-plane
plane	closed half-plane
segment	angle
half-line	triangle

Which figures are
(a) curves?   (b) closed curves?
(c) simple curves?   (d) simple closed curves?

2. Which of the curves shown in Figure 10.24 is a simple curve? A closed curve? A simple closed curve?

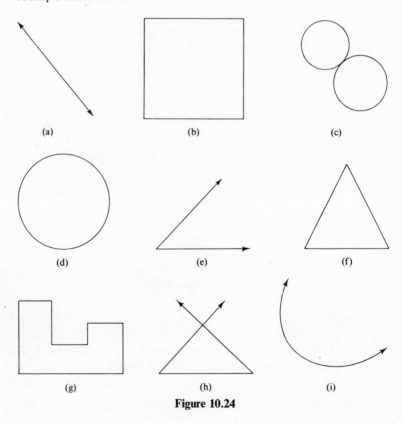

(a)   (b)   (c)

(d)   (e)   (f)

(g)   (h)   (i)

Figure 10.24

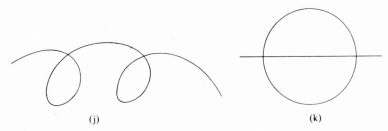

(j)   (k)

Figure 10.24 (continued)

3. Draw pictures to show:

   (a) Three coplanar half-planes whose intersection is the interior of a triangle.
   (b) Three coplanar half-planes whose intersection is not the interior of a triangle.
   (c) Four half-planes whose intersection is the interior of a triangle.

4. Prove Theorem 10.16.

5. (a) Review the definitions of *angle* and *triangle* carefully. Would you say that a triangle contains an angle?
   (b) In $\triangle ABC$ is the union of sides $\overline{AB}$ and $\overline{BC}$ contained in an angle? Name the angle.

   Note: Owing to the relationship between any two sides of a triangle and an angle, we say the triangle *has three angles*, or we speak of the *angles of a triangle*. This does *not* mean that a triangle *contains* any angles.
   (c) Give an alternative definition of the *interior of a triangle* in terms of the interiors of angles.

6. Draw triangle ABC and lines $\overleftrightarrow{AB}, \overleftrightarrow{AC}, \overleftrightarrow{BC}$.

   (a) Mark a point P on $\overleftrightarrow{AB}$ so that (ABP); mark a point Q on $\overleftrightarrow{BC}$ so that (BCQ); mark a point R on $\overleftrightarrow{AC}$ so that (RAC).
   (b) Shade $\overleftrightarrow{AB}|R, \overleftrightarrow{AC}|Q$, and $\overleftrightarrow{BC}|P$.
   (c) Consider the union of $\overleftrightarrow{AB}|R, \overleftrightarrow{AC}|Q$, and $\overleftrightarrow{BC}|P$ and state an alternative definition of the exterior of a triangle.

10.8 POLYGONS

The student doubtless thinks of a *polygon* as a closed figure consisting entirely of line segments. We now define polygons a bit more formally: A polygon is a union of line segments $\overline{A_1A_2}, \overline{A_2A_3}, \overline{A_3A_4}, \ldots, \overline{A_{n-1}A_n}$, and $\overline{A_nA_1}$ where $n \geq 3$ (read "n is greater than or equal to 3"), no three successive points $A_1, A_2, A_3, \ldots, A_n$ are collinear, and these points are the only points of intersection of the given line segments. The segments $\overline{A_1A_2}, \overline{A_2A_3}, \overline{A_3A_4}, \ldots, \overline{A_nA_1}$ are the **sides** of the polygon. If $n = 3$, we have a triangle; if $n = 4$, we have a four-sided figure called a **quadrilateral**. The following chart shows the names of the polygons having $3, 4, 5, 6, 7, 8, 9, 10,$ and 12 sides, that is, $n = 3, 4, 5, 6, 7, 8, 9, 10, 12$.

n	Name of polygon
3	triangle
4	quadrilateral
5	pentagon
6	hexagon
7	heptagon
8	octagon
9	nonagon
10	decagon
12	dodecagon

A quadrilateral is a polygon studied extensively in plane geometry. At this point we can define two sorts of quadrilaterals having special characteristics: A **trapezoid** is a quadrilateral having at least one pair of parallel sides. A **parallelogram** is a quadrilateral having two pairs of parallel sides. In a later chapter we shall consider other special quadrilaterals: squares, rhombuses, and rectangles. The concept of congruence is needed to define these figures.

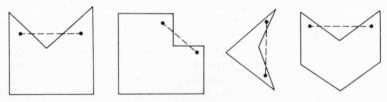

Figure 10.25

In the consideration of polygons according to the number of sides, once we begin to study polygons with four or more sides, we recognize intuitively that the union of the polygon and its interior need not be a convex set. For example, consider the polygons in Figure 10.25. Here we see that we may have polygons of four or more sides whose interiors are not convex sets. We call such polygons **concave** polygons, while the polygon whose interior is a convex set is called a **convex** polygon. The reader is probably aware of the imprecise nature of these descriptions of concave and convex polygons. We know what a convex set is; this term has been carefully defined. But we have no careful definition of the *interior* of any polygon other than that of the triangle. We therefore proceed to define the term *convex polygon* more formally, avoiding the use of an undefined term such as "interior." If for each side of a polygon, one of the two closed half-planes that has that side as an edge contains each of the other sides of the polygon, then this polygon is called a **convex polygon**. In Figure 10.26 the quadrilateral $ABCD$ is a convex polygon as indicated by the four closed half-planes.

A **concave polygon** is a polygon that is not convex. In Figure 10.27 the quadrilateral $ABCD$ is concave since neither of the closed half-planes whose edge contains $\overline{AB}$, contains each of the other sides of the quadrilateral. We observe that neither

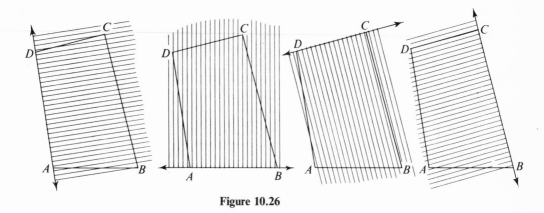

Figure 10.26

$\overleftrightarrow{CD}$ nor $\overline{BC}$ are entirely contained in the closed half-plane that is the union of $\overleftrightarrow{AB}$ and $\overleftrightarrow{AB}|D$, while the other closed half-plane, whose edge contains $\overleftrightarrow{AB}$, does not entirely contain $\overline{CD}$ and $\overline{AD}$.

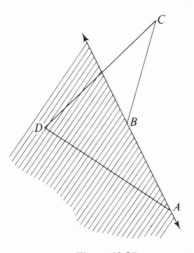

Figure 10.27

We are now ready to consider the idea of the *interior of a convex polygon*. Recall that each of the sides of a convex polygon is in the boundary of a closed half-plane that contains each of the other sides of the polygon. Consider the *half-plane* contained in this closed half-plane. The intersection of all such half-planes is defined as the **interior of the convex polygon**. The union of a polygon and its interior is called a **polygonal region**. At this point, we can prove (Exercise 6) the following theorem about the interior of a convex polygon.

Theorem 10.18:   The interior of every convex polygon is a convex set.

When considering polygons of more than three sides, we find that some of the vertices, taken in pairs, are endpoints of line segments that are not sides of the polygons. These segments are called **diagonals**. Several of the exercises are concerned with diagonals.

EXERCISES 10.8

1. Complete the table in Figure 10.28 and thereby classify the figures on the left by placing a check mark in each column that represents a class of figures to which the given figure belongs.

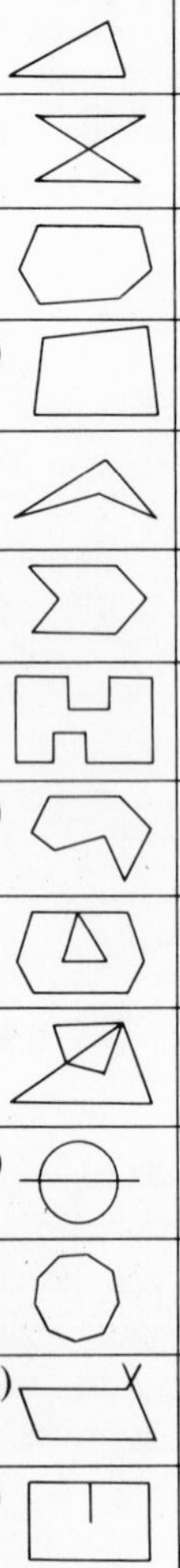

	Curve	Simple Curve	Closed Curve	Polygon	Convex Polygon	Concave Polygon
(a)			X	X	X	
(b)			X	X		X
(c)			X	X	X	
(d)			X	X	X	
(e)			X	X		X
(f)			X	X		X
(g)			X	X		X
(h)			X	X		X
(i)			X	X		X
(j)			X	X		X
(k)	X		X			X
(l)			X	X	X	
(m)			X	X	X	X
(n)			X	X		X

Figure 10.28

2. (a) Draw convex polygons having 3, 4, 5, 6, and 7 sides.
   (b) Draw all the diagonals in each of these convex polygons.
   (c) Draw concave polygons having 4, 5, 6, 7, and 8 sides.
   (d) Draw all the diagonals for each of these concave polygons.
   (e) Suggest an alternative definition for a convex polygon.

3. (a) Is a convex polygon a convex set?
   (b) Is the interior of a convex polygon a convex set?
   (c) Is a polygonal region consisting of the union of a convex polygon and its interior a convex set?

4. (a) Can the interior of an angle contain a line? A ray? A segment?
   (b) Can the interior of a triangle contain a line? A ray? A segment?
   (c) Can the interior of a convex polygon contain a line? A ray? A segment?

5. Discuss the similarities and differences among the ways in which the following figures separate the plane:

   (a) The line.   (b) The angle.
   (c) The nonsimple closed curve.   (d) The simple closed curve.
   (e) The triangle.   (f) The polygon.
   (g) The concave polygon.   (h) The convex polygon.

6. Prove Theorem 10.18.

7. Refer to the definition of "closed half-plane" in Section 10.5.

   (a) Sketch three closed half-planes that intersect to form the union of a triangle and its interior.
   (b) Sketch four closed half-planes that intersect to form the union of a quadrilateral and its interior.
   (c) Sketch five closed half-planes that intersect to form the union of a pentagon and its interior.
   (d) Make a sketch showing that the intersection of four closed half-planes may not be a polygonal region.

Polygon	Number of sides	Number of diagonals
Triangle	3	0
Quadrilateral	4	2
Pentagon	5	5
Hexagon	6	9
Heptagon	7	14
Octagon	8	20
Nonagon	9	27

Figure 10.29

8. (a) Draw convex polygons having 8 and 9 sides and draw all the diagonals of each.

(b) Complete the table in Figure 10.29.

(c) Project how many diagonals a 15-sided convex polygon will have.

(d) In terms of n, express the number of diagonals an n-sided convex polygon will have.

CHAPTER 11

Measurement and Congruence

11.1 THE TRADITIONAL TREATMENT OF CONGRUENCE

The statement, "Two figures are congruent if they have the same size and shape," is one that has been commonly used to explain the concept of **congruence**. This statement suggests that one figure must be an exact copy of the other if the two figures are to be considered congruent. In the typical traditional textbook used to introduce Euclidean geometry in the high school a decade or more ago, the classic definition was: "Congruent figures are figures that can be made to coincide." This definition was generally followed by a discussion of testing figures for congruence by superimposing one figure on the other to see if they "fit exactly," in which case they do "coincide" and are congruent.

"Superposition of figures" is a phrase that seems to suggest a physical process; thus, the phrase is hardly suitable for the explanation of a mathematical concept. Modern treatments of *congruence*, therefore, avoid the use of the terms "coincide" or "superposition" for different figures. The statement included in the first sentence of this section indicates that congruence might be definable in terms of such notions as measures of segments and angles. For example, two segments could be considered congruent if their lengths are equal, or two angles could be considered congruent if their measures (to be discussed in more detail in the following pages) are equal. This is the approach we shall use to develop the concept of

205

congruence. We shall postulate some metrical notions involving the measure of a line segment and the measure of an angle, and we shall then define congruence of several types of geometric figures in terms of these notions.

The student would do well, before proceeding to the next section, to review Sections 8.2 and 8.3 in Chapter 8.

11.2 NUMBERS AND MEASUREMENT

In previous chapters, we have used the idea of a correspondence between numbers and points on a line to help us understand *properties of numbers*. In this chapter, we are using this correspondence to help us develop *properties of the line and properties of subsets of the line*. In Chapter 8, we stated a property called the *completeness property of real numbers*. This property is essential to our development of the concepts of length and congruence of line segments and is implicit in the first assumption we make relative to these concepts.

> **Assumption 11.1:** Any line may be used as a real number line; that is,
> 1. Any point of the line may be selected as the **origin** with coordinate 0.
> 2. Any second point of the line may be selected as the **unit point** with coordinate 1.
> 3. After the origin and unit point have been selected, each point of the line has a unique real number as its **coordinate** and each real number has a point of the line as its **graph**.

For any two points A and B of a line, the **distance** between A and B, that is, the **length** of the line segment $\overline{AB}$ [symbolized $m(\overline{AB})$], is the absolute value of the difference of the coordinates of A and B. Once we have selected a point on the line as the origin and a second point as the unit point, we have determined a **coordinate system** for the line, that is, the coordinates of the remaining points on the line have been uniquely determined. Theorem 11.1 follows directly from Assumption 11.1 and the definition of the length of a line segment.

> **Theorem 11.1** (the point location theorem): Let $\overrightarrow{OB}$ be a ray such that the number zero is the coordinate of O and the number one is the co-ordinate of a second point P on $\overrightarrow{OB}$, and let x be any positive real number. Then there is exactly one point X contained in $\overrightarrow{OB}$ such that $m(\overline{OX}) = x$.

EXERCISES 11.2

1. Draw a ray $\overrightarrow{AB}$ similar to the one illustrated in Figure 11.1.

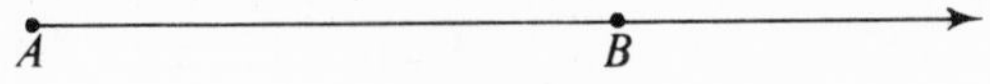

Figure 11.1

(a) Locate two points P and Q on $\overrightarrow{AB}$ such that (APQ). Let zero be the coordinate of A and locate the point on $\overrightarrow{AB}$ which is one inch from A and assign it the coordinate 1.

(b) Use your ruler and the coordinate system established in part (a), and find the coordinate of P. Of Q. Find $m(\overline{PQ})$.

(c) Is $m(\overline{PQ})$ a real number? Explain your answer.

(d) Can you locate a segment $\overline{AS}$ on $\overrightarrow{AB}$ which will have the same length as $\overline{PQ}$? Explain your answer.

(e) If you choose a point M on $\overline{AB}$, such that (PQM), can you locate a segment $\overline{MN}$ such that $m(\overline{MN}) = m(\overline{PQ})$? Is it unique? Justify your answer.

2. Find the distance between each pair of points whose coordinates are listed below.

(a) $0, 9$ (b) $9, 0$ (c) $0, \,^-9$

(d) $3, 14$ (e) $3, \,^-14$ (f) $2\frac{1}{2}, 3\frac{4}{5}$

(g) $^-7.2, \,^-3.7$ (h) $^-7.2, 7.2$ (i) $7.2, 7.2$

(j) $3x, \,^-3x$ (k) $3x, 3x$ (l) x_1, x_2

(m) $x + y, x - y$ (n) $\sqrt{2}, 3\sqrt{2}$ (o) $\sqrt{2}, \sqrt{3}$

3. In Figure 11.2, two different coordinate systems have been chosen for the line, both having the same unit.

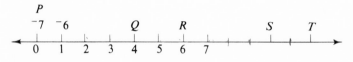

Figure 11.2

(a) What is the coordinate of the point Q in the coordinate system below the line? The point R?

(b) Find $m(\overline{QR})$ using the coordinate system below the line. Show that no change in $m(\overline{QR})$ would result from using the coordinate system above the line.

(c) What seems to be the coordinate of the point S in the scale above the line? Of the point T? Find the coordinates of these points in the scale below the line.

(d) Using the coordinate system below the line, find $m(\overline{PQ}), m(\overline{PR}), m(\overline{PS}), m(\overline{PT}), m(\overline{ST})$.

(e) Find the lengths of the segments given in part (d) using the scale above the line. Are your results different from the ones found when you used the scale below the line?

4. In the illustration (Fig. 11.3), two different coordinate systems are shown for the line.

Figure 11.3

(a) What is the coordinate of the point P in the upper scale? Of the point Q? Of the point R?

(b) Find $m(\overline{PQ})$ using coordinates in the lower scale. Find $m(\overline{PQ})$ using coordinates in the upper system.

(c) Do you get the same result for $m(\overline{PQ})$ regardless of which of the given coordinate systems you use?

(d) Explain the results of part (b).

(e) The integral coordinates of the upper scale can be listed as follows:

$$^{-}3, \quad ^{-}2, \quad ^{-}1, \quad 0, \quad 1, \quad 2, \quad 3, \quad \cdots, \quad n \quad \cdots,$$
$$\updownarrow \quad \updownarrow \quad \updownarrow \quad \updownarrow \quad \updownarrow \quad \updownarrow \quad \updownarrow \qquad \quad \updownarrow$$

Copy this and complete the chart showing the correspondence between these coordinates and a subset of the integral coordinates (for the same points) of the lower scale.

5. The first numbering of the points on the line in Figure 11.4 represents a coordinate system. Which of the other numberings are not coordinate systems according to Assumptions 11.1 and the definition of a coordinate system?

	$^{-}6$	$^{-}5$	$^{-}4$	$^{-}3$	$^{-}2$	$^{-}1$	0	1	2	3	4	5	6	7
(a)	$^{-}8$	$^{-}7$	$^{-}6$	$^{-}5$	$^{-}4$	$^{-}3$	$^{-}2$	$^{-}1$	0	1	2	3	4	5
(b)	0	1	2	3	4	5	6	5	4	3	2	1	0	1
(c)	11	12	13	14	15	16	17	18	19	20	21	22	23	24
(d)	$^{-}11$	$^{-}12$	$^{-}13$	$^{-}14$	$^{-}15$	$^{-}16$	$^{-}17$	$^{-}18$	$^{-}19$	$^{-}20$	$^{-}21$	$^{-}22$	$^{-}23$	$^{-}24$
(e)	$^{-}3$	$^{-}2$	$^{-}1$	0	$^{-}1$	2	3	4	5	6	7	8	9	10

Figure 11.4

6. Consider the points on a line whose coordinates are described below.

(a) $x < 3$ (b) $x = 1$
(c) $x > 2$ (d) $x \leq 1$
(e) $x = {}^{-}3$ (f) $|x| \leq 3$
(g) $|x| > 2$ (h) $|x| \geq 0$

Which of the above sets is a ray? A half-line? A set containing a single point? A line? A segment? Something other than any of these?

11.3   MORE ABOUT THE REAL NUMBERS AND THE LINE

Using the idea of the length of a segment, which we have introduced in the preceding section, and a new concept, that of *nested intervals*, we can give the reader an intuitive verification of the *completeness property of real numbers*; that is, we can help the reader to feel more assured that this statement of a one-to-one correspondence between real numbers and points on a line does make sense.

Consider any two real numbers p and q that are coordinates of points P and Q on a number line. The points with coordinates x where $p \leq x \leq q$ form a **closed interval** $[P, Q]$ (read "the interval P, Q"). We indicate the interval $[P, Q]$ on the number line as shown in Figure 11.5.

$$P \qquad\qquad Q$$

Figure 11.5

A sequence of nested intervals is an ordered set of closed intervals such that each interval is a subset of the previous interval. If, in an infinite sequence of nested intervals, the sequence of the lengths of the intervals approaches zero, then we say the sequence "closes down" on a single point P. In this kind of sequence, the point P is contained in every interval of the sequence. An example of such a sequence is shown in Figure 11.6. The Interval I_1 is $[P_1, Q_1]$; I_2 is $[P_2, Q_2]$ where P_2 and Q_2 are chosen as points that divide $\overline{P_1 Q_1}$ into three equal segments; I_3 is $[P_3, Q_3]$ where P_3 and Q_3 are points that divide $\overline{P_2 Q_2}$ into three equal segments, etc. Clearly *this* sequence of nested intervals closes down on a point P that is the midpoint of $\overline{P_1 Q_1}$.

Figure 11.6

Now consider the set D of points, on the line, whose coordinates are terminating decimal fractions. We shall indicate the reasonableness of a one-to-one correspondence between the real numbers and the points on the line by using sequences of nested intervals whose endpoints belong to this set D, and which do close down on a single point. *First*, we shall indicate that given a particular real number, we can always find a point on the line corresponding to it; *second*, we shall indicate that, given a point on the line, we can always find a real number coordinate for it.

1. **Assignment of a point to a real number.** To give the reader an intuitive notion that we can always assign a point to *any* real number, we shall use the real number $\sqrt{2} = 1.4142135\ldots$. The sequence of number pairs $(1, 2)$, $(1.4, 1.5)$, $(1.41, 1.42)$, $(1.414, 1.415)$, $\ldots$ corresponds to the sequence of nested intervals $[P_1, Q_1]$, $[P_2, Q_2]$, $[P_3, Q_3]$, $[P_4, Q_4]$, $\ldots$ as pictured on the line in Figure 11.7.

Notice that the point corresponding to $\sqrt{2}$ is in each of these intervals. For each digit of $\sqrt{2}$ beyond the fourth, we get an additional interval in this sequence. The first interval has length 1; the second interval has length $\frac{1}{10}$; the third interval has length $\frac{1}{100}$, and the nth interval has length $\frac{1}{10}^{n-1}$. The sequence is closing down on a point that is the graph of the real number $\sqrt{2}$.

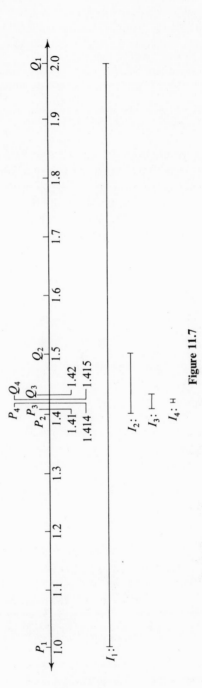

Figure 11.7

2. Assignment of a real number to a point. To give the notion that we can always assign a real number to a point, we imagine the line divided (by geometric construction) into intervals of length $\frac{1}{10}$. To these we assign coordinates

$$\ldots, \ ^{-}0.4, \ ^{-}0.3, \ ^{-}0.2, \ ^{-}0.1, 0, 0.1, 0.2, 0.3, \ldots$$

Suppose, then, there is a given point on the line whose coordinate we wish to determine, as indicated by the point P in Figure 11.8.

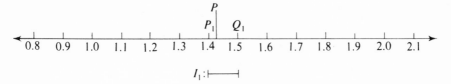

Figure 11.8

We see that the point P is in the interval I_1, $[P_1, Q_1]$, corresponding to the coordinate pair (1.4, 1.5); thus we know that the first two digits of the coordinate of P are 1.4. Now it is possible to divide I_1 into intervals of length $\frac{1}{100}$ (by geometric construction). Figure 11.9 is drawn on a larger scale and shows P to be in the interval I_2, $[P_2, Q_2]$, corresponding to the coordinate pair (1.41, 1.42).

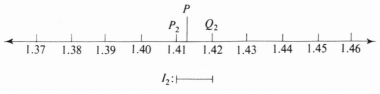

Figure 11.9

The interval I_2 can be divided into intervals of length $\frac{1}{1000}$, shown on a still larger scale. In Figure 11.10 we see that P lies in the interval I_3, $[P_3, Q_3]$, corresponding to the coordinate pair (1.414, 1.415); hence the first four digits of its coordinate are 1.414.

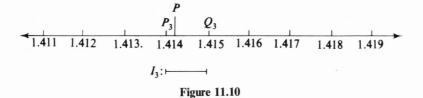

Figure 11.10

By continuing this process, we can find nested intervals $I_4, I_5, \ldots, I_n, \ldots$, and we can assign a real number to P having as many decimal places as desired.

The real number that is the coordinate of the point P that we have chosen can be represented as $\sqrt{2}$. For a point P chosen arbitrarily, however, there would very frequently be no simpler way to express its coordinate than by an infinite non-terminating decimal fraction.

EXERCISES 11.3

1. (a) Locate on a line the first five intervals of the sequence corresponding to the following coordinate pairs.

$$(^-1, 1), (^-1/2, 1/2), (^-1/3, 1/3), \ldots, (^-1/n, 1/n), \ldots$$

   (b) Is this a sequence of nested intervals?
   (c) Does it close down on a point?
   (d) If so, what is the coordinate of the point?

2. (a) Locate on a line the first five intervals of the sequence corresponding to the following coordinate pairs.

$$(^-1, 2), (^-1/2, 3/2), (^-1/3, 4/3), \ldots, (^-1/n, [n + 1]/n), \ldots$$

   (b) Is this a sequence of nested intervals?
   (c) Does it close down on a point?
   (d) If so, what is the coordinate of the point?

3. A sequence of nested intervals, $I_1, I_2, I_3, \ldots$ begins with intervals corresponding to the following coordinate pairs: (0, 1), (0.1, 0.2), (0.11, 0.12), (0.111, 0.112).

   (a) Give the coordinate pairs corresponding to the next three intervals.
   (b) What is the coordinate of the point on which this sequence closes down?

4. Consider the points (on a line) whose coordinates are the real numbers listed below. For each point, begin with the interval whose coordinate pair contains the nearest integers, and identify the first five intervals of a sequence of nested intervals that closes down on each point.

   (a) $\frac{1}{3} = 0.\overline{3}$ (b) $\frac{1}{7} = 0.\overline{142857}$

   (c) $\sqrt{3} = 1.73205\ldots$ (d) $\sqrt[3]{3} = 1.44225\ldots$

   (e) $\sqrt{7} = 2.64575\ldots$ (f) $\sqrt[3]{9} = 2.08008\ldots$

5. Suppose a point P lies exactly $\frac{1}{6}$ of the distance from P_1 to Q_1 as shown in Figure 11.11. .

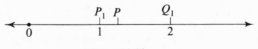

Figure 11.11

Begin with $[P_1, Q_1]$, and write the coordinate pairs of the first five intervals of a sequence of nested intervals that closes down on P.

11.4 DISTANCE AND ORDER OF POINTS

If, in addition to Assumption 11.1, we make a second assumption about the distance, or measurement, of a segment, we can state and prove four theorems involving the ordering of measurements. This assumption is frequently called the "addition postulate."

Assumption 11.2 (the addition postulate): If (ABC), then

$$m(\overline{AB}) + m(\overline{BC}) = m(\overline{AC})$$

Proofs of three of these theorems are left to the student as Exercises in Exercises 11.5.

Theorem 11.2: If (APQ), then $m(\overline{AP}) < m(\overline{AQ})$.

Theorem 11.3: Given points P and Q on a ray $\overrightarrow{AB}$, if $m(\overline{AP}) < m(\overline{AQ})$, then (APQ).

A proof of Theorem 11.3 might proceed as follows: If P is on $\overrightarrow{AB}$ then, by definition of a ray, $\overrightarrow{AB} = \overrightarrow{AP}$. Then Q lies on $\overrightarrow{AP}$. This means that if P and Q are distinct (Assumption 10.6), then (AQP) or (APQ); if P and Q are not distinct, then $P = Q$. If (AQP), then by Theorem 11.2, $m(\overline{AQ}) < m(\overline{AP})$. If $Q = P$, then by Theorem 11.1, $m(\overline{AQ}) = m(\overline{AP})$. The conclusion of each of these last two statements is contrary to the hypothesis of the theorem. Hence we must conclude that (APQ).

Theorem 11.4: If $m(\overline{AB}) < m(\overline{CD})$, then there exists exactly one point P in $\overrightarrow{CD}$ such that $m(\overline{AB}) = m(\overline{CP})$ and (CPD).

Theorem 11.5: If A, B, and C lie on a line and if $m(\overline{AB}) + m(\overline{BC}) = m(\overline{AC})$, then (ABC).

11.5 CONGRUENCE OF SEGMENTS

With the notion of measurement, or length, of segments established by the definitions, assumptions, and theorems given in Section 11.2 and 11.4, we are ready to define **congruence of segments**. If segments $\overline{AB}$ and $\overline{CD}$ have the same length, that is, $m(\overline{AB}) = m(\overline{CD})$, then $\overline{AB}$ is defined to be **congruent** to $\overline{CD}$. The statement "$\overline{AB}$ is congruent to $\overline{CD}$" is symbolized by $\overline{AB} \cong \overline{CD}$.

Congruence of segments is an equivalence relation, a fact that follows immediately from the fact that equality of real numbers is an equivalence relation (review Section 5.12). Also, two other important properties of congruence of segments follow directly from Theorem 11.1, Assumption 11.2, and the addition property of equality of real numbers. In Theorem 11.6, we list the reflexive, symmetric, transitive, and two additional properties of congruence.

Theorem 11.6: Congruence of segments has the following properties:

1. $\overline{AB} \cong \overline{AB}$; thus congruence of segments is reflexive.
2. If $\overline{AB} \cong \overline{CD}$, then $\overline{CD} \cong \overline{AB}$; thus congruence of segments is symmetric.
3. If $\overline{AB} \cong \overline{CD}$ and $\overline{CD} \cong \overline{EF}$, then $\overline{AB} \cong \overline{EF}$; thus congruence of segments is transitive.
4. Given a ray $\overrightarrow{AB}$ and a segment $\overline{CD}$, there is exactly one point P in $\overrightarrow{AB}$ such that $\overline{AP} \cong \overline{CD}$.
5. If $\overline{AB} \cong \overline{A_1B_1}$, and $\overline{BC} \cong \overline{B_1C_1}$, and if (ABC) and $(A_1B_1C_1)$, then $\overline{AC} \cong \overline{A_1C_1}$.

Using the concept of congruent segments, we define the expression **midpoint of a segment**. If the point M is in $\overline{AB}$ and $\overline{AM} \cong \overline{MB}$, then M is the **midpoint** of $\overline{AB}$. A proof of Theorem 11.7 will be considered in Exercise 8.

Theorem 11.7: A segment has one and only one midpoint.

The midpoint of a segment is said to **bisect** the segment. In general, any figure whose intersection with a segment is the midpoint of the segment is said to **bisect** the segment.

EXERCISES 11.5

1. Suppose the point P lies on a line and n is a positive real number. How many points exist on the line and at a distance n from P?

2. The letter pairs AB, BC, AC, and DB in the next three sentences are intended to denote either numbers, lines, segments, or rays. The sentences are a description of a single figure.

   (1) $AB + BC = AC$
   (2) DB contains A and C, but DB contains neither point A nor point C.
   (3) The point A belongs to DB, but point C does not.

(a) Use appropriate symbols to establish a correct identity for each letter pair in the previous three sentences.

(b) Draw a picture of the figure described in the three given sentences.

3. Four points A, B, C, and D lie on a line so that $m(\overline{AC}) > m(\overline{AB})$ and $m(\overline{BD}) < m(\overline{BC})$.

(a) Draw the line and locate the four points properly.

(b) Is there more than one possible order? Explain.

4. Refer to Figure 11.12 and tell which of the following are correct statements.

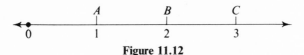

Figure 11.12

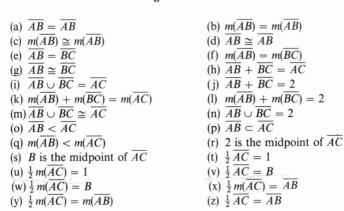

(a) $\overline{AB} = \overline{AB}$

(b) $m(\overline{AB}) = m(\overline{AB})$

(c) $m(\overline{AB}) \cong m(\overline{AB})$

(d) $\overline{AB} \cong \overline{AB}$

(e) $\overline{AB} = \overline{BC}$

(f) $m(\overline{AB}) = m(\overline{BC})$

(g) $\overline{AB} \cong \overline{BC}$

(h) $\overline{AB} + \overline{BC} = \overline{AC}$

(i) $\overline{AB} \cup \overline{BC} = \overline{AC}$

(j) $\overline{AB} + \overline{BC} = 2$

(k) $m(\overline{AB}) + m(\overline{BC}) = m(\overline{AC})$

(l) $m(\overline{AB}) + m(\overline{BC}) = 2$

(m) $\overline{AB} \cup \overline{BC} \cong \overline{AC}$

(n) $\overline{AB} \cup \overline{BC} = 2$

(o) $\overline{AB} < \overline{AC}$

(p) $\overline{AB} \subset \overline{AC}$

(q) $m(\overline{AB}) < m(\overline{AC})$

(r) 2 is the midpoint of $\overline{AC}$

(s) B is the midpoint of $\overline{AC}$

(t) $\frac{1}{2}\overline{AC} = 1$

(u) $\frac{1}{2} m(\overline{AC}) = 1$

(v) $\frac{1}{2}\overline{AC} = B$

(w) $\frac{1}{2} m(\overline{AC}) = B$

(x) $\frac{1}{2} m(\overline{AC}) = \overline{AB}$

(y) $\frac{1}{2} m(\overline{AC}) = m(\overline{AB})$

(z) $\frac{1}{2}\overline{AC} = \overline{AB}$

5. Prove Theorem 11.2: If (APQ), then $m(\overline{AP}) < m(\overline{AQ})$.

6. Your answers to the questions in this exercise will constitute a proof of Theorem 11.4: If $m(\overline{AB}) < m(\overline{CD})$, then there exists exactly one point P in $\overrightarrow{CD}$ such that $m(\overline{AB}) = m(\overline{CP})$ and (CPD).

    PROOF: There exists exactly one point P in $\overrightarrow{CD}$ such that $m(\overline{CP}) = m(\overline{AB})$. (a) Why is there just one such point? Then (CPD), $P = D$, or (CDP). (b) Why must just one of these statements be true? If $P = D$, then $m(\overline{AB}) = m(\overline{CD})$, contrary to the hypothesis of the theorem. If (CDP), then $m(\overline{CD}) < m(\overline{CP})$. (c) Why? And $m(\overline{CD}) < m(\overline{AB})$, contrary to hypothesis. (d) Why? Thus we are forced to conclude that (CPD).

7. Prove Theorem 11.5: If $A, B,$ and C lie on a line and if $m(\overline{AB}) + m(\overline{BC}) = m(\overline{AC})$, then (ABC).

8. Your answers to the questions in this exercise will constitute a proof of Theorem 11.7: A segment has one and only one midpoint.

    PROOF: I. *Existence* Let any segment be represented by $\overline{AB}$. Then there is a point N in $\overline{AB}$ such that $m(\overline{AN}) = \frac{1}{2}m(\overline{AB})$. (a) Why does such a point N

exist? Thus $m(\overline{AN}) < m(\overline{AB})$. (b) Why? It follows that (ANB) and N is in $\overline{AB}$. (c) Why? Then $m(\overline{AN}) + m(\overline{NB}) = m(\overline{AB})$. (d) How can we justify the previous statement? Since $m(\overline{AN}) = \frac{1}{2}m(AB), m(\overline{AB}) = 2m(\overline{AN})$. (e) Why? Then $m(\overline{AN}) + m(\overline{NB}) = 2m(\overline{AN})$ and $m(\overline{NB}) = m(\overline{AN})$. (f) Why? This means that N is the midpoint of $\overline{AB}$ by definition. Thus we have proved that the midpoint of a segment exists.

II. _Uniqueness_ Suppose a point M is also the midpoint of $\overline{AB}$; then M is in $\overline{AB}$ and $m(\overline{AM}) = m(\overline{MB})$. (g) Why is the previous statement true? This means (AMB) and M is in $\overline{AB}$, which tells us that $m(\overline{AM}) + m(\overline{MB}) = m(\overline{AB})$. (h) Why is $m(\overline{AM}) + m(\overline{MB}) = m(\overline{AB})$? Thus $m(\overline{AM}) = \frac{1}{2}m(\overline{AB})$ by properties of real numbers and $m(\overline{AM}) = m(\overline{AN})$. (i) Why? We must therefore conclude that $M = N$; that is, the midpoint of a segment is unique. (j) Explain.

11.6 ANGLES AND MEASUREMENT

Doubtless the student is familiar with the common _degree_ system of angle measurement and perhaps has studied _radian_ measure also. In our discussion of angle measurement, we shall make assumptions that will be in agreement with the degree system of angle measurement and will retain its properties. Also, in this chapter, we are interested in _angles that have interiors_. The student will recall that the interior of an angle ABC is defined to be the intersection of half-planes $\overleftrightarrow{BA}|C$ and $\overleftrightarrow{BC}|A$. The first assumption given (often called "the angle measurement postulate") reflects the fact that we are limiting our discussion to angles that have interiors. The name for rays that are not contained in the same line is **noncollinear rays**.

Assumption 11.3 (the angle measurement postulate): If $\overrightarrow{BA}$ and $\overrightarrow{BC}$ are noncollinear rays, then to each angle $\angle ABC$ there corresponds a unique real number between 0 and 180 called the measure of the angle and symbolized as $m(\angle ABC)$.

It is important that the student notice that the symbol $m(\angle ABC)$ represents a real number. You are accustomed to statements such as $\angle ABC = 30°$ (30 degrees). This statement means: The _number_ of degrees in $\angle ABC$ is 30. Since we are attempting to develop properties of angles in terms of properties of real numbers, we are not interested in the _degrees_ (units of measurement) specified in this statement, but rather in the _number_ that is specified. The angle measurement postulate (Assumption 11.3) is related to the measurement system with degrees as units only inasmuch as Assumption 11.3 specifies that the measure of a non-straight angle is a real number between 0 and 180. Suppose, on the other hand, that we had stated the postulate as follows: To each angle $\angle ABC$ there corresponds a real number between 0 and π. Then the measure of an angle would have been related to the system of angle measurement that has radians as units. Other conceivable units of angle measure could be used and would alter the postulate only by changing the *set of real numbers* that are used here as measures of angles.

> **Assumption 11.4** (the angle construction postulate): Given a half-plane H, a ray $\overrightarrow{AB}$ on the boundary of H and a real number r between 0 and 180, there exists exactly one ray $\overrightarrow{AP}$, where P is in H, such that $m(\angle PAB) = r$. (See illustration, Fig. 11.13.)

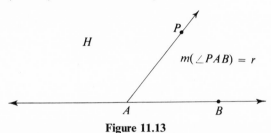

Figure 11.13

11.7 ANGLE MEASURE
AND ORDER OF RAYS

You will recall that the relationship between distance and order of points involved concepts about betweenness for points on a line. In dealing with angles, we need some ideas about betweenness for rays. We define **betweenness for rays** as follows: Suppose we have rays $\overrightarrow{OA}$, $\overrightarrow{OB}$, and $\overrightarrow{OC}$ with O as a common endpoint such that $\overrightarrow{OA}$ and $\overrightarrow{OC}$ are distinct and not opposite. If there exist, in $\overrightarrow{OA}$, $\overrightarrow{OB}$, and $\overrightarrow{OC}$ respectively, points A_1, B_1, and C_1 so that $(A_1 B_1 C_1)$, then we say that $\overrightarrow{OB}$ is **between** $\overrightarrow{OA}$ and $\overrightarrow{OC}$; this relation is written symbolically as $(\overrightarrow{OA}\,\overrightarrow{OB}\,\overrightarrow{OC})$. The method of showing that there is a natural link between the order of points and the order of rays is simple: Suppose we have $\angle ABC$; consider a point $A_1 \neq B$ lying on $\overrightarrow{BA}$ and a point $C_1 \neq B$ on $\overrightarrow{BC}$. Now think of each point P such that $(A_1 P C_1)$ as associated with a ray $\overrightarrow{BP}$ (Fig. 11.14) . This establishes a one-to-one

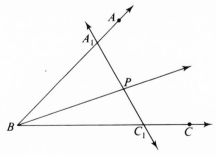

Figure 11.14

correspondence between all the points on $\overleftrightarrow{A_1 C_1}$ between A_1 and C_1 and all the rays between $\overrightarrow{BA}$ and $\overrightarrow{BC}$. This indicates that much of what we have developed about betweenness for points can be associated with the concept of betweenness for rays.

We shall list those properties of betweenness for rays that are essential to our work with angles as assumptions, although they can be proved as theorems. The proofs would be time-consuming and are not necessary for an understanding of the properties of angle measure.

Assumption 11.5: If $(\overrightarrow{OA}\ \overrightarrow{OB}\ \overrightarrow{OC})$, then $(\overrightarrow{OC}\ \overrightarrow{OB}\ \overrightarrow{OA})$.

Assumption 11.6: If $(\overrightarrow{OA}\ \overrightarrow{OB}\ \overrightarrow{OC})$, then each pair of $\overrightarrow{OA}$, $\overrightarrow{OB}$, and $\overrightarrow{OC}$ are distinct and not opposite.

In Section 10.5 a *half-plane* is defined as the set of points on one side of a line. In the next several assumptions we find it convenient to refer to a half-plane with a particular line as its edge as a *side of that line*. The reader may find Figure 11.15 useful for visualizing several of the following assumptions.

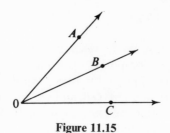

Figure 11.15

Assumption 11.7: If $(\overrightarrow{OA}\ \overrightarrow{OB}\ \overrightarrow{OC})$, then
1. A and B are on the same side of $\overleftrightarrow{OC}$.
2. B and C are on the same side of $\overleftrightarrow{OA}$.
3. A and C are on opposite sides of $\overleftrightarrow{OB}$.

The next assumption is closely related to Assumption 11.7.

Assumption 11.8: If $(\overrightarrow{OA}\ \overrightarrow{OB}\ \overrightarrow{OC})$, then
1. Each point of $\overrightarrow{OA}$ and each point of $\overrightarrow{OB}$ are on the same side of $\overleftrightarrow{OC}$.
2. Each point of $\overrightarrow{OB}$ and each point of $\overrightarrow{OC}$ are on the same side of $\overleftrightarrow{OA}$.
3. Each point of $\overrightarrow{OA}$ and each point of $\overrightarrow{OC}$ are on opposite sides of $\overleftrightarrow{OB}$.

Assumption 11.9:   If $(\overrightarrow{OA}\ \overrightarrow{OB}\ \overrightarrow{OC})$ and $A_1 \neq O$ is on $\overrightarrow{OA}$ and $C_1 \neq O$ is on $\overrightarrow{OC}$, then $\overrightarrow{OB}$ intersects $\overline{A_1 C_1}$ in exactly one point.

Assumption 11.10: If $(\overrightarrow{OA}\ \overrightarrow{OB}\ \overrightarrow{OC})$, then $(\overrightarrow{OB}\ \overrightarrow{OC}\ \overrightarrow{OA})$ is not true.

Assumption 11.11: If B and C are on the same side of $\overleftrightarrow{OA}$, then $(\overrightarrow{OA}\ \overrightarrow{OB}\ \overrightarrow{OC})$, $(\overrightarrow{OA}\ \overrightarrow{OC}\ \overrightarrow{OB})$ or $\overrightarrow{OB} = \overrightarrow{OC}$.

Assumption 11.12: If A and B are on the same side of $\overleftrightarrow{OC}$, and B and C are on the same side of $\overleftrightarrow{OA}$, then $(\overrightarrow{OA}\ \overrightarrow{OB}\ \overrightarrow{OC})$

This completes the list of properties of betweenness for rays that are needed for our work with angles. We are now ready to make an important assumption about angle measure that makes use of the idea of betweenness for rays. It is frequently called the angle addition postulate.

Assumption 11.13 (the angle addition postulate):   If $(\overrightarrow{BA}\ \overrightarrow{BP}\ \overrightarrow{BC})$, then

$$m(\angle ABP) + m(\angle PBC) = m(\angle ABC).$$

The angle addition postulate is illustrated in Figure 11.16.

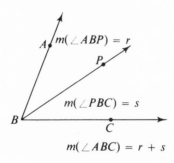

$$m(\angle ABC) = r + s$$

Figure 11.16

We can now state some theorems about order and angle measure. Proofs of these theorems will be considered in Exercises 3, 4, and 5.

Theorem 11.8: If $(\overrightarrow{BA}\ \overrightarrow{BP}\ \overrightarrow{BC})$, then $m(\angle ABP) < m(\angle ABC)$.

Theorem 11.9: (Converse of Theorem 11.7).   If P and C are on the same side of $\overleftrightarrow{AB}$ and $m(\angle ABP) < m(\angle ABC)$, then $(\overrightarrow{BA}\ \overrightarrow{BP}\ \overrightarrow{BC})$.

Theorem 11.10:   If $m(\angle ABC) < m(\angle DEF)$, then there exists a ray $\overrightarrow{EP}$ such that $m(\angle ABC) = m(\angle DEP)$ and $(\overrightarrow{ED}\ \overrightarrow{EP}\ \overrightarrow{EF})$.

EXERCISES 11.7

1. The instrument we use to measure segments is the ruler, which usually has one edge divided into units we call "inches," and the opposite edge is often divided into units we call "centimeters." The instrument that we use to measure angles is called the **protractor**. We have pictured a protractor in Figure 11.17.

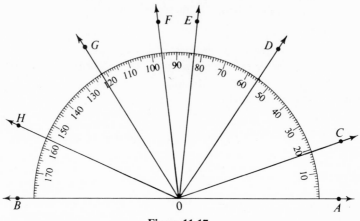

Figure 11.17

The curved edge of the protractor is divided into 180 units, which we call "degrees." We have superimposed several angles on the picture. Find the value of each of the following:

(a) $m(\angle AOC)$

(b) $m(\angle AOB)$

(c) $m(\angle BOH)$

(d) $m(\angle BOF)$

(e) $m(\angle BOE)$

(f) $m(\angle DOC)$

(g) $m(\angle FOD)$

(h) $m(\angle HOC)$

(i) $m(\angle AOC) + m(\angle COE)$

(j) $m(\angle EOF) + m(\angle FOB)$

(k) $m(\angle AOD) + m(\angle GOH)$

(l) $m(\angle GOB) + m(\angle AOC)$

(m) $m(\angle AOC) + m(\angle COB)$

(n) $m(\angle AOD) + m(\angle COB)$

(o) $m(\angle AOD) - m(\angle AOC)$

(p) $m(\angle BOH) - m(\angle COA)$

(q) $m(\angle HOC) - m(\angle FOD)$

(r) $m(\angle FOC) - m(\angle HOF)$

2. (a) Given a ray $\overrightarrow{AC}$ lying on the edge of a half-plane H and a number n between 0 and 180, in how many ways can you construct a ray $\overrightarrow{AB}$ in H so that $m(\angle BAC) = n$? Why?

(b) Given a ray $\overrightarrow{AC}$ in a plane α and a number n between 0 and 180, in how many ways can you construct a ray $\overrightarrow{AB}$ in α so that $m(\angle BAC) = n$? Explain.

3. Prove Theorem 11.8: If $(\overrightarrow{BA}\ \overrightarrow{BP}\ \overrightarrow{BC})$, then $m(\angle ABP) < m(\angle ABC)$. *Hint:* Use the angle addition postulate and the definition of order for real numbers.

4. This exercise will involve the student in a proof of Theorem 11.9 (converse of Theorem 11.8): If P and C are on the same side of $\overleftrightarrow{AB}$ and $m(\angle ABP) < m(\angle ABC)$, then $(\overrightarrow{BA}\ \overrightarrow{BP}\ \overrightarrow{BC})$.

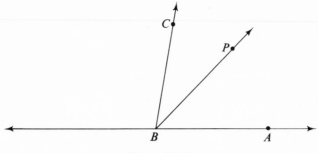

Figure 11.18

PROOF: If P and C are on the same side of $\overleftrightarrow{AB}$, then $(\overrightarrow{BA}\ \overrightarrow{BP}\ \overrightarrow{BC})$, $(\overrightarrow{BA}\ \overrightarrow{BC}\ \overrightarrow{BP})$ or $\overrightarrow{BP} = \overrightarrow{BC}$. (a) Why? If $(\overrightarrow{BA}\ \overrightarrow{BC}\ \overrightarrow{BP})$, then, contrary to hypothesis, $m(\angle ABC) < m(\angle ABP)$. (b) Explain why $m(\angle ABC) < m(\angle ABP)$. If $\overrightarrow{BP} = \overrightarrow{BC}$, then, also contrary to hypothesis, $\angle ABP = \angle ABC$ and $m(\angle ABP) = m(\angle ABC)$. (c) Why? Therefore we must conclude that $(\overrightarrow{BA}\ \overrightarrow{BP}\ \overrightarrow{BC})$.

5. This exercise will constitute a proof of Theorem 11.10: If $m(\angle ABC) < m(\angle DEF)$, then there exists a ray $\overrightarrow{EP}$ such that $m(\angle ABC) = m(\angle DEP)$ and $(\overrightarrow{ED}\ \overrightarrow{EP}\ \overrightarrow{EF})$.

Figure 11.19

PROOF: There exists a ray $\overrightarrow{EP}$ with P in the half-plane $\overleftrightarrow{ED}|F$ such that $m(\angle DEP) = m(\angle ABC)$. (a) Why is this statement true? Since P and F are on the same side of $\overleftrightarrow{ED}$, $(\overrightarrow{ED}\ \overrightarrow{EF}\ \overrightarrow{EP})$, $(\overrightarrow{ED}\ \overrightarrow{EP}\ \overrightarrow{EF})$, or $\overrightarrow{EP} = \overrightarrow{EF}$. (b) Why must just one of these statements be true? If $(\overrightarrow{ED}\ \overrightarrow{EF}\ \overrightarrow{EP})$, then $m(\angle DEF) < m(\angle DEP)$. (c) Why? It follows that, contrary to hypothesis, $m(\angle DEF) < m(\angle ABC)$. (d) Explain. If $\overrightarrow{EP} = \overrightarrow{EF}$, then $m(\angle DEF) = m(\angle DEP)$ and, contrary to hypothesis, $m(\angle DEF) = m(\angle ABC)$. (e) Why? Thus we must conclude that $(\overrightarrow{ED}\ \overrightarrow{EP}\ \overrightarrow{EF})$.

11.8 CONGRUENCE OF ANGLES

Once we had established the concept of measurement of segments, we were ready to define congruence of segments in terms of this concept. Similarly, now that we have postulated notions and developed some theorems for angle measurement, we shall use the concept of measurement to define **congruence of angles**.

If $m(\angle ABC) = m(\angle DEF)$, then we say $\angle ABC$ is **congruent** to $\angle DEF$. This relation is written symbolically as $\angle ABC \cong \angle DEF$.

Continuing the analogy between our treatment of congruence of segments and congruence of angles, we state a theorem that shows congruence of angles to be an equivalence relation and that includes two other properties of congruence of angles that follow from Assumptions 11.4 and 11.13 and from the addition property of equality of real numbers.

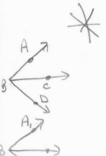

Theorem 11.11:   Congruence of angles has the following properties:

1. $\angle ABC \cong \angle ABC$; thus congruence of angles is reflexive.

2. If $\angle ABC \cong \angle DEF$, then $\angle DEF \cong \angle ABC$; thus congruence of angles is symmetric.

3. If $\angle ABC \cong \angle DEF$ and $\angle DEF \cong \angle GHI$, then $\angle ABC \cong \angle GHI$; thus congruence of angles is transitive.

4. Given a half-plane H, a ray $\overrightarrow{AB}$ on the boundary of H and an angle $\angle PQR$, there exists exactly one ray $\overrightarrow{AC}$ with C in H such that $\angle ABC \cong \angle PQR$.

5. If $\angle ABC \cong \angle A_1B_1C_1$, $\angle CBD \cong \angle C_1B_1D_1$, $(\overrightarrow{BA}\,\overrightarrow{BC}\,\overrightarrow{BD})$, and $(\overrightarrow{B_1A_1}\,\overrightarrow{B_1C_1}\,\overrightarrow{B_1D_1})$, then $\angle ABD \cong \angle A_1B_1D_1$.

Analogous to the midpoint of a segment is the **bisector of an angle**. We define the **bisector** of an angle $\angle ABC$ to be the ray $\overrightarrow{BP}$ such that $(\overrightarrow{BA}\,\overrightarrow{BP}\,\overrightarrow{BC})$ and $\angle ABP \cong \angle PBC$. A proof of the following theorem will be considered in Exercise 4.

Theorem 11.12:   An angle has one and only one bisector.

In order to state and prove several other theorems about angle measurement, we need to introduce some new terms. **Vertical angles** are defined as a pair of angles, having a common vertex, such that the sides of one angle are opposite to the sides of the other angle. Two angles are said to form a **linear pair** if they have a side in common and the other two sides are opposite rays. Two angles are **supplementary** and each is said to be the **supplement** of the other if they are respectively congruent to the angles of a linear pair. The following assumption is often called the "supplement postulate."

Assumption 11.14 (the supplement postulate):   If $\overrightarrow{OA}$ is opposite $\overrightarrow{OB}$ and the point P is not in $\overleftrightarrow{OA}$, then $m(\angle AOP) + m(\angle POB) = 180$.

We are now ready for more theorems.

Theorem 11.13:   Vertical angles are congruent.

A proof of Theorem 11.13 might be given as follows: Given $\angle ABC$ and $\angle DBE$ such that $\overrightarrow{BA}$ is opposite $\overrightarrow{BE}$ and $\overrightarrow{BC}$ is opposite $\overrightarrow{BD}$ as shown in Figure 11.20,

$$m(\angle ABC) + m(\angle CBE) = 180$$

and

$$m(\angle DBE) + m(\angle CBE) = 180$$

by the supplement postulate. Thus we can equate the left members of the equalities: $m(\angle ABC) + m(\angle CBE) = m(\angle DBE) + m(\angle CBE)$. By the addition property of equality of real numbers, then, $m(\angle ABC) = m(\angle DBE)$ and, by the definition of congruence of angles, $\angle ABC \cong \angle DBE$. We can prove vertical angles $\angle ABD$ and $\angle CBE$ congruent by a similar argument.

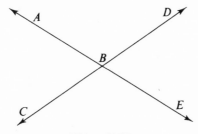

Figure 11.20

Proofs of the following two theorems will be considered in Exercises 5 and 6.

Theorem 11.14:   Supplements of congruent angles are congruent.

Theorem 11.15:   If $\angle AOB$ and $\angle AOC$ are supplementary and B and C are on opposite sides of $\overleftrightarrow{AO}$, then $\overrightarrow{OB}$ is opposite $\overrightarrow{OC}$ and $\angle AOB$ and $\angle AOC$ form a linear pair.

At this point we introduce some additional terms. If two angles of a linear pair are congruent, then each of the angles is a **right angle**. It follows from Assumption 11.14 that the sum of the measures of the angles in a linear pair is 180. If these angles are congruent, then it can easily be shown that the measure of each is 90. Thus **the measure of a right angle is 90**, a statement that can be used as an alternate definition of a right angle. In a drawing, it is customary to identify an angle as a right angle by forming a small square as shown in Figure 11.21.

Figure 11.21

If two lines intersect, their point of intersection separates each line into two opposite rays, and these rays form four angles. Thus we say that the intersection of two lines determines (or forms) four angles. Consider two lines containing two intersecting sets either of which is a line, a ray, or a segment. If these two lines determine at least one right angle, we say the intersecting sets are **perpendicular**. If l is perpendicular to m, we symbolize this relation as $l \perp m$. Hence, in Figure 11.22 we can write $l \perp m$, $\overline{AB} \perp \overline{CD}$, $\overrightarrow{AB} \perp \overrightarrow{CD}$, $\overrightarrow{AB} \perp \overleftrightarrow{CD}$, etc.

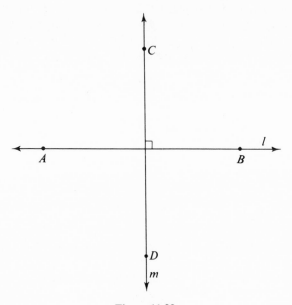

Figure 11.22

Two angles the sum of whose measures is 90 are said to be **complementary angles**, and each is said to be the **complement** of the other. An angle with measure less than 90 is called an **acute angle**; an angle with measure greater than 90 is called an **obtuse angle**. The following statements can be deduced from the definition of a right angle with such little difficulty that they almost constitute restatements of the definition:

1. An angle is a right angle iff it is congruent to one of its supplements.
2. An angle is a right angle iff its measure is 90.
3. Any supplement of a right angle is a right angle.
4. Any two right angles are congruent.
5. Any angle congruent to a right angle is a right angle.
6. No line is perpendicular to (or forms a right angle with) itself.

We resume the list of theorems that follow from the assumptions we have made.

Theorem 11.16: If two lines are perpendicular, the four angles they form are right angles.

We offer a proof of this theorem: Suppose $\overleftrightarrow{AB}$ and $\overleftrightarrow{BC}$ intersect to form a right angle $\angle ABC$ (Fig. 11.23). Choose a point D on $\overleftrightarrow{AB}$ such that $\overrightarrow{BA}$ is opposite $\overrightarrow{BD}$ and let E be a point on $\overleftrightarrow{BC}$ such that $\overrightarrow{BC}$ is opposite to $\overrightarrow{BE}$. Then $\angle ABC$ and $\angle BDE$ are vertical angles and are congruent by Theorem 11.13. Thus $\angle BDE$ is a right angle by statement 5 in the listed statements about right angles. Since linear pairs are formed by $\angle ABC$ and $\angle ABE$, and by $\angle ABC$ and $\angle CBD$, $\angle ABE$ is supplementary to $\angle ABC$ and $\angle CBD$ is supplementary to $\angle ABC$. By statement 3 in the listed statements about right angles, then, $\angle ABE$ and $\angle CBD$ are both right angles, and the theorem is proved.

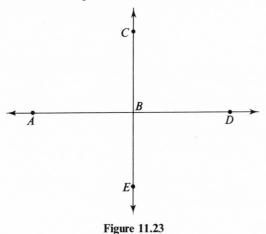

Figure 11.23

A proof of the following theorem will be considered in Exercise 7.

Theorem 11.17: Given a plane α containing a line l, if P is on l, then there is one and only one line m on α such that P is on m and $m \perp l$.

Theorem 11.18: Complements of congruent angles are congruent.

The proof of Theorem 11.18 is so closely analogous to the proof of Theorem 11.14 that it will neither be given nor assigned as an exercise.

Theorem 11.19: If two angles are complementary, then both are acute.

Theorem 11.19 follows directly from the definitions of complementary angles and acute angles.

EXERCISES 11.8

1. Referring to the picture in Figure 11.24, which of the following are correct statements?

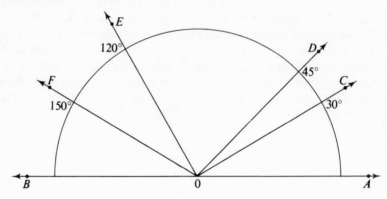

Figure 11.24

(a) $\angle AOC = \angle BOF$

(b) $m(\angle AOC) = m(\angle BOF)$

(c) $\angle AOC \cong \angle BOF$

(d) $\angle AOC = \angle AOC$

(e) $\angle AOC \cong \angle AOC$

(f) $m(\angle AOC) \cong m(\angle AOC)$

(g) $\angle AOC + \angle COD = \angle AOD$

(h) $\angle AOC \cup \angle COD = \angle AOD$

(i) $m(\angle AOC) + m(\angle COD) = m(\angle AOD)$

(j) $\angle AOC \cup \angle BOF = \angle BOE$

(k) $\angle AOC \cup \angle BOF \cong \angle BOE$

(l) $m(\angle AOC) + m(\angle BOF) = m(\angle BOE)$

(m) $m(\angle AOC) \cup m(\angle BOF) = m(\angle BOE)$

(n) $\angle AOC = \frac{1}{2}\angle BOE$

(o) $m(\angle AOC) = \frac{1}{2}m(\angle BOE)$

(p) $\angle AOC = 30$

(q) $m(\angle AOC) = 30$

(r) $m(\angle AOC) = 30°$

(s) $\angle AOC = 30°$

(t) $\angle AOD + \angle BOD = 180$

(u) $m(\angle AOD) + m(\angle BOD) = 180$

(v) $\angle AOD$ and $\angle BOD$ are supplementary

(w) $m(\angle AOE)$ is a supplement of $m(\angle BOE)$

(x) $\angle COD + \angle DOE = 90$

(y) $m(\angle COD) + m(\angle DOE) = 90$

(z) $\angle AOC$ is a complement of $\angle BOE$

2. Referring to Figure 11.24,

    (a) Name a pair of perpendicular rays.
    (b) Name three acute angles.
    (c) Name three obtuse angles.
    (d) Name a pair of vertical angles.
    (e) Name the angles in a linear pair.
    (f) Indicate the betweenness relationship for rays $\overrightarrow{OA}, \overrightarrow{OF},$ and $\overrightarrow{OD}$.
    (g) Indicate the order between $m(\angle COD)$ and $m(\angle COE)$; between $m(\angle COD)$ and $m(\angle EOF)$; between $m(\angle COD)$ and $m(\angle COB)$.

3. If two lines intersect, how many pairs of vertical angles are formed? If $x =$ the measure of one of the angles formed, write a formula for each of the others.

4. This exercise will constitute a proof of Theorem 11.12: An angle has one and only one bisector.

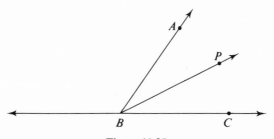

Figure 11.25

PROOF: I. *Existence* Given $\angle ABC$ there exists a ray $\overrightarrow{BP}$ with P in $\overleftrightarrow{BC}|A$ such that $m(PBC) = \frac{1}{2}m(\angle ABC)$. (a) Why? Then $(\overrightarrow{BC}\ \overrightarrow{BP}\ \overrightarrow{BA})$. (b) Explain why $(\overrightarrow{BC}\ \overrightarrow{BP}\ \overrightarrow{BA})$. Also, $m(\angle PBA) + m(\angle PBC) = m(\angle ABC)$. (c) Why? That is, $m(\angle PBA) + \frac{1}{2}m(\angle ABC) = m(\angle ABC)$. (d) Why? Thus $m(\angle PBA) = m(\angle ABC) - \frac{1}{2}m(\angle ABC)$. (e) Why? That is, $m(\angle PBA) = \frac{1}{2}m(\angle ABC)$, and $m(\angle PBC) = m(\angle PBA)$; thus $\angle PBC \cong \angle PBA$. (f) Explain. Hence $\angle ABC$ has one bisector, the ray $\overrightarrow{BP}$.

II. *Uniqueness* Suppose a second ray $\overrightarrow{BP_1}$ exists such that $\overrightarrow{BP_1}$ is also a bisector of $\angle ABC$. Then $(\overrightarrow{BC}\ \overrightarrow{BP_1}\ \overrightarrow{BA})$, and $\angle P_1BC \cong \angle P_1BA$. (g) Justify this statement. And $m(\angle P_1BC) + m(\angle P_1BA) = m(\angle ABC)$. (h) Why? Also $m(\angle P_1BC) + m(\angle P_1BC) = m(\angle ABC)$, that is, $2m(\angle P_1BC) = m(\angle ABC)$. (i) Why? Thus $m(\angle P_1BC) = \frac{1}{2}m(\angle ABC)$. (j) Why? But only one ray $\overrightarrow{BP}$ exists with P in $\overleftrightarrow{BC}|A$ such that $m(\angle PBC) = \frac{1}{2}m(\angle ABC)$. (k) Why does just one such ray exist? Thus we must conclude that $\overrightarrow{BP} = \overrightarrow{BP_1}$; hence $\angle ABC$ has only one bisector.

5. Prove Theorem 11.14: Supplements of congruent angles are congruent.

6. This exercise will involve the student in a proof of Theorem 11.15: If $\angle AOB$ and $\angle AOC$ are supplementary and B and C are on opposite sides of $\overleftrightarrow{OA}$, then $\overrightarrow{OB}$ is opposite $\overrightarrow{OC}$ and $\angle AOB$ and $\angle AOC$ form a linear pair. Figure 11.26 may help the student to visualize this theorem.

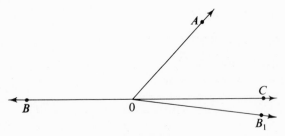

Figure 11.26

PROOF: Let $\overrightarrow{OB_1}$ be opposite $\overrightarrow{OB}$. Then B_1 and C are on the same side of $\overleftrightarrow{OA}$. $\angle AOB$ is supplementary to $\angle AOC$. (a) Why is $\angle AOB$ supplementary to $\angle AOC$? But $\angle AOB_1$ is supplementary to $\angle AOC$. (b) Explain. Thus $\angle AOC \cong \angle AOB_1$. (c) Why? Only one ray $\overrightarrow{OX}$ exists with O in $\overleftrightarrow{OA}$ and X in $\overleftrightarrow{OA}|C$ such that $\angle AOX \cong \angle AOC$. (d) Why does just one such ray exist? Thus $\overrightarrow{OB_1} = \overrightarrow{OC}$ and $\overrightarrow{OB}$ and $\overrightarrow{OC}$ are opposite rays. Also, $\angle AOB$ and $\angle AOC$ form a linear pair. (e) Why?

7. This exercise will constitute a proof of Theorem 11.17. Given a plane containing a line l, if P is on l, then there is one and only one line m such that P is on m and $m \perp l$ (Fig. 11.27).

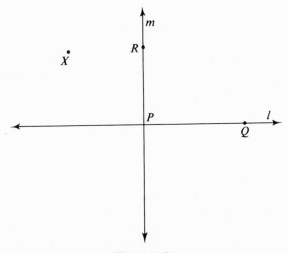

Figure 11.27

PROOF: I. *Existence* Given the line l containing the point P, we identify a second point Q on l and a point X not in l, as shown in the drawing. Then there exists exactly one ray $\overrightarrow{PR}$ with P in $\overrightarrow{PQ}$ (or l) and R in $\overrightarrow{PQ}|X$ such that $m(\angle RPQ) = 90$. (a) Why does just one such ray exist? Thus $\angle RPQ$ is a right angle. (b) Why? And the line m containing P and R is perpendicular to l.

II. *Uniqueness* $\overrightarrow{PR}$ with R in $\overleftrightarrow{PQ}|X$ is the only ray containing P such that $m(\angle RPQ) = 90$. (c) Why is $\overrightarrow{PR}$ the only such ray? Thus $\overrightarrow{PR}$ is the only ray in $\overleftrightarrow{PQ}|X$ that is perpendicular to $\overrightarrow{PQ}$. Since $\overrightarrow{PR}$ lies on only one line m, that line is the only line perpendicular to $\overleftrightarrow{PQ}$ (or l). (d) Why?

11.9 CONGRUENT TRIANGLES

The classic definition of *congruent triangles* reads something like this: Two triangles are said to be congruent if their corresponding parts are equal. This statement contains the basic elements of the concept of congruence of triangles, but it is not clear because certain of its terms are vague in meaning. A satisfactory definition of congruent triangles must be in terms of primitives and/or previously defined terms. First of all, we may remark that the word *equal* in the given classic definition must be replaced by the term *congruent*, since *equal*, for us, means *identical*. Certainly parts of two different triangles cannot be identical.

Next we need to define the term *corresponding*, and we must understand that the *parts* of a triangle are its three vertices, its three sides and its three angles. A one-to-one correspondence between the vertices of two triangles is called a **correspondence between the vertices of the triangles**. For example, if ABC and DEF are triangles, then

$$
\begin{array}{ccc}
A & B & C \\
\updownarrow & \updownarrow & \updownarrow \\
D & E & F
\end{array}
$$

is a correspondence between the vertices of $\triangle ABC$ and the vertices of $\triangle DEF$. It is important that the reader recognize that this is not the only correspondence between the vertices of these two triangles. Another correspondence would be:

$$
\begin{array}{ccc}
A & B & C \\
\updownarrow & \updownarrow & \updownarrow \\
F & E & D
\end{array}
$$

We use a notation to symbolize a correspondence between vertices that identifies the specific correspondence we are considering. The first correspondence displayed is indicated symbolically by $ABC \leftrightarrow DEF$, and the second one is indicated by $ABC \leftrightarrow FED$.

A correspondence between vertices of two triangles leads to a *correspondence between the sides of the triangles*. If the endpoints of two sides correspond, for example, if A corresponds to D and B corresponds to E, then $\overline{AB}$ corresponds to $\overline{DE}$. If $ABC \leftrightarrow DEF$, then the **correspondence between the sides** is

$$
\begin{array}{ccc}
\overline{AB} & \overline{BC} & \overline{AC} \\
\updownarrow & \updownarrow & \updownarrow \\
\overline{DE} & \overline{EF} & \overline{DF}
\end{array}
$$

Pairs of corresponding sides are often indicated in a horizontal notation: $\overline{AB} \leftrightarrow \overline{DE}$, $\overline{BC} \leftrightarrow \overline{EF}$, $\overline{AC} \leftrightarrow \overline{DF}$.

In $\triangle ABC$ and $\triangle DEF$, the correspondence between vertices $ABC \leftrightarrow DEF$ also leads to a *correspondence between the angles* of the triangles. If $ABC \leftrightarrow DEF$, then the **correspondence between angles** is

$$\angle ABC \quad \angle BCA \quad \angle CAB$$
$$\updownarrow \qquad \updownarrow \qquad \updownarrow$$
$$\angle DEF \quad \angle EFD \quad \angle FDE$$

Pairs of corresponding angles may also be indicated in horizontal notation: $\angle ABC \leftrightarrow \angle DEF$, $\angle BCA \leftrightarrow \angle EFD$, $\angle CAB \leftrightarrow \angle FDE$.

We are now ready to define **congruent triangles** precisely: Two triangles $\triangle ABC$ and $\triangle DEF$ are said to be **congruent** if there exists a correspondence between their vertices such that the corresponding sides are congruent and the corresponding angles are congruent. Thus, under the correspondence $ABC \leftrightarrow DEF$, $\triangle ABC \cong \triangle DEF$ if

$$\overline{AB} \cong \overline{DE}, \quad \overline{BC} \cong \overline{EF}, \quad \overline{AC} \cong \overline{DF}$$

$$\angle ABC \cong \angle DEF, \quad \angle BCA \cong \angle EFD, \quad \text{and} \quad \angle CAB \cong \angle FDE$$

The reader will recall, from his high school geometry, the term *opposite side*, *opposite angle*, *included side*, and *included angle*. We shall define these terms precisely:

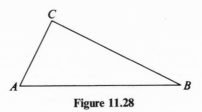

Figure 11.28

In $\triangle ABC$ (Fig. 11.28), side $\overline{AC}$ is **opposite** $\angle B$ and $\angle B$ is **opposite** $\overline{AC}$. The same relation is said to exist between $\overline{AB}$ and $\angle C$, and between $\overline{BC}$ and $\angle A$, that is, $\overline{AB}$ and $\angle C$ are **opposite** each other, and $\overline{BC}$ and $\angle A$ are **opposite** each other. We define $\angle ABC$ to be **included** by sides $\overline{AB}$ and $\overline{BC}$, $\angle BAC$ to be **included** by sides $\overline{AB}$ and $\overline{AC}$, and $\angle ACB$ to be **included** by sides $\overline{AC}$ and $\overline{BC}$. Side $\overline{AB}$ is said to be **included** by $\angle BAC$ and $\angle ABC$, side $\overline{BC}$ is said to be **included** by $\angle ABC$ and $\angle ACB$, and side $\overline{AC}$ is **included** by $\angle BAC$ and $\angle ACB$.

Other terms that need precise definition are *isosceles triangle*, *equilateral triangle*, *equiangular triangle*, *acute triangle*, *obtuse triangle*, *right triangle*, and *median of a triangle.* If two or more sides of a triangle are congruent, the triangle is said to be **isosceles**. If all three sides of a triangle are congruent, the triangle is said to be **equilateral**. A triangle is said to be **equiangular** if all three of its angles are congruent. An **acute triangle** is a triangle each of whose angles is acute. A triangle is said to be an **obtuse triangle** if it has an obtuse angle. If a triangle has a right angle, it is said to be a **right triangle**. The side opposite the right angle of a right triangle is called the **hypotenuse**. The other two sides of a right triangle are called its **legs**. A **median of a triangle** is a segment one of whose endpoints is a vertex of the triangle; the other endpoint is the midpoint of the opposite side.

We have now defined enough terms to develop some important theorems about congruent triangles. However, we need one new assumption in order to proceed. This assumption is often called the *side-angle-side* postulate, or simply the *SAS* postulate. Notice that this assumption involves the concepts we have developed about *congruence of segments* and *congruence of angles*.

> **Assumption 11.15** (the *SAS* postulate): If there exists a correspondence between the vertices of $\triangle ABC$ and $\triangle DEF$ such that two sides and the included angles of $\triangle ABC$ are congruent respectively to the corresponding two sides and included angle of $\triangle DEF$, then $\triangle ABC \cong \triangle DEF$.

Several theorems follow from the definitions and the assumption of this section.

> **Theorem 11.20:** If a triangle is isosceles, the angles opposite the congruent sides are congruent.

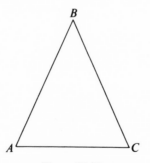

Figure 11.29

A proof of this theorem might be given as follows: In $\triangle ABC$ (Fig. 11.29) let $\overline{AB} \cong \overline{BC}$. We shall examine the correspondence $ABC \leftrightarrow CBA$ between the vertices of $\triangle ABC$ and itself. Since $\overline{AB} \cong \overline{BC}$, $\overline{BC} \cong \overline{AB}$, and $\angle ABC = \angle CBA$, then $\triangle ABC \cong \triangle CBA$ by the *SAS* postulate. By the definition of congruent triangles, corresponding angles $\angle ACB$ and $\angle BAC$ (which are opposite congruent sides $\overline{AB}$ and $\overline{BC}$ respectively) are congruent.

A **corollary** of Theorem 11.20, that is, a proposition that arises directly from the theorem, follows:

Corollary: An equilateral triangle is equiangular.

A proof of the following theorems will be considered in Exercises 4 and 5.

> **Theorem 11.21** (the ASA theorem):   If there exists a correspondence between the vertices of $\triangle ABC$ and $\triangle DEF$ such that two angles and the included side of $\triangle ABC$ are congruent, respectively, to the corresponding two angles and included side of $\triangle DEF$, then $\triangle ABC \cong \triangle DEF$.

> **Theorem 11.22:**   If two angles of a triangle are congruent, then the sides opposite these angles are congruent.

In order to continue stating theorems that arise out of the definitions, assumptions, and previously stated theorems of this section, we need to define the set of points on a plane that are *equidistant* from two other of its points. First, we say a point P is **equidistant** from distinct points A and B on a plane iff $\overline{PA} \cong \overline{PB}$. Next we define the **perpendicular bisector** of a segment $\overline{AB}$ to be a line l that is perpendicular to $\overline{AB}$ and contains the midpoint of $\overline{AB}$. The *perpendicular bisector* of a segment $\overline{AB}$ is *unique*. This fact follows from Theorem 11.7 which attests to the uniqueness of the midpoint of $\overline{AB}$ and from Theorem 11.17 which confirms that l, a line perpendicular to $\overline{AB}$ through a point M, is unique.

We may now state two more familiar and important theorems, proofs of which will be considered in Exercises 6 and 7.

> **Theorem 11.23:**   On a plane, a point is equidistant from the endpoints of a segment if and only if it is on the perpendicular bisector of the segment.

The expression *if and only if* in Theorem 11.23 means that we are actually stating a theorem and its converse as a single theorem. We must prove

1. In a plane, if a point is on the perpendicular bisector of a segment, then it is equidistant from the endpoints of the segment.
2. In a plane, if a point is equidistant from the endpoints of a segment, then it is on the perpendicular bisector of the segment.

Theorem 11.23 may be used to prove another familiar triangle congruence theorem.

> **Theorem 11.24** (the SSS theorem):   If there exists a correspondence between the vertices of $\triangle ABC$ and $\triangle DEF$ such that the sides of $\triangle ABC$

are congruent, respectively, to the corresponding sides of $\triangle DEF$, then $\triangle ABC \cong \triangle DEF$.

We define two angles in a plane to be **adjacent** iff they have a common side and the intersection of their interiors is empty. We have defined *adjacent angles* so that we can use the term to define the term *exterior angle*: An angle that is adjacent and supplementary to an angle of a triangle is called an **exterior angle** of the triangle. In discussions involving *exterior angles*, we shall call the angles of a triangle **interior angles**. An interior angle that is not adjacent to a given exterior angle is called (relative to the given exterior angle) a **nonadjacent interior angle**, or a **remote interior angle**. Any triangle has six exterior angles. The six exterior angles of $\triangle ABC$, as pictured in Figure 11.30, are identified by drawing arcs in their interiors, terminating on their sides; we have numbered these angles.

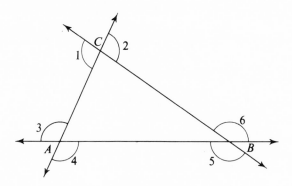

Figure 11.30

Assumption 11.16: If a point D is in the interior of $\angle BAC$, then $\overrightarrow{AD}$ meets $\overline{BC}$.

Assumption 11.16 has been introduced because it is essential to a proof of Theorem 11.25.

Theorem 11.25 (the exterior angle theorem): The measure of any exterior angle of a triangle is greater than the measure of either of the remote interior angles of the triangle. (A proof of this theorem will be considered in Exercises 11.9.)

In Theorem 11.17, we proved that in a plane there exists one and only one line m perpendicular to a given line l through a point P in l. The next theorem involves the case where the point P is not on the given line.

Theorem 11.26: If a point P is not in a line l, then there exists one and only one line m that contains P and is perpendicular to l.

A proof of Theorem 11.26 may be given as follows:

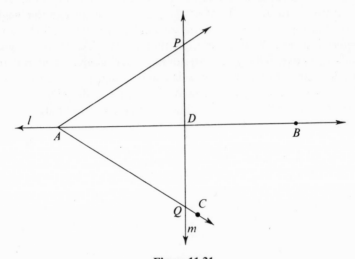

Figure 11.31

I. *Existence*   Let A and B be distinct points in l (Fig. 11.31); let P be a point not in l. Consider ray $\overrightarrow{AP}$, whose union with $\overrightarrow{AB}$ forms $\angle PAB$. In the half-plane (with edge l) that does not contain P, there exists exactly one ray $\overrightarrow{AC}$ such that $m(\angle PAB) = m(\angle CAB)$, by the angle construction postulate. We select a point Q on $\overrightarrow{AC}$ such that $\overline{AQ} \cong \overline{AP}$. Consider line $\overleftrightarrow{PQ}$ and let D be the intersection of $\overleftrightarrow{PQ}$ and l. (Why does this intersection exist?) In $\triangle PAD$ and $\triangle QAD$, consider the correspondence between vertices $PAD \leftrightarrow QAD$: $\overline{AP} \cong \overline{AQ}$ by a previous statement in the proof; $\overline{AD} \cong \overline{AD}$ because these segments are identical; and $\angle PAD \cong \angle QAD$ by a previous statement in the proof. Therefore, by the SAS postulate, $\triangle PAD \cong \triangle QAD$. Then by definition of congruent triangles, $\angle PDA \cong \angle QDA$ and $\overleftrightarrow{PQ}$ is perpendicular to l. We may call $\overleftrightarrow{PQ}$ line m; thus the existence of the desired perpendicular is proved.

II. *Uniqueness*   Assume there is more than one line through P that is perpendicular to l. For example, suppose $\overleftrightarrow{PF}$, as well as $\overleftrightarrow{PD}$, is perpendicular to l as shown in Figure 11.32. The ray $\overrightarrow{DP}$ intersects l at D; $\overleftrightarrow{PF}$ intersects l at F. By the point location postulate, there exists on the ray opposite $\overrightarrow{DP}$ exactly one point R such that $m(\overline{DR}) = m(\overline{PD})$. In $\triangle PDF$ and $\triangle RDF$, consider the correspondence between vertices $PDF \leftrightarrow RDF$. Since $m(\overline{DR}) = m(\overline{PD})$, $\overline{DR} \cong \overline{PD}$. Also, $\angle PDF \cong$

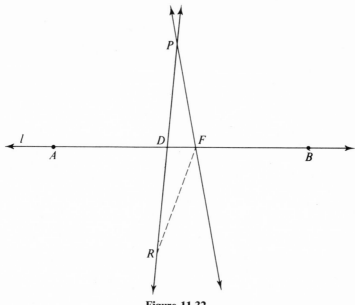

Figure 11.32

$\angle RDF$ because all right angles are congruent, and $\overline{DF} \cong \overline{DF}$ because these segments are identical. By the SAS postulate, then, $\triangle PDF \cong \triangle RDF$. By definition of congruent triangles, $\angle PFD \cong \angle RFD$. Since $\angle PFD$ is a right angle, then $\angle RFD$ is also a right angle. Therefore $\overrightarrow{RF}$ is perpendicular to l at F; but $\overleftrightarrow{PF}$ is perpendicular to l at F. This contradicts Theorem 11.17 (Section 11.8). Thus our assumption that there is more than one line through P that is perpendicular to l is false. Hence uniqueness has been proved.

EXERCISES 11.9

1. Consider $\triangle ABC$ and $\triangle DEF$. Which of the listed correspondences between vertices are the same?

   (a) $ABC \leftrightarrow DEF$ (b) $BAC \leftrightarrow DEF$ (c) $BAC \leftrightarrow EFD$
   (d) $BAC \leftrightarrow EDF$ (e) $ABC \leftrightarrow EDF$ (f) $ACB \leftrightarrow DFE$
   (g) $BCA \leftrightarrow DFE$ (h) $ACB \leftrightarrow FDE$ (i) $CBA \leftrightarrow DFE$

2. Given $\triangle ABC$ and $\triangle DEF$, how many different correspondences exist between their vertices? List these correspondences.

3. If each of the pairs of triangles in Figure 11.33 are congruent, as they appear to be, give the correspondence between vertices, in each case, that leads to the congruence; that is, give the correspondence between vertices that results in congruence of the corresponding sides and angles.

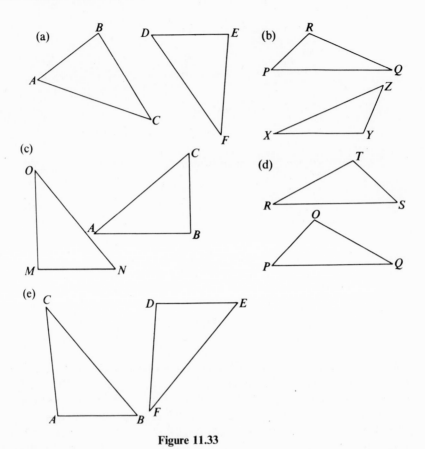

Figure 11.33

4. This exercise will involve the student in a proof of Theorem 11.21 (the ASA theorem): If there exists a correspondence between △ABC and △DEF such that two angles and the included side of △ABC are congruent, respectively, to the corresponding two angles and included side of △DEF, then △ABC ≅ △DEF.

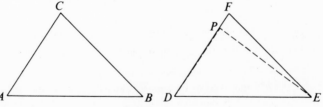

Figure 11.34

Given: △ABC and △DEF where ABC ↔ DEF results in the following congruences: ∠A ≅ ∠D, ∠B ≅ ∠DEF, and $\overline{AB} \cong \overline{DE}$.

Prove: $\triangle ABC \cong \triangle DEF$.

PROOF

Statements	*Reasons*
1. In $\triangle ABC$ and $\triangle DEF \angle A \cong \angle D$,	
$\angle B \cong \angle DEF$, and $\overline{AB} \cong \overline{DE}$.	1. Why?
2. There exists a point P in $\overrightarrow{DF}$	2. Why?
such that $\overline{DP} \cong \overline{AC}$.	
3. If $\overline{DP} \cong \overline{AC}$, $\triangle ABC \cong \triangle DEP$.	3. Why?
4. Then $\angle DEP \cong \angle B$.	4. Why?
5. But $\angle B \cong \angle DEF$.	5. Why?
6. Thus $\angle DEP \cong \angle DEF$.	6. Why?
7. And $\overrightarrow{EP} = \overrightarrow{EF}$.	7. Why?
8. Hence $P = F$ and $\overline{BC} \cong \overline{EF}$.	8. Why?
9. Therefore, $\triangle ABC \cong \triangle DEF$.	9. Why?

5. Prove Theorem 11.22: If two angles of a triangle are congruent, then the sides opposite these angles are congruent.

6. This exercise will involve the student in a proof of Theorem 11.23: In a plane, a point is equidistant from the endpoints of a segment if and only if it is on the perpendicular bisector of the segment. Recall that this actually constitutes two theorems in one.

    I. In a plane, if a point is on the perpendicular bisector of a segment, then it is equidistant from the endpoints of the segment.

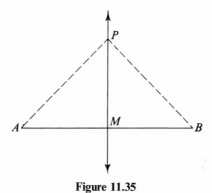

Figure 11.35

Given: Segment $\overline{AB}$ with its perpendicular bisector $\overleftrightarrow{PM}$; M is the midpoint of $\overline{AB}$; point P is in $\overleftrightarrow{PM}$. (If $P = M$, then P is equidistant from the endpoints of $\overline{AB}$ by definition of a midpoint. Therefore, in the following proof, we shall assume the case where $P \neq M$.)
Prove: $\overline{PA} \cong \overline{PB}$.

PROOF

Statements	Reasons
1. $\overleftrightarrow{PM} \perp \overline{AB}$ and bisects $\overline{AB}$ at M; $P \neq M$.	1. Why?
2. M is the midpoint of $\overline{AB}$.	2. Why?
3. $\overline{AM} \cong \overline{MB}$.	3. Why?
4. $\angle AMB$ is a right angle; $\angle BMP$ is a right angle.	4. Why?
5. $\angle AMB \cong \angle BMP$.	5. Why?
6. $\overline{PM} \cong \overline{PM}$.	6. Why?
7. $\triangle AMP \cong \triangle BMP$.	7. Why?
8. $\overline{PA} \cong \overline{PB}$.	8. Why?

II. In a plane, if a point is equidistant from the endpoints of a segment, then it is on the perpendicular bisector of the segment. We have two cases here that we must consider:

1. Assume P is in $\overline{AB}$. Since, by hypothesis, P is equidistant from A and B, $\overline{AP} \cong \overline{PB}$. (a) Why? Then P is the midpoint of $\overline{AB}$. (b) Why? And P is on the perpendicular bisector of segment $\overline{AB}$. (c) Why?

2. Assume P is not on $\overline{AB}$, recall that M is the midpoint of $\overline{AB}$, and consider $\triangle AMP$ and $\triangle BMP$ with correspondence between vertices $AMP \leftrightarrow BMP$.

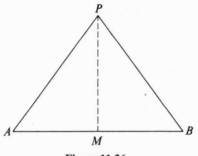

Figure 11.36

Given: Segment $\overline{AB}$ with midpoint M and point P not in $\overline{AB}$; $\overline{AP} \cong \overline{BP}$.
Prove: $\overleftrightarrow{PM}$ is the perpendicular bisector of $\overline{AB}$.

PROOF

Statements	Reasons
1. M is the midpoint of $\overline{AB}$; $P \notin \overline{AB}$; $\overline{AP} \cong \overline{BP}$.	1. Why?
2. $\overline{AM} \cong \overline{MB}$.	2. Why?
3. $\angle PBA \cong \angle PAB$	3. Why?
4. $\triangle AMP \cong \triangle BMP$.	4. Why?
5. $\angle AMP \cong \angle BMP$.	5. Why?
6. Since they form a linear pair, $\angle AMP$ and $\angle PMB$ are right angles.	6. Why?
7. $\overleftrightarrow{PM} \perp \overline{AB}$ at M.	7. Why?
8. $\overleftrightarrow{PM}$ is the perpendicular bisector of $\overline{AB}$.	8. Why?

7. This exercise will involve the student in a proof of Theorem 11.24 (the SSS theorem): If there exists a correspondence between the vertices of $\triangle ABC$ and $\triangle DEF$ such that the sides of $\triangle ABC$ are congruent, respectively, to the corresponding sides of $\triangle DEF$, then $\triangle ABC \cong \triangle DEF$.

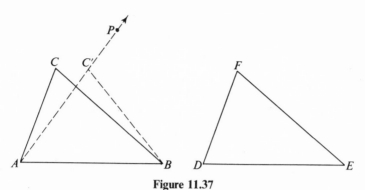

Figure 11.37

Given: $\triangle ABC$ and $\triangle DEF$ where correspondence between vertices $ABC \leftrightarrow DEF$ is such that $\overline{AB} \cong \overline{DE}, \overline{BC} \cong \overline{EF}$, and $\overline{AC} \cong \overline{DF}$.
Prove: $\triangle ABC \cong \triangle DEF$.

PROOF

Statements	*Reasons*
1. In $\triangle ABC$ and $\triangle DEF$, $\overline{AB} \cong \overline{DE}$, $\overline{BC} \cong \overline{EF}$, and $\overline{AC} \cong \overline{DF}$.	1. Why?
2. There exists a ray $\overrightarrow{AP}$ in $\overleftrightarrow{AB}|C$ such that $\angle BAP \cong \angle EDF$.	2. Why?
3. There is a point C' in $\overrightarrow{AP}$ such that $\overline{AC'} \cong \overline{DF}$.	3. Why?
4. $\angle BAP = \angle BAC'$.	4. Definition of a ray and definition of an angle.
5. $\angle BAC' \cong \angle EDF$.	5. Why?
6. $\triangle ABC' \cong \triangle DEF$.	6. Why?
7. Thus $\overline{BC'} \cong \overline{EF}$.	7. Why?
8. But $\overline{AC} \cong \overline{DF}$ and $\overline{BC} \cong \overline{EF}$.	8. Given
9. Therefore, $\overline{AC} \cong \overline{AC'}$ and $\overline{BC} \cong \overline{BC'}$, or A is equidistant from C and C' and B is equidistant from C and C'.	9. Why?
10. Assume C and C' are distinct. Then A is on the perpendicular bisector of $\overline{CC'}$ and B is on the perpendicular bisector of $\overline{CC'}$.	10. Why?
11. $\overleftrightarrow{AB}$ is the perpendicular bisector of $\overline{CC'}$.	11. Why?
12. Thus $\overleftrightarrow{AB}$ meets $\overline{CC'}$ and C and C' are on opposite sides of $\overleftrightarrow{AB}$.	12. Why?
13. But C' is in $\overrightarrow{AP}$ which is in $\overleftrightarrow{AB}|C$.	13. Statements 2 and 3.

Since statement 12 is a contradiction, we must abandon the assumption which leads to it: the assumption that C and C' are distinct. Hence $C = C'$ and $\triangle ABC \cong \triangle DEF$.

8. This exercise will involve the student in a proof of part of Theorem 11.25 (the exterior angle theorem): The measure of any exterior angle is greater than the measure of either of the remote interior angles of the triangle.

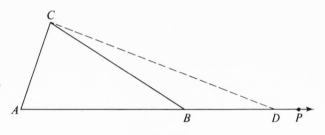

Figure 11.38

Given: $\triangle ABC$ with exterior angle $\angle CBP$, where (ABP).
Prove: $m(\angle CBP) > m(\angle ACB)$.

PROOF: (Note that we are proving only a part of the theorem—that the measure of only *one* of the *six* exterior angles of $\triangle ABC$ is greater than the measure of only *one* of *two* remote interior angles of the triangle. The proofs of the other parts of the theorem, however, are similar.)

Statements	*Reasons*

We know that $m(\angle CBP) \gtreqless m(\angle ACB)$. Let us assume that $m(\angle CBP) = m(\angle ACB)$.

1. $\angle CBP$ is an exterior angle of $\triangle ABC$; (ABP).   1. Why?
2. On $\overrightarrow{BP}$ there exists a point D such that $\overline{BD} \cong \overline{AC}$.   2. Why?

In $\triangle ABC$ and $\triangle BCD$, consider the correspondence $ABC \leftrightarrow DCB$:

3. $\angle CBD \cong \angle ACB$.   3. Why?
4. $\overline{BC} \cong \overline{BC}$.   4. Why?
5. $\triangle ABC \cong \triangle BCD$.   5. Why?
6. $\angle ABC$ and $\angle ACB$ of $\triangle ABC$ are congruent, respectively, to corresponding angles $\angle BCD$ and $\angle CBD$ of $\triangle BCD$.   6. Why?
7. $\angle ABC$ and $\angle CBD$ are supplementary.   7. Why?
8. Then $\angle ACB$ and $\angle BCD$ are supplementary.   8. Why?
9. Also, $\angle ACD$ is a straight angle.   9. Why?
10. And A, C, and D are collinear points.   10. Why?

11. This means that two distinct lines, $\overleftrightarrow{AB}$ and $\overleftrightarrow{AC}$,   11. Why?
    intersect in *two* distinct points, A and D, which
    is impossible.

Thus, the assumption that $m(\angle CBP) = m(\angle ACB)$ is false.
Let us assume that $m(\angle CBP) < m(\angle ACB)$:

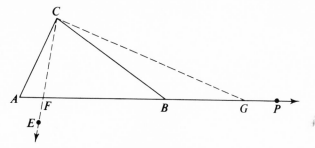

Figure 11.39

12. There exists a ray $\overrightarrow{CE}$ where $(\overrightarrow{CA}\ \overrightarrow{CE}\ \overrightarrow{CB})$ such   12. Why?
    that $m(\angle ECB) = m(\angle CBP)$.
13. $\overrightarrow{CE}$ meets $\overline{AB}$ in point F.   13. Why?
14. There exists a point G in $\overrightarrow{BP}$ such that $\overline{BG} \cong \overline{CF}$.   14. Why?
15. $\overline{BC} \cong \overline{BC}$.   15. Why?
16. $\triangle BCF \cong \triangle BCG$ 16. Why?
    (under correspondence $BCF \leftrightarrow CBG$).

By arguments similar to steps 4 through 9, we have two distinct lines $\overleftrightarrow{FC}$ and $\overleftrightarrow{FB}$
intersecting in two distinct points F and G, which is impossible. Therefore the
assumption that $m(\angle CBP) < m(\angle ACB)$ is false.

17. If $m(\angle CBP) \neq m(\angle ACB)$ and   17. Why?
    $m(\angle CBP) \not< m(\angle ACB)$, then
    $m(\angle CBP) > m(\angle ACB)$.

11.10 PARALLEL LINES AND CONGRUENCE

In a plane, if a line n meets two distinct lines l and m at distinct points P and Q,
line n is called a **transversal** of lines l and m. A ray or segment may also be called
a **transversal** if it meets two distinct coplanar lines at distinct points.
    Consider points A and B on l such that (APB) and points C and D on m such that
(CQD), with A and C in the same half-plane with edge $\overleftrightarrow{PQ}$, as shown in Figure 11.40.
The pair of angles, $\angle APQ$ and $\angle PQD$, are called **alternate interior angles**; a
second pair of alternate interior angles formed by lines l and m and transversal n
are $\angle BPQ$ and $\angle CQP$. Consider points E and F on n such that (QPE) and (PQF).
The pair of angles, $\angle EPB$ and $\angle PQD$, are called **corresponding angles**. There are
three other pairs of corresponding angles formed by lines l and m and transversal n.

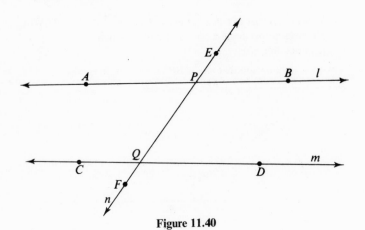

Figure 11.40

They are

$$\angle BPQ \text{ and } \angle DQF$$
$$\angle APE \text{ and } \angle CQP$$
$$\angle APQ \text{ and } \angle CQF$$

In Section 10.4 *parallel lines* are defined as distinct coplanar lines that do not intersect. The *existence* of parallel lines is established by Assumption 10.13 (the parallel postulate). However, the reader will notice that the *existence of parallel lines could be proved* by using the angle construction postulate and the following theorem.

Theorem 11.27: If two lines in a plane are cut by a transversal so that a pair of alternate interior angles are congruent, then the lines are parallel.

A proof of Theorem 11.27 may be given as follows:

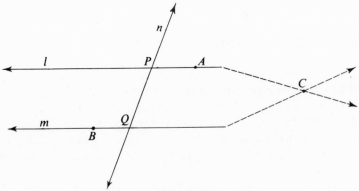

Figure 11.41

Given: Lines *l* and *m* cut by transversal *n* where $\angle APQ \cong \angle BQP$.

Prove: Line *l* is parallel to *m*.

Proof

Statements	*Reasons*
1. Transversal n meets l and m in points P and Q respectively; $\angle APQ \cong \angle BQP$.	1. Given.
2. If l and m meet, the point C of this intersection will be on one side of n or the other; we shall suppose the lines l and m do meet in point C on the side of n that contains A.	2. Assumption 10.15: In a plane, for any line l, the set of all points not in l consists of two disjoint sets, called the sides of the plane.
3. $\overline{PC} \cup \overline{PQ} \cup \overline{QC}$, then, forms $\triangle PQC$.	3. Definition of a triangle.
4. $\angle BQP$ is an exterior angle of $\triangle PQC$.	4. Definition of an exterior angle.
5. Then $\angle BQP > \angle APQ$.	5. Theorem 11.25. The exterior angle theorem.
6. But $\angle APQ \cong \angle BQP$.	6. Given.

Thus the assumption that l meets m results in a contradiction of our hypothesis. Therefore l does not meet m on the side of n that contains A. Had we assumed that l and m meet in a point D on the side of n containing B, a similar argument, involving $\triangle PQD$ and exterior angle $\angle APQ$, would have been possible. Therefore l does not meet m on the side of n that contains B. Hence l is parallel to m. The converse of Theorem 11.27 is also very useful:

Theorem 11.28: If lines l and m are parallel, then a pair of alternate interior angles formed by l, m and a transversal n are congruent.

A proof of Theorem 11.28 is to be considered in Exercise 4.

We are now prepared to state and prove an interesting and important theorem about the angles of a triangle.

Theorem 11.29:   The sum of the measures of the angles of a triangle is 180.

A proof of this theorem might be given as follows:

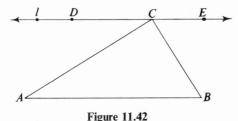

Figure 11.42

Given: $\triangle ABC$. Let l be a line that contains C and is parallel to $\overleftrightarrow{AB}$. Let E be a point on l in $\overleftrightarrow{AC}|B$; let D be a point in the ray opposite $\overrightarrow{CE}$.

Prove: $m(\angle ABC) + m(\angle BAC) + m(\angle ACB) = 180$.

PROOF

Statements	*Reasons*
1. Vertex C of $\triangle ABC$ is on the line l; l is parallel to $\overline{AB}$; E is on l in $\overleftrightarrow{AC}|B$; and D is on the ray opposite $\overrightarrow{CE}$.	1. Given.
2. A and B are in the same side of l.	2. Assumption 10.15: If a point A is in one side of a line l and B is in the other, $\overline{AB}$ intersects l in exactly one point.
3. Then $(\overrightarrow{CA}\ \overrightarrow{CB}\ \overrightarrow{CE})$.	3. Assumption 11.12. Since A and B are on the same side of $\overleftrightarrow{CE}$, and B and E are on the same side of $\overleftrightarrow{AC}$, then $(\overrightarrow{CA}\ \overrightarrow{CB}\ \overrightarrow{CE})$.
4. $m(\angle ACB) + m(\angle BCE) = m(\angle ACE)$.	4. Assumption 11.13. The angle addition postulate.
5. $\angle ACE$ and $\angle DCA$ form a linear pair.	5. Definition of a linear pair.
6. $\angle ACE$ and $\angle DCA$ are supplementary.	6. Definition of supplementary angles.
7. $m(\angle ACE) + m(\angle DCA) = 180$.	7. Assumption 11.14. The supplement postulate.
8. $\angle ABC \cong \angle BCE$.	8. Theorem 11.28.
9. $m(\angle ABC) = m(\angle BCE)$.	9. Definition of congruent angles.
10. $m(\angle ACB) + m(\angle ABC) = m(\angle ACE)$.	10. Statements 4 and 9.
11. $\angle CAB \cong \angle DCA$.	11. Theorem 11.28.
12. $m(\angle CAB) = m(\angle DCA)$.	12. Definition of congruent angles.
13. $m(\angle ACB) + m(\angle ABC) + m(\angle CAB) = 180$.	13. Statements 7, 10, and 12.

Several corollaries follow from Theorem 11.29. Two of these will be considered in Exercise 5. A third corollary follows so directly from Theorem 11.29 that we shall not write its proof.

Corollary: The acute angles of a right triangle are complementary.

We offer the next theorem because it is useful in problems and proofs. We shall assume Theorem 11.30 without proof.

Theorem 11.30. The length of the median to the hypotenuse of a right triangle is one-half the length of the hypotenuse.

Several classic theorems involving parallelism and congruence are introduced in Exercises 11.10. These theorems may, of course, be used to justify statements in proofs that follow them. Therefore, they must be read even though their proofs are not assigned as exercises.

In Chapter 10 a **quadrilateral** was defined as the union of line segments $\overline{A_1A_2}$, $\overline{A_2A_3}$, $\overline{A_3A_4}$, and $\overline{A_4A_1}$ where $\overline{A_1A_2}$, $\overline{A_2A_3}$, $\overline{A_3A_4}$, and $\overline{A_4A_1}$ are coplanar and intersect only at the vertices A_1, A_2, A_3, and A_4. **Consecutive sides** of a quadrilateral are two sides that have a common vertex. **Opposite sides** of a quadrilateral are two sides that do not have a point in common. In Exercises 11.10 there are theorems about quadrilaterals that we have not defined. Therefore we offer these definitions: A parallelogram (defined in Section 10.8) with at least one right angle is a **rectangle**. A rectangle with at least two consecutive sides congruent is a **square**. A parallelogram with at least two consecutive sides congruent is a **rhombus**.

The concept of the distance from a point to a line is also needed. We define the **distance from a point P to a line l** as the length of the segment $\overline{PQ}$ where $\overline{PQ}$ is perpendicular to l and meets l at Q.

EXERCISES 11.10

1. Which of the following implications are true?

   (a) If a figure is a square, then it is a quadrilateral.
   (b) If a figure is a square, then it is a parallelogram.
   (c) If a figure is a parallelogram, then it is a rhombus.
   (d) If a figure is a rhombus, then it is a parallelogram.
   (e) If a figure is a square, then it is a rhombus.
   (f) If a figure is a parallelogram, then it is a rectangle.
   (g) If a figure is a square, then it is a rectangle.
   (h) If a figure is a square, then it is a trapezoid.
   (i) If a figure is a trapezoid, then it is a quadrilateral.
   (j) If a figure is a trapezoid, then it is a parallelogram.
   (k) If a figure is a rectangle, then it is a square.
   (l) If a figure is a rhombus, then it is a square.
   (m) If a figure is a rectangle, then it is a parallelogram.

2. Name the pairs of alternate interior angles to be found in Figure 11.43.

3. Use the illustration in Exercise 3 and name the pairs of corresponding angles.

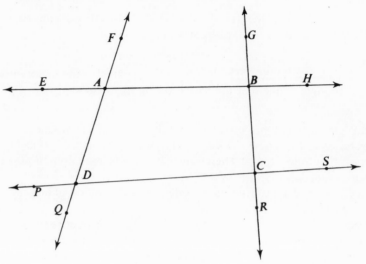

Figure 11.43

4. This exercise will involve the student in a proof of Theorem 11.28: If lines *l* and *m* are parallel, then a pair of alternate interior angles formed by *l*, *m*, and a transversal *n* are congruent.

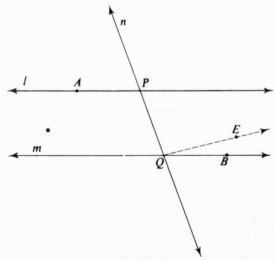

Figure 11.44

Given: Line *l* is parallel to line *m*, and they are cut by transversal *n*. Let *P* and *Q* be the points of intersection of *n* with *l* and *m* respectively. Let *A* and *B* be points on *l* and *m* respectively such that *A* and *B* are on opposite sides of *n*.

Prove: $\angle APQ \cong \angle PQB$.

PROOF

Statements	Reasons

1. $l \parallel m$ and they are cut by transversal n in points P and Q respectively. Points A and B are on l and m respectively and are on opposite sides of n.

1. Why?

2. There exists a ray $\overrightarrow{QE}$ in $\overrightarrow{nB}$ such that $\angle APQ \cong \angle PQE$.

2. Why?

3. Then $\overleftrightarrow{AP}$ is parallel to $\overleftrightarrow{QE}$.

3. Why?

4. $\overleftrightarrow{QB} = m$ and $\overleftrightarrow{AP} = l$.

4. Assumption 10.2: Two distinct points are contained in one and only one line.

5. $\overleftrightarrow{QB}$ is parallel to $\overleftrightarrow{AP}$.

5. Why?

6. $\overleftrightarrow{QE} = \overleftrightarrow{QB}$.

6. Why?

7. $\angle PQE = \angle PQB$.

7. Why?

8. Hence $\angle APQ \cong \angle PQB$.

8. Why?

5. Prove the following corollaries of Theorem 11.29:

(a) Given a correspondence between the vertices of two triangles, if two pairs of corresponding angles are congruent, then the third pair of corresponding angles are congruent.

(b) For any triangle the measure of an exterior angle is equal to the sum of the measures of the two remote interior angles.

6. Given the angle measures indicated in the illustration in Figure 11.45,

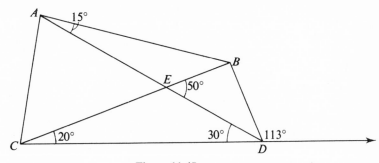

Figure 11.45

find:

(a) $m(\angle CDB)$
(b) $m(\angle CBD)$
(c) $m(\angle ADB)$
(d) $m(\angle ABD)$
(e) $m(\angle ABC)$
(f) $m(\angle AEB)$
(g) $m(\angle DAC) + m(\angle BCA)$

7. In $\triangle ABC$ (Fig. 11.46) and $\triangle PQR$, $\angle A \cong \angle Q$, and $\angle B \cong \angle P$.

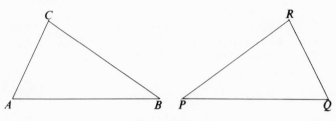

Figure 11.46

Is it true that

(a) $\angle C \cong \angle R$?   (b) $\overline{AB} \cong \overline{PQ}$?

8. This exercise will involve the student in a proof of a new theorem: If lines l and m are cut by a transversal n such that one pair of alternate interior angles are congruent, then the other pair of alternate interior angles are also congruent.

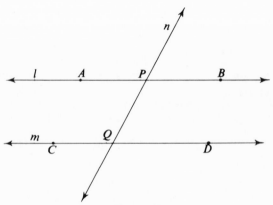

Figure 11.47

Given: Lines l and m cut by transversal n. The points of intersection of n with l and m respectively are P and Q. Points A and B are in l and (APB). Points C and D are in m and (CQD). Points C and A are on the same side of n. $\angle APQ \cong \angle PQD$.
Prove: $\angle BPQ \cong \angle PQC$.

PROOF

Statements	Reasons
1. Transversal n intersects l and m in points P and Q respectively; points A and B are in l and (APB); points C and D are in m and (CQD); points C and A are on the same side of n; $\angle APQ \cong \angle PQD$.	1. Why?

2. $\angle APQ$ and $\angle BPQ$ form a linear pair; $\angle PQC$ 2. Why?
and $\angle PQD$ form a linear pair.
3. $\angle APQ$ and $\angle BPQ$ are supplementary; $\angle PQC$ 3. Why?
and $\angle PQD$ are supplementary.
4. $\angle PQC \cong \angle BPQ$.   4. Why?

9. Complete the proof of the following theorem.

Theorem: If lines l and m are cut by a transversal n so that a pair of corresponding angles are congruent, then $l \parallel m$.

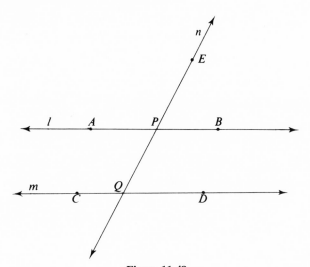

Figure 11.48

Given: Lines l and m cut by transversal n. Points P and Q are the points of intersection of n with l and m respectively. Points A and B are in l and (APB); points C and D are in m and (CQD). Points A and C are on the same side of n. Point E is in n and (EPQ). $\angle EPB \cong \angle PQD$.
Prove: Line l is parallel to m.

PROOF

Statements	Reasons
1. Transversal n intersects l and m in points P and Q respectively; points A and B are in l and (APB); points C and D are in m and (CQD); points A and C are on the same side of n; E is in n and (EPQ); $\angle EPB \cong \angle PQD$.	1. Why?
2. $\angle EPB \cong \angle APQ$.	2. Why?
3. $\angle APQ \cong \angle PQD$.	3. Why?
4. $l \parallel m$.	4. Why?

10. Complete the proof of the following theorem: If lines *l* and *m* are cut by a transversal *n* so that a pair of corresponding angles are congruent, then the other three pairs of corresponding angles are also congruent.

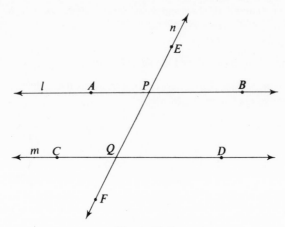

Figure 11.49

Given: Lines *l* and *m* and transversal *n*. The points of intersection of *n* with *l* and *m* are *P* and *Q* respectively. Points *A* and *B* are in *l* so that (*APB*); points *E* and *F* are in *n* so that (*EPQ*) and (*PQF*); points *C* and *D* are in *m* so that (*CQD*); points *A* and *C* are on the same side of *n*; $\angle EPB \cong \angle PQD$.
Prove: $\angle BPQ \cong \angle DQF$, $\angle APE \cong \angle CQP$, and $\angle APQ \cong CQF$.

PROOF

Statements	*Reasons*
1. Transversal *n* intersects *l* and *m* in points *P* and *Q* respectively; points *A* and *B* are in *l* and (*APB*); points *E* and *F* are in *n*, (*EPQ*); and (*PQF*); points *C* and *D* are in *m* and (*CQD*); points *A* and *C* are on the same side of *n*; $\angle EPB \cong \angle PQD$.	1. Why?
2. $\angle EPB$ and $\angle BPQ$ form a linear pair; $\angle PQD$ and $\angle DQF$ form a linear pair.	2. Why?
3. $\angle EPB$ is supplementary to $\angle BPQ$; $\angle PQD$ is supplementary to $\angle DQF$.	3. Why?
4. $\angle BPQ \cong \angle DQF$.	4. Why?
5. $\angle EPB \cong \angle APQ$; $\angle PQD \cong \angle CQF$.	5. Why?
6. $\angle APQ \cong \angle CQF$.	6. Why?
7. $\angle BPQ \cong \angle APE$; $\angle DQF \cong \angle CQP$.	7. Why?
8. $\angle APE \cong \angle CQP$.	8. Why?

11. Prove this theorem: If parallel lines *l* and *m* are cut by a transversal *n*, then a pair of corresponding angles are congruent.

12. Prove this theorem: In a plane, if two distinct lines *l* and *m* are perpendicular to the same line *n*, then *l* is parallel to *m*.

13. Complete the following proof:

Theorem: A parallelogram and one of its diagonals form two congruent triangles.

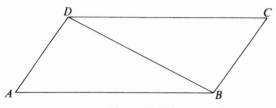

Figure 11.50

Given: Parallelogram *ABCD* with diagonal $\overline{BD}$.
Prove: $\triangle ABD \cong \triangle BCD$.
Hint: Use theorem 11.28 and the ASA theorem.

14. Prove this theorem: In a parallelogram, any two opposite sides are congruent.
    Hint: Use the theorem in Exercise 13.

15. Prove this theorem: If a line *l* is parallel to a line *m*, then the distances of any points *P* and *Q* on *l* from *m* are equal. This is an important property of parallel lines; it is often stated as: Parallel lines are everywhere equidistant.

16. Prove this theorem: In a parallelogram, two opposite angles are congruent.
    Hint: Use the theorem in Exercise 13.

17. Complete the proof of the following theorem: If two sides of a quadrilateral are parallel and congruent, then the quadrilateral is a parallelogram.

Figure 11.51

Given: Quadrilateral *ABCD* with $\overline{AB} \parallel \overline{CD}$ and $\overline{AB} \cong \overline{CD}$. We shall use diagonal $\overline{BD}$ in our proof.
Prove: *ABCD* is a parallelogram.

PROOF

Statements	*Reasons*
1. In *ABCD*, $\overline{AB} \parallel \overline{CD}$, $\overline{AB} \cong \overline{DC}$. $\overline{BD}$ is a diagonal of *ABCD*.	1. Given
2. $\angle BDC \cong \angle ABD$.	2. Why?
3. $\overline{BD} \cong \overline{BD}$.	3. Why?
4. $\triangle ABD \cong \triangle BCD$.	4. Why?
5. $\angle ADB \cong \angle CBD$.	5. Why?
6. $\overline{AD}$ is parallel to $\overline{BC}$.	6. Why?
7. *ABCD* is a parallelogram.	7. Why?

18. Complete the proof of this theorem: The segment whose endpoints are the midpoints of two sides of a triangle is parallel to the third side and its length is one-half the length of the third side.

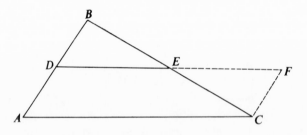

Figure 11.52

Given: $\triangle ABC$ with *D* the midpoint of $\overline{AB}$ and *E* the midpoint of $\overline{BC}$.
Prove: $m(\overline{DE}) = \frac{1}{2}m(\overline{AC})$ and $\overline{DE} \parallel \overline{AC}$. We shall use segments $\overline{EF}$ and $\overline{FC}$ in our proof. $\overline{EF}$ is a segment congruent to $\overline{DE}$ and it is on the ray $\overrightarrow{EF}$ opposite the ray $\overrightarrow{ED}$. $\overline{FC}$ is the segment with endpoints *F* and *C*.

PROOF

Statements	*Reasons*
1. In $\triangle ABC$, *D* is the midpoint of $\overline{AB}$ and *E* is the midpoint of $\overline{BC}$.	1. Why?
2. $\overline{EF} \cong \overline{DE}$.	2. Construction of $\overline{EF}$ is justified by what assumption?
3. $\overline{BE} \cong \overline{EC}$.	3. Why?
4. $\angle BED \cong \angle CEF$.	4. Why?
5. $\triangle BDE \cong \triangle CEF$.	5. Why?
6. $\angle DBE \cong \angle ECF$.	6. Why?
7. $\overline{AB}$ is parallel to $\overline{CF}$.	7. Why?

Statements	*Reasons*
8. $\overline{CF} \cong \overline{BD}$.	8. Why?
9. $\overline{BD} \cong \overline{AD}$.	9. Why?
10. $\overline{CF} \cong \overline{AD}$.	10. Why?
11. *ADFC* is a parallelogram.	11. Why?
12. $\overline{DE} \parallel \overline{AC}$.	12. Why?
13. $\overline{DE} = \frac{1}{2}\overline{DF}$.	13. Why?
14. $\overline{DF} = \overline{AC}$.	14. Why?
15. $\overline{DE} = \frac{1}{2}\overline{AC}$.	15. Why?

19. Prove this theorem: All four angles of a rectangle are right angles.

Figure 11.53

Given: Parallelogram *ABCD* with $\angle A$ a right angle (a rectangle is a parallelogram with at least one right angle).

Prove: $\angle ABC$, $\angle C$, and $\angle ADC$ are right angles. In our proof, we shall use points *E* and *F*: *E* is a point on $\overleftrightarrow{DC}$ such that (*EDC*) and *F* is a point on $\overleftrightarrow{AB}$ such that (*ABF*).

PROOF

Statements	*Reasons*
1. *ABCD* is a parallelogram and $\angle A$ is a right angle.	1. Why?
2. $\angle C$ is a right angle.	2. Why?
3. $\overleftrightarrow{DC}$ is parallel to $\overleftrightarrow{AB}$.	3. Why?
4. $\angle EDA \cong \angle A$; $\angle CBF \cong \angle C$.	4. Why?
5. Thus $\angle EDA$ and $\angle CBF$ are right angles.	5. Why?
6. $\angle EDC$ and $\angle ADC$ form a linear pair and are supplementary; $\angle ABC$ and $\angle CBF$ form a linear pair and are supplementary.	6. Why?
7. $\angle ADC$ is a right angle; $\angle ABC$ is a right angle.	7. Why?

20. Complete the proof of this theorem: In a rhombus, the diagonals are perpendicular to each other.

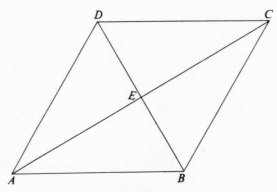

Figure 11.54

Given: Rhombus *ABCD* with diagonals $\overline{AC}$ and $\overline{BD}$.
Prove: Diagonals $\overline{AC}$ and $\overline{BD}$ are perpendicular to each other at their point of intersection *E*.

PROOF

Statements	*Reasons*
1. *ABCD* is a rhombus with diagonals $\overline{AC}$ and $\overline{BD}$.	1. Given.
In △*AED* and △*BEC*:	
2. $\overline{AD} \cong \overline{BC}$.	2. Why?
3. $\angle ADE \cong \angle EBC$.	3. Why?
4. $\angle DAE \cong \angle BCE$.	4. Why?
5. △*ADE* ≅ △*BCE*.	5. Why?
6. In △*ADC*, $\angle DAC \cong \angle ACD$.	6. Why?
In △*ADE* and △*CDE*:	
7. $\angle DAC \cong \angle ACD$.	7. Statement 6.
8. $\overline{AE} \cong \overline{EC}$.	8. Why?
9. $\overline{AD} \cong \overline{BC}$.	9. Why?
10. △*ADE* ≅ △*CDE*.	10. Why?
11. $\angle AED \cong \angle CED$.	11. Why?
12. $\angle AED$ is supplementary to $\angle CED$.	12. Why?
13. $\angle AED$ and $\angle CED$ are right angles.	13. Why?
14. Hence $\overline{AC} \perp \overline{BD}$.	14. Why?

21. Prove this theorem: If the hypotenuse and an acute angle of a right triangle are congruent to the corresponding parts of a second right triangle, then the triangles are congruent.

22. Complete the proof of the following theorem: If the hypotenuse and one leg of a right triangle are congruent to the corresponding parts of a second right triangle, then the triangles are congruent.

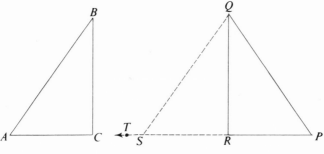

Figure 11.55

Given: $\triangle ABC$ and $\triangle PQR$. $\angle C$ and $\angle QRP$ are right angles; $\overline{AB} \cong \overline{PQ}$ and $\overline{BC} \cong \overline{QR}$.
Prove: $\triangle ABC \cong \triangle PQR$.

PROOF

Statements	*Reasons*
1. In $\triangle ABC$ and $\triangle PQR$, $\angle C$ and $\angle QRP$ are right angles; $\overline{AB} \cong \overline{PQ}$ and $\overline{BC} \cong \overline{QR}$.	1. Why?
2. On $\overrightarrow{RT}$, opposite $\overrightarrow{RP}$, there exists a point S such that $\overline{RS} \cong \overline{AC}$.	2. Why?
3. $\overline{QR} \perp \overleftrightarrow{SP}$.	3. Why?
4. $\angle QRS$ is a right angle.	4. Why?
5. $\angle QRS \cong \angle C$.	5. Why?
6. $\triangle ABC \cong \triangle QRS$.	6. Why?
7. $\overline{QS} \cong \overline{AB}$.	7. Why?
8. $\overline{AB} \cong \overline{PQ}$.	8. Given.
9. Then $\overline{QS} \cong \overline{PQ}$.	9. Why?
10. $\angle P \cong \angle QSR$.	10. Why?
11. $\angle QSR \cong \angle A$.	11. Why?
12. Then $\angle P \cong \angle A$.	12. Why?
13. $\triangle ABC \cong \triangle PQR$.	13. Why?

23. Complete the proof of the following theorem: If the measure of an acute angle of a right triangle is 30, then the length of the side opposite this angle is one-half the length of the hypotenuse.

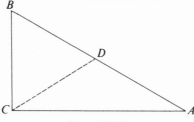

Figure 11.56

Given: $\triangle ABC$ with $\angle C$ a right angle and $m(\angle A) = 30$. Let D be the midpoint of hypotenuse $\overline{AB}$.

Prove: $m(\overline{BC}) = \frac{1}{2}m(\overline{AB})$.

PROOF

Statements	*Reasons*
1. In ABC, $\angle C$ is a right angle and $m(\angle A) = 30$; D is the midpoint of hypotenuse $\overline{AB}$.	1. Why?
2. $m(\angle C) = 90$.	2. Why?
3. $\angle B$ is complementary to $\angle A$.	3. Why?
4. $m(\angle B) + m(\angle A) = 90$.	4. Why?
5. $m(\angle B) = 60$.	5. Why?
6. $m(\overline{CD}) = \frac{1}{2}m(\overline{AB})$	6. Why?
7. $\overline{CD} \cong \overline{BD}$.	7. Why?
8. $m(\angle BCD) = 60$.	8. Why?
9. Then $m(\angle B) + m(\angle BCD) = 120$.	9. Why?
10. And $m(\angle BDC) = 60$.	10. Why?
11. Then $\overline{BC} \cong \overline{BD}$.	11. Why?
12. $m(\overline{BC}) = \frac{1}{2}m(\overline{AB})$.	12. Why?

24. Consider a trapezoid with just two sides parallel. The parallel sides of such a trapezoid are its only bases; the median of the trapezoid is the segment with the midpoints of the non-parallel sides as endpoints. Complete the proof of the following theorem: The median of a trapezoid with just two parallel sides is parallel to the bases and equal in length to one-half the sum of the lengths of the bases.

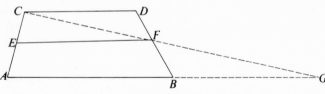

Figure 11.57

Given: Trapezoid $ABCD$ with $\overline{AB} \parallel \overline{CD}$ and median $\overline{EF}$. Since $\overleftrightarrow{CF}$ is not parallel to $\overleftrightarrow{AB}$ (why?), we can conclude that it will meet $\overleftrightarrow{AB}$ at a point G.

Prove: $m(\overline{EF}) = \frac{1}{2}[m(\overline{AB}) + m(\overline{CD})]$ and $\overline{EF} \parallel \overline{CD}$ and $\overline{AB}$.

PROOF

(a) Prove $\triangle CDF \cong \triangle BFG$.
(b) Show that $m(\overline{AB}) + m(\overline{BG}) = m(\overline{AB}) + m(\overline{CD})$.
(c) In $\triangle ACG$, show that $m(\overline{EF}) = \frac{1}{2}m(\overline{AG})$ and $\overline{EF} \parallel \overline{AG}$.
(d) Show that the proof of part c leads to the proof of the given theorem.

25. Complete the proof of the following theorem: The diagonals of a parallelogram bisect each other.

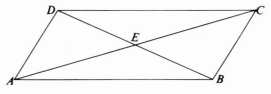

Figure 11.58

Given: Parallelogram $ABCD$ with diagonals $\overline{AC}$ and $\overline{BD}$ intersecting at E.
Prove: $\overline{AE} \cong \overline{EC}$ and $\overline{DE} \cong \overline{BE}$. (Is this equivalent to saying the diagonals bisect each other?)

PROOF

(a) Prove $\triangle ADE \cong \triangle BCE$.
(b) Show that it follows that $\overline{DE} \cong \overline{EB}$ and $\overline{AE} \cong \overline{ED}$.

CHAPTER 12

Areas of Polygonal Regions

12.1 POLYGONAL REGIONS

In Chapter 10 we stated that the union of a polygon and its interior is called a *polygonal region*. The reader will find that the set of polygonal regions described in Chapter 10 is a proper subset of the set of *polygonal regions* as we are about to define the term in this chapter. The first polygonal region that we shall introduce is a *triangular region*. A **triangular region** is defined to be the union of a triangle and its interior. Our reason for defining the term *triangular region* first becomes apparent when we define the term *polygonal region*: A **polygonal region** is defined to be the union of a finite number of coplanar triangular regions such that the intersection of any two of these regions is a point, a segment, or the empty set. Examples of polygonal regions, where the broken lines show how each of the figures may be divided into triangular regions, are shown in Figure 12.1.

Figure 12.1

258

Notice that the intersection of any two of the triangular regions shown in Figure 12.1 meets the requirements of the definition of a polygonal region. The one restriction that is placed on the polygonal region by the definition is that the triangular regions involved do not "overlap," that is, any points they share must be in the triangles, but not in the interiors of the triangles. The polygonal region *ABCDEFG*, pictured in Figure 12.2, is not divided into triangular regions in an acceptable way according to the requirements of the definition. It is a polygonal region, however, as shown in Figure 12.3.

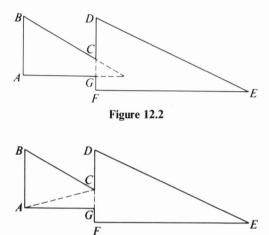

Figure 12.2

Figure 12.3

Further, if a figure is a polygonal region, it can be divided into triangular regions in many ways. Consider the divisions of a trapezoid in Figure 12.4, for example.

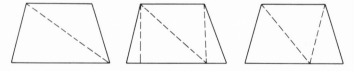

Figure 12.4

In our study of measures of segments and measures of angles, we began by introducing appropriate assumptions, relating these measures to the real number system. Since the *area* of a polygonal region is a *measure* of that region, we shall begin the study of areas of polygonal regions in the same way.

> **Assumption 12.1:** To every polygonal region there corresponds a unique positive real number.

We define the **area of a polygonal region** to be the unique real number assigned to it by Assumption 12.1. A polygonal region will be designated by the symbol **R**. Its area will be designated by $A(\mathbf{R})$. Since we shall be dealing only with regions that are *polygonal* regions, we shall refer to a polygonal region simply as a *region* for the remainder of the chapter. Thus the term *region* is understood to mean *polygonal region*.

In our study of the measures of segments and angles, it was made clear that the unique number associated with a segment or an angle was *unique for a particular unit segment or unit angle*. Similarly, the unique number associated with a region is *unique for a particular unit of area measure*. A *unit of area* for the measure of polygonal regions is customarily related to a unit of measure for segments; that is, we define a *unit of area* in terms of a *unit of length*. Some of the units of length commonly used are the inch, the foot, the yard, and the mile. A unit of area is a *square unit* that corresponds to the unit of length we are using. A **square unit** is the union of a square and its interior where a side of the square is one unit in length. Thus we may measure area in *square inches, square feet, square yards, square miles*, etc.

The next assumption is motivated by our intuition. We feel that two regions having the same size and shape should have the same area.

> **Assumption 12.2:**   The regions associated with congruent triangles have equal areas.

It is important that the student recognize that the area of a region is not a figure (a set of points), but *a number* associated with a figure. This means that certain properties of numbers are also properties of area.

If a given region is divided into two regions, it is intuitively clear to us that the area of the given region is equal to the sum of the areas of the two regions. We state this fact as an assumption.

> **Assumption 12.3:**   If a region R is the union of regions R_1 and R_2, and if the intersection of R_1 and R_2 is at most a finite number of points and segments, then $A(R) = A(R_1) + A(R_2)$.

In Figure 12.5 each region R has been divided into two regions R_1 and R_2. The intersection of R_1 and R_2 has been marked heavily in each case. Since, in each case, the intersection of R_1 and R_2 is a finite number of points and segments, $A(R_1) + A(R_2) = A(R)$.

One more assumption must be made before we can develop formulas for the area of various polygonal regions. This assumption will establish a way to compute the area of one particular region. The simplest region whose area we might postulate is the square. We would assume that the area of a square region

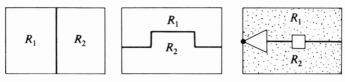

Figure 12.5

with sides of length s is s^2. This is not, however, the assumption we shall choose to make; it is a simple statement, but great difficulty is encountered when we try to develop formulas for the areas of other regions from it. It seems that the best region to use in our assumption is the rectangle. In this assumption and in other statements we shall refer to the *base* and *altitude* of a polygon. Therefore, we need a clear understanding of these terms.

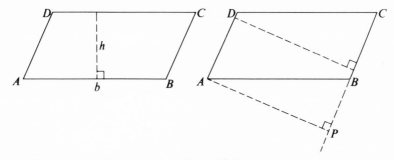

Figure 12.6

Any side of a parallelogram may be considered to be a **base**. A corresponding **altitude** is a segment, perpendicular to this *base*, that has one endpoint on the opposite side and the other endpoint on the line containing the base. This definition is often stated: An **altitude** of a parallelogram is a perpendicular segment from any point on the opposite side to the base. Consider a parallelogram $ABCD$ (Fig. 12.6). If we choose to consider $\overline{AB}$ as base b, then an altitude h may be chosen as pictured in the drawing on the left in Figure 12.6. If we choose $\overline{BC}$ as base, the altitude from vertex A to the line containing $\overline{BC}$ will meet this line at a point P so that (PBC) as shown in the other drawing in Figure 12.6; a second altitude from a point between A and D may meet $\overleftrightarrow{BC}$ in a point between B and C.

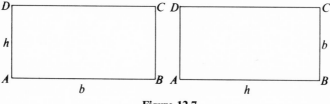

Figure 12.7

A rectangle is a parallelogram. Therefore, the base and altitude of a rectangle have been defined in the preceding paragraph. However, since the rectangle is a specialized parallelogram where each side is perpendicular to each of its *consecutive* sides, we can consider either of the sides that is consecutive to a base as an altitude of the rectangle to that base. Two possibilities are pictured in Figure 12.7 for rectangle *ABCD*.

Any side of a triangle may be considered as a base. A corresponding **altitude** is a segment, perpendicular to the base, and with the opposite vertex as one endpoint and a point of the line containing the base as the other endpoint. A triangle has three altitudes, then: one altitude for each side as a base. We picture these altitudes for a triangle *ABC* in Figure 12.8.

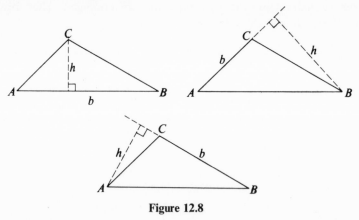

Figure 12.8

A trapezoid has at least two bases: its parallel sides. An **altitude** is a segment perpendicular to the bases and with endpoints on the lines containing the bases (see Fig. 12.9).

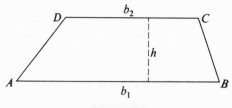

Figure 12.9

Having clarified the meaning of the terms *base* and *altitude*, we state the assumption previously mentioned about the area of a rectangle.

Assumption 12.4:   The area A of a rectangular region is the product of the length of a base and the length of an altitude to that base. Symbolically, this is stated $A = bh$.

For the sake of convenience we shall henceforward speak of the area of a parti-cular polygonal region, for example, a triangular region, a square region, the region consisting of the union of a parallelogram and its interior, etc., as simply the area of the polygon. Thus, when we speak of the area of any polygon (a triangle, a square, a parallelogram, etc.), it shall be understood that this means the area of the polygonal region determined by that polygon.

> **Theorem 12.1:** The area of a square is the square of the length of one side. If the length of a side $= s$, this statement may be symbolized as $A = s^2$.

Theorem 12.1 follows directly from Assumption 12.4, because a square is a rectangle with its base equal to its altitude for any choice of base.

EXERCISES 12.1

1. Show that each of the given figures (Fig. 12.10) is a polygonal region by dividing it into triangular regions so that the intersection of any two of these triangular regions is a segment, a point, or the empty set. Try to find the smallest possible number of triangular regions in each case.

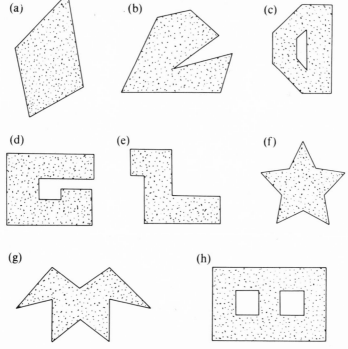

(a) (b) (c)

(d) (e) (f)

(g) (h)

Figure 12.10

2. Find the area of a square with a side having length

(a) 15 ft   (b) $10\frac{1}{2}$ in.   (c) 3.2 mi
(d) 30 ft   (e) 16.04 cm   (f) 4.01 cm

3. Refer to parts (a) and (d) of Exercise 2. Notice that the length of a side of the square in part (d) is twice the length of a side of the square in part (a).

(a) Compare the areas of these squares.
(b) What effect upon the area of a square do you expect when a side of the square is doubled?

Refer to parts (e) and (f) of Exercise 2. Notice that the length of a side of the square in part (f) is $\frac{1}{4}$ the length of a side of the square in part (e).

(c) Compare the area of these squares.
(d) What effect upon the area of a square do you expect when a side of the square is divided by 4?
(e) What effect upon the area of a square do you expect when a side of the square is multiplied by a positive number k?
(f) What effect upon the area of a square do you expect when a side of the square is divided by a positive number k?

4. Find the area of each of the following rectangles; the given lengths of base and altitude are understood to be in inches.

(a) $b = 5, h = 4$ (b) $b = 12, h = 3\frac{1}{2}$ (c) $b = 3\frac{1}{2}, h = 12$

(d) $b = 5.5, h = 3.2$ (e) $b = 3\sqrt{2}, h = \sqrt{2}$ (f) $b = 2\sqrt{3}, h = \sqrt{2}$

(g) $b = 4.3, h = \sqrt{5}$ (h) $b = x, h = y$ (i) $b = x, h = 3y$

(j) $b = 5x, h = y$ (k) $b = 2x, h = 2y$ (l) $b = 3x, h = x + y$

5. Refer to parts (h), (i), (j), and (k) in Exercise 4.

(a) What change in the area of the rectangle occurs when the altitude is multiplied by 3?
(b) What change in the area of the rectangle takes place when the base is multiplied by 5?
(c) What change in the area of the rectangle occurs when both base and altitude are multiplied by 2?
(d) What change in the area of a rectangle would you expect if either the base or the altitude (but not both) are multiplied by some positive factor n?
(e) What change in the area of a rectangle would you expect if both the base and altitude are multiplied by some positive factor n?

6. Find the area of each of the pictured regions in Figure 12.11. Assume lengths of segments to be in inches; assume that angles appearing to be right angles are right angles.

7. How many square regions with sides having length $\frac{1}{2}$ in. can be constructed as subsets of a rectangular region with base 6 in. and altitude $4\frac{1}{2}$ in. so that the intersection of any two squares is either a segment, a point, or the empty set?

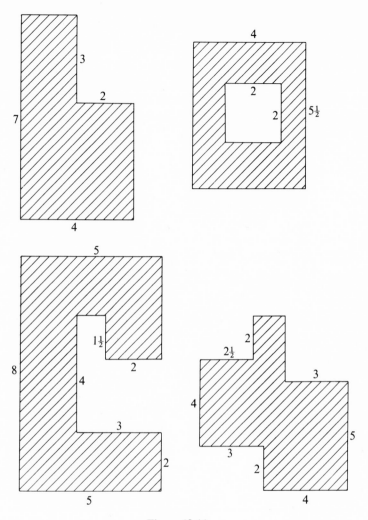

Figure 12.11

12.2   AREAS OF TRIANGLES AND CERTAIN QUADRILATERALS

The assumptions we have made in Section 12.1 make it possible for us to develop theorems about areas of some polygonal regions.

Theorem 12.2:   The area of a right triangle is one-half the product of the lengths of its legs.

A proof of this theorem follows:

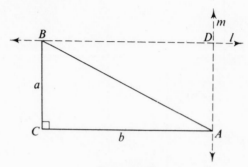

Figure 12.12

Given: $\triangle ABC$ with $m(\angle C) = 90$, $m(\overline{BC}) = a$, and $m(\overline{AC}) = b$. Let l be a line through B that is parallel to $\overleftrightarrow{AC}$; let m be a line through A that is parallel to $\overleftrightarrow{BC}$; l and m intersect at D.

Prove: Area($\triangle ABC$) $= \frac{1}{2}ab$.

PROOF

Statements	*Reasons*
1. In $\triangle ABC$ $m(\angle C) = 90$, $m(\overline{BC}) = a$, $m(\overline{AC}) = b$; $l \| \overleftrightarrow{AC}$ and contains B; $m \| \overleftrightarrow{BC}$ and contains A; l and m intersect at D.	1. Given.
2. $l \perp \overleftrightarrow{BC}$; $m \perp \overleftrightarrow{AC}$.	2. When two parallel lines are cut by a transversal, the alternate interior angles are congruent.
3. Quadrilateral $CADB$ is a parallelogram.	3. A quadrilateral with both pairs of opposite sides parallel is a parallelogram (definition).
4. Parallelogram $CADB$ is a rectangle.	4. A parallelogram having one right angle is a rectangle (definition).
5. $A(CADB) = ab$.	5. Assumption 12.4.
6. $\triangle ABC \cong \triangle ABD$.	6. A diagonal of a parallelogram forms two congruent triangles (theorem).
7. $A(\triangle ABC) + A(\triangle ABD) = A(CADB)$.	7. Assumption 12.3.
8. $A(\triangle ABC) = A(\triangle ABD)$.	8. Assumption 12.2.
9. $2A(\triangle ABC) = A(CADB)$.	9. Statements 7 and 8.

Statements	*Reasons*
10. $A(\triangle ABC) = \frac{1}{2}A(CADB)$.	10. Multiplication property of equality of real numbers.
11. $A(\triangle ABC) = \frac{1}{2}ab$.	11. Transitive property of equality of real numbers.

Theorem 12.2 provides the basis for a second theorem about the area of any triangle.

Theorem 12.3: The area of a triangle is one-half the product of the length of a base and the length of the altitude corresponding to that base.

A proof of this theorem follows:

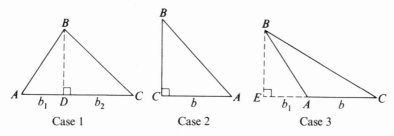

| Case 1 | Case 2 | Case 3 |

Figure 12.13

Given: $\triangle ABC$; $m(\overline{AC}) = b$. There are three cases that must be considered as indicated by the drawings in Figure 12.13:

Case 1. $\overline{AC}$ has a corresponding altitude $\overline{BD}$ where D is an interior point of $\overline{AC}$; $m(\overline{BD}) = h$.

Case 2. $\overline{AC}$ has a corresponding altitude $\overline{BC}$ where C is an endpoint of $\overline{AC}$; $m(\overline{BC}) = h$.

Case 3. $\overline{AC}$ has a corresponding altitude $\overline{BE}$ where E is on $\overleftrightarrow{AC}$ and not on $\overline{AC}$; $m(\overline{BE}) = h$.

Prove: Area($\triangle ABC$) = $\frac{1}{2}bh$.

PROOF

Statements	*Reasons*
Case 1. Let $m(\overline{AD}) = b_1$, $m(\overline{DC}) = b_2$.	
1. Since (ADC), $b_1 + b_2 = b$.	1. The addition postulate. Assumption 11.2.

Statements	*Reasons*
2. Since $\overline{BD} \perp \overline{AC}$, $\triangle ADB$ and $\triangle BDC$ are right triangles,	2. Definition of an altitude of a triangle and definition of a right triangle.
3. $A(\triangle ADB) = \frac{1}{2}b_1h$; $A(\triangle BDC) = \frac{1}{2}b_2h$.	3. Theorem 12.2.
4. $A(\triangle ABC) = A(\triangle ADB) + A(\triangle BDC) = \frac{1}{2}b_1h + \frac{1}{2}b_2h$.	4. Assumption 12.3.
5. $\frac{1}{2}b_1h + \frac{1}{2}b_2h = \frac{1}{2}h(b_1 + b_2)$.	5. Distributive property of multiplication over addition for real numbers.
6. $\frac{1}{2}h(b_1 + b_2) = \frac{1}{2}hb$.	6. Statement 1.
7. $A(\triangle ABC) = \frac{1}{2}bh$.	7. Transitive property of equality of real numbers.

Case 2. If $C = D$, Area$(\triangle ABC) = \frac{1}{2}bh$ by Theorem 12.2.

Case 3. Let $m(\overline{AE}) = b_1$, $m(\overline{AC}) = b$.

1. Since (EAC), $b_1 + b = m(\overline{EC})$.	1. Assumption 11.2.
2. Since $BE \perp AC$, $\triangle BEC$ and $\triangle BEA$ are right triangles.	2. Definition of an altitude of a triangle and definition of a right triangle.
3. $A(\triangle BEC) = \frac{1}{2}h(b_1 + b)$.	3. Theorem 12.2.
4. $A(\triangle BEA) = \frac{1}{2}hb_1$.	4. Theorem 12.2.
5. $A(\triangle BEC) = A(\triangle BEA) + A(\triangle BAC)$.	5. Assumption 12.3.
6. $A(\triangle BAC) = A(\triangle BEC) - A(\triangle BEA) = \frac{1}{2}h(b_1 + b) - \frac{1}{2}b_1h$.	6. Addition property of equality for real numbers.
7. $\frac{1}{2}h(b_1 + b) = \frac{1}{2}hb_1 + \frac{1}{2}hb$.	7. Distributive property of multiplication over addition for real numbers.
8. $A(\triangle BAC) = (\frac{1}{2}b_1h + \frac{1}{2}bh) - \frac{1}{2}b_1h$.	8. Statements 6 and 7.
9. $A(\triangle BAC) = \frac{1}{2}bh$.	9. Subtraction performed.

As Assumption 12.2 and the statement of Theorem 12.3 imply, the area of a triangle is independent of the selection of a side as base.

We have learned that a polygonal region can be divided into triangular regions in many ways (Section 12.1). Further, by Assumption 12.3, the area of a polygonal region is equal to the sum of the areas of these triangular regions. The next two theorems involve the areas of a parallelogram and a trapezoid. The proof of each theorem (Exercises 9 and 10) involves dividing the given region into triangular regions. Since the area of the given region is unique (Assumption 12.1), it does not depend upon the manner in which the region is divided into triangular regions.

Theorem 12.4:   The area of a parallelogram is the product of the length of any base and the length of a corresponding altitude.

Theorem 12.5:   The area of a trapezoid is one-half the product of the length of an altitude and the sum of the lengths of its corresponding bases.

The reader is probably familiar with the term *ratio*. A **ratio** may be defined as the indicated quotient of two numbers. It is often useful for purposes of comparison of two measures. In the following theorem we consider the *ratio* of the *areas* of two triangles.

Theorem 12.6:   If two triangles have altitudes that are equal in length, then the ratio of the areas of the triangles is equal to the ratio of the lengths of the corresponding bases.

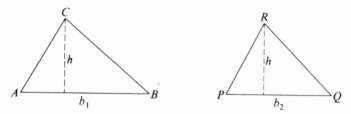

Figure 12.14

This theorem is easily proved. Suppose the two triangles are $\triangle ABC$ with altitude of length h and base $\overline{AB}$ of length b_1, and $\triangle PQR$ with altitude of length h and base $\overline{PQ}$ of length b_2. Then $A(\triangle ABC) = \frac{1}{2}b_1 h$; $A(\triangle PQR) = \frac{1}{2}b_2 h$. The ratio of their areas is

$$\frac{A(\triangle ABC)}{A(\triangle PQR)} = \frac{\frac{1}{2}b_1 h}{\frac{1}{2}b_2 h}$$

$$= \frac{\frac{1}{2}h \cdot b_1}{\frac{1}{2}h \cdot b_2}$$

$$= \frac{\frac{1}{2}h}{\frac{1}{2}h} \cdot \frac{b_1}{b_2}$$

$$= \frac{b_1}{b_2}$$

The student may identify the properties of real numbers that justify the steps in this proof of Theorem 12.6.

Theorem 12.7: If two triangles have altitudes of equal length and corresponding bases of equal length, they have equal areas.

Theorem 12.7 follows directly from Theorem 12.6, since use of the formula, area $= \frac{1}{2}bh$, will yield identical results for both triangles.

EXERCISES 12.2

1. Find the area of $\triangle ABC$, given the information that $\overline{CD}$ is an altitude to $\overline{AB}$, $\overline{CD} \perp \overline{AB}$, and

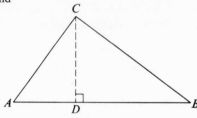

Figure 12.15

(a) $m(\overline{AB}) = 10$ and $m(\overline{CD}) = 4$.
(b) $m(\overline{AB}) = 3\frac{1}{2}$, $m(\overline{CD}) = 6$.
(c) $m(\overline{AB}) = 8$, $m(\overline{BC}) = 7$, $m(\angle B) = 30$.
(d) $m(\overline{AB}) = 11$, $m(\overline{AC}) = 4$, $m(\angle A) = 30$.
(e) $m(\angle A) = 60$, $m(\angle B) = 30$, $m(\overline{BC}) = (7\sqrt{3})/2$, $m(\overline{AD}) = 3\frac{1}{2}$.
(f) $m(\angle A) = 45$, $m(\overline{AC}) = 5$, $m(\angle B) = 45$.
(g) $m(\angle ACB) = 90$, $m(\overline{AD}) = m(\overline{DB})$, $m(\overline{CD}) = 2.5$.

2. The hypotenuse of a triangle is 30 units long, one leg is 24 units, and the area of the triangle is 432 square units. (a) Find the length of the other leg. (b) Find the length of the altitude to the hypotenuse.

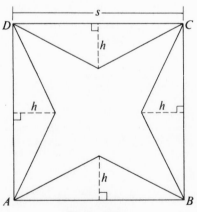

Figure 12.16

3. If *ABCD* is a square, find the area of the star in the illustration (Fig. 12.16) in terms of *s* and *h*. The segments forming the sides of the star are congruent.

4. In parallelogram *ABCD*, $\overline{DE} \perp \overline{BC}$, and $\overline{FG} \perp \overline{AB}$ as shown in Figure 12.17.

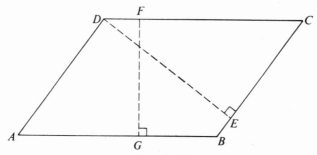

Figure 12.17

(a) If $m(\overline{AB}) = 10$, $m(\overline{FG}) = 5$, and $m(\overline{DE}) = 8$, find $m(\overline{AD})$.
(b) If $m(\overline{AB}) = 16$, $m(\overline{BC}) = 12$, and $m(\overline{FG}) = 9$, find $m(\overline{DE})$.
(c) If $m(\overline{EC}) = 4$, $m(\angle C) = 60$, and $m(\overline{FG}) = 5$, find $A(ABCD)$.

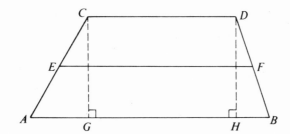

Figure 12.18

5. Find the area of a trapezoid *ABCD* if

(a) $m(\overline{AB}) = 15$, $m(\overline{CD}) = 6$, and $m(\overline{CG}) = 5$.
(b) $m(\overline{AB}) = 20$, $m(\overline{AG}) = 3$, $m(\overline{GH}) = m(\overline{HB})$, and $m(\overline{DH}) = 5$.
(c) $m(\angle B) = 30$, $m(\overline{AB}) = 16$, $m(\overline{CD}) = 7$, and $m(\overline{BD}) = 8$.
(d) $\overline{EF}$ is the median of *ABCD*, $m(\overline{EF}) = 9$, $m(\overline{CG}) = 6$.

6. Compare the areas of

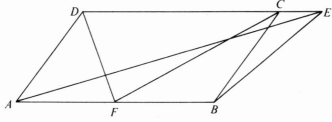

Figure 12.19

(a) $\triangle ABE$ and $\triangle DCF$ ($ABCD$ is a parallelogram).

(b) Parallelogram $ABCD$ and $\triangle ABE$.

(c) $\triangle ADF$ and $\triangle BCF$ if F is the midpoint of $\overline{AB}$.

(d) Parallelogram $ABCD$ and $\triangle AFD$ if F is the midpoint of $\overline{AB}$.

(e) $\triangle BCF$ and $\triangle ABE$ if F is the midpoint of $\overline{AB}$.

7. Prove that the area of a rhombus is equal to one-half the product of the lengths of its diagonals.

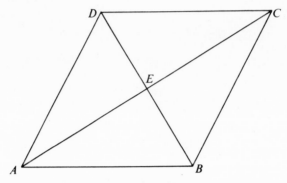

Figure 12.20

Hint: Remember, the diagonals of a rhombus are perpendicular to each other (Exercise 20, Exercises 11.10).

8. If the diagonals of a rhombus have lengths 12 and 16, find its area. If an altitude is of length 7, find the length of a side.

9. Complete this proof of Theorem 12.4: The area of a parallelogram is the product of the length of any base multiplied by the length of a corresponding altitude.

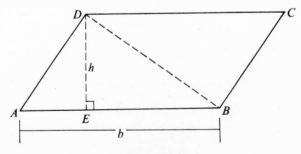

Figure 12.21

Given: Parallelogram $ABCD$ with altitude $\overline{DE} = h$, diagonal $\overline{BD}$; $m(\overline{AB}) = b$.
Prove: $A(ABCD) = bh$.

Proof

Statements	*Reasons*
1. $ABCD$ is a parallelogram with altitude $\overline{DE}$ of length h; $m(\overline{AB}) = b$.	1. Why?
2. Diagonal $\overline{BD}$ divides $ABCD$ into two congruent triangles, $\triangle ABD$ and $\triangle BCD$.	2. Why?
3. $A(\triangle ABD) = A(\triangle BCD)$.	3. Why?
4. $A(\triangle ABD) = \frac{1}{2}bh$.	4. Why?
5. $A(\triangle BCD) = \frac{1}{2}bh$.	5. Why?
6. $A(ABCD) = A(\triangle ABD) + A(\triangle BCD)$.	6. Why?
7. $A(ABCD) = \frac{1}{2}bh + \frac{1}{2}bh$.	7. Why?
8. $A(ABCD) = bh$.	8. Why?

10. Complete this proof of Theorem 12.5: The area of a trapezoid is one-half the product of the length of an altitude multiplied by the sum of the lengths of its corresponding bases.

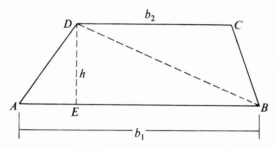

Figure 12.22

Given: Trapezoid $ABCD$ with $\overline{AB} \parallel \overline{CD}$. Let $m(\overline{AB}) = b_1$, $m(\overline{CD}) = b_2$, and $m(\overline{DE}) = h$. Either diagonal divides the trapezoid into two triangles. We shall consider $\triangle ABD$ and $\triangle BCD$ formed when $\overline{BD}$ divides the trapezoid.
   Prove: $A(ABCD) = \frac{1}{2}h(b_1 + b_2)$.

Proof

(a) Show that $A(\triangle ABD) = \frac{1}{2}hb_1$.
(b) Show that $A(\triangle BCD) = \frac{1}{2}hb_2$.
(c) Show that parts (a) and (b) lead to the conclusion we are to prove.

12.3 THE PYTHAGOREAN THEOREM

A theorem involving a relationship between the legs of a right triangle and its hypotenuse is probably the most famous theorem in mathematics. The theorem is credited to Pythagoras, a Greek geometer who is said to have proved it around 525 B.C. There are many ways to prove the Pythagorean theorem. There is one method that was discovered by President Garfield.† The proof we shall present, however, is not President Garfield's proof.

Theorem 12.8 (the Pythagorean theorem):   In a right triangle, the square of the length of the hypotenuse is equal to the sum of the squares of the lengths of the legs.

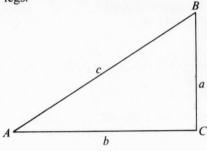

Figure 12.23

Given: Right triangle $\triangle ABC$ with $m(\overline{AB}) = c$, $m(\overline{AC}) = b$ and $m(\overline{BC}) = a$.

Prove: $c^2 = a^2 + b^2$.

PROOF: We use a square $ABCD$ the length of whose sides is $a + b$. Four right triangles may be formed with legs of length a and b as shown in Figure 12.24.

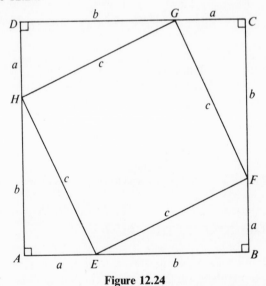

Figure 12.24

† The reader will find this proof in a text book by Herman Hyatt and Charles C. Carico entitled *Modern Plane Geometry for College Students* (New York: The Macmillan Company, 1967), pp. 228–9.

By the SAS postulate, each of these four right triangles is congruent to given $\triangle ABC$. Therefore, the hypotenuse of each is of length c. The quadrilateral $EFGH$ formed by the four hypotenuses is a square, which we now show:

The acute angles of each right triangle are complementary. Thus $m(\angle AEH) + m(\angle AHE) = 90$. Now, $\angle BEF \cong \angle AHE$ because $\triangle AEH \cong \triangle EBF$. Therefore $m(\angle AEH) + m(\angle BEF) = 90$. Since $\angle AEB$ is a straight angle, $[m(\angle AEH) + m(\angle BEF)] + m(\angle HEF) = 180$. By the addition property of equality of real numbers, $m(\angle HEF) = 90$; hence $\angle HEF$ is a right angle. A similar proof may be presented showing each of the other angles of $EFGH$ to be a right angle. Since the sides of $EFGH$ are congruent, and each of its angles is a right angle, then $EFGH$ is a square.

By repeated applications of Assumption 12.3, we can show that area($ABCD$) = area($EFGH$) + area($\triangle AEH$) + area($\triangle BEF$) + area($\triangle CFG$) + area($\triangle DGH$). But

$$\text{Area}(ABCD) = (a + b)^2 \text{ by Theorem 12.1}$$

$$\text{Area}(EFGH) = c^2 \text{ by Theorem 12.1}$$

$$\text{Area}(\triangle AEH) = \text{area}(\triangle BEF)$$

$$= \text{area}(\triangle CFG)$$

$$= \text{area}(\triangle DGH)$$

$$= \tfrac{1}{2}ab \text{ by Theorem 12.3}$$

Thus,

$$(a + b)^2 = c^2 + 4(\tfrac{1}{2}ab)$$

$a^2 + 2ab + b^2 = c^2 + 2ab$ by performing multiplication of real numbers

$a^2 + b^2 = c^2$ by the addition property of equality of real numbers

This proves the theorem.

The converse of the Pythagorean theorem is also true.

Theorem 12.9: If the square of the length of one side of a triangle is equal to the sum of the squares of the lengths of the other two sides, then the triangle is a right triangle, with the right angle opposite the largest side.

We offer a proof of this theorem.

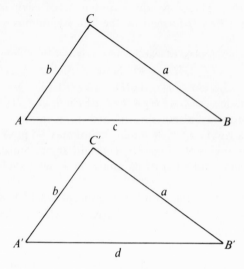

Figure 12.25

Given: $\triangle ABC$ with $m(\overline{AB}) = c$, $m(\overline{AC}) = b$ and $m(\overline{BC}) = a$. $c^2 = a^2 + b^2$.

Prove: $\angle C$ is a right angle.

PROOF

Statements	*Reasons*
1. In $\triangle ABC$, $m(\overline{AB}) = c$, $m(\overline{AC}) = b$, and $m(\overline{BC}) = a$; $c^2 = a^2 + b^2$.	1. Given.
2. We construct $\triangle A'B'C'$ with $\angle C'$ a right angle; $m(\overline{A'C'}) = m(\overline{AC}) = b$, $m(\overline{B'C'}) = m(\overline{BC}) = a$; let $m(\overline{A'B'}) = d$.	2. The angle construction postulate (Section 11.6) and the point location theorem (Section 11.1).
3. $\triangle A'B'C'$ is a right triangle with right angle $\angle C$	3. Step 2.
4. $d^2 = a^2 + b^2$.	4. Theorem 12.8. The Pythagorean theorem.
5. $c^2 = a^2 + b^2$.	5. Given.
6. Therefore $d^2 = c^2$.	6. Transitive property of equality of real numbers.
7. Since c and d are positive real numbers, $d = c$.	7. For positive real numbers a and b, if $a^2 = b^2$, then $a = b$.

Statements	*Reasons*
8. Then $\triangle ABC \cong \triangle A'B'C'$.	8. The SSS theorem.
9. $\angle C = \angle C'$ and therefore is a right angle.	9. Definition of congruent triangles.

These theorems will be especially important in the next chapter when we introduce some elements of trigonometry.

EXERCISES 12.3

1. Find the length of the side whose measure has been omitted in each of the following right triangles. The symbols a and b will denote measures of legs, while c will denote the measure of an hypotenuse. Table I in the Appendix may be used to find square roots. You will find that some of the results are irrational numbers. In these cases, the results may be left as indicated square roots; for example, $\sqrt{23}$.

    (a) $a = 6, b = 5$ (b) $b = 3, c = 5$ (c) $a = 5, c = 13$

    (d) $a = 2, b = 4$ (e) $a = 2, c = 5$ (f) $a = 6, b = 7$

2. Which of the following sets of numbers could be the lengths of the sides of a right triangle?

    (a) 5, 7, 8   (b) 15, 20, 25   (c) 11, 60, 61

    (d) 6, 8, 11   (e) 9, 40, 41   (f) 5, 12, 13

    (g) $1, 1, \sqrt{2}$ (h) $\sqrt{2}, \sqrt{2}, 4$ (i) $\sqrt{3}, \sqrt{5}, 2\sqrt{2}$

    (j) $4\frac{1}{2}, 20, 20\frac{1}{2}$ (k) $\frac{1}{3}, \frac{1}{4}, \frac{1}{5}$ (l) 1.8, 8.0, 8.2

3. With right angles and lengths marked as in Figure 12.26, find:

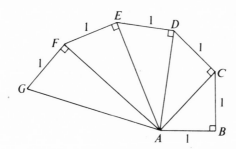

Figure 12.26

    (a) $m(\overline{AC})$ (b) $m(\overline{AD})$ (c) $m(\overline{AE})$ (d) $m(\overline{AG})$

4. (a) In Exercise 3, imagine adding another segment $\overline{GH}$ where $m(\overline{GH}) = 1$ and $\overline{GH} \perp \overline{AG}$. Without computing, what do you think $m(\overline{AH})$ would be?

    (b) What is $m(\angle DAE)$?

5. In rhombus $ABCD$, $m(\overline{AB}) = 8$, $m(\overline{AC}) = 10$.

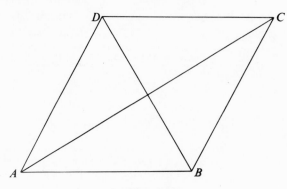

Figure 12.27

Find $m(\overline{BD})$.

6. In the accompanying figure $\overline{BC} \perp \overline{CA}$, $m(\overline{BC}) = 5$, $m(\overline{CA}) = 12$, $\overline{CD} \perp \overline{AB}$.

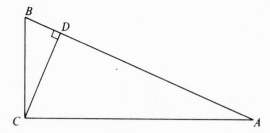

Figure 12.28

Find $m(\overline{CD})$.

7. The lengths of the legs of a right triangle are 9 and 12. Find the length of the hypotenuse; find the length of the altitude to the hypotenuse.

8. The lengths of the legs of a right triangle are a and b. Find the length h of the altitude to the hypotenuse.

CHAPTER 13

Similarity
and Trigonometry

13.1 AN INFORMAL DISCUSSION OF
SIMILARITY

The idea of *similarity* is a very important one in geometry. Many significant conclusions arise from the similarity concept. As with other terms we have used, we shall need to define *similar figures* carefully, but we shall preface formal definition with an informal discussion.

The reader will recall that *congruent figures* can be described informally as figures having the same size and shape. *Similar figures* are often described as figures having the same shape though not necessarily the same size. If a slide is projected onto a screen, so that the plane of the slide is parallel to the plane of the

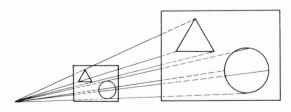

Figure 13.1

279

screen, then the shapes of figures in the slide are the same as the shapes of the projections of these figures on the screen, although the size of the figures has been enlarged. A segment in the slide projects into a segment on the screen; a circle in the slide projects into a circle on the screen, an equilateral triangle projects into an equilateral triangle—any figure in the slide projects into a larger, *similar*, figure on the screen. Thus the slide projector provides us with familiar examples of *similar figures*.

13.2 RATIO AND PROPORTION

The concepts of *ratio* and *proportion* provide a basis for a formal definition of similar figures. The reader will recall that a **ratio** is the indicated quotient of two numbers. If two numbers are symbolized as a and b, their ratio may be written either as a/b or $a:b$. The usual purpose of a ratio is to compare two measures. Since we have worked with measures that are real numbers, the numbers in the ratios we shall consider may be any real numbers.

A **proportion** is a statement that two or more ratios are equal; for example, $\frac{2}{3} = \frac{4}{6} = \frac{6}{9}$. We say two sequences of numbers, $a, b, c, \ldots$ and $p, q, r, \ldots$, are **proportional** iff

$$a/p = b/q = c/r = \ldots, \quad \text{that is,} \quad p/a = q/b = r/c = \ldots$$

The simplest proportion, one that involves only four numbers (not necessarily distinct), has algebraic properties that are very useful. If $a/b = c/d$ and a, b, c, and d are all different from zero,

1. $ad = bc$
2. $a/c = b/d$
3. $b/a = d/c$
4. $\dfrac{a + b}{b} = \dfrac{c + d}{d}$
5. $\dfrac{a - b}{b} = \dfrac{c - d}{d}$

These properties arise from the fact that the proportion $a/b = c/d$ is a statement expressing equality of two real numbers. Therefore the properties of equality of real numbers may be applied to the given proportion to obtain the listed special properties. Consider the original proportion $a/b = c/d$:

1. By the multiplication property of equality of real numbers, we can multiply both ratios by bd to get $ad = bc$.

2. By the multiplication property of equality of real numbers, we can multiply both ratios by b/c to get $a/c = b/d$.

3. By the multiplication property of equality of real numbers, we can multiply both ratios by bd/ac to get $b/a = d/c$.

4. By the addition property of equality of real numbers, we can add 1 (represented as b/b and d/d) to both ratios to get

$$\frac{a + b}{b} = \frac{c + d}{d}$$

5. By the addition property of equality, we can add $^-1$ (represented as $^-b/b$ and $^-d/d$) to both ratios to get

$$\frac{a - b}{b} = \frac{c - d}{d}$$

Consider a proportion $a/b = c/d$ where $b = c$, that is,

$$\frac{a}{b} = \frac{b}{d}$$

If a, b, and d are positive numbers, then b is said to be the **geometric mean** between a and d iff $a/b = b/d$. Property 1 may be used to find that $b = \sqrt{ad}$. Thus the **geometric mean** between **a** and **d** is $\sqrt{ad}$.

EXERCISES 13.2

1. Find x in each proportion:

(a) $3/4 = x/8$ (b) $21/7 = 3/x$ (c) $x/48 = 5/12$ (d) $13/x = 52/8$

2. Complete each statement:

(a) If $3/8 = a/b$, then $3b =$ ____.
(b) If $x/5 = 12/60$, then $60x =$ ____.
(c) If $4/9 = x/2$, then $9x =$ ____.
(d) If $x/3 = 1/11$, then $11x =$ ____.
(e) If $3/7 = 12/x$, then $3x =$ ____.
(f) If $3/8 = a/b$, then $3/b =$ ____.
(g) If $x/5 = 12/60$, then $x/12 =$ ____.
(h) If $x/3 = 1/11$, then $x/1 =$ ____.
(i) If $4/9 = x/2$, then $4/x =$ ____.
(j) If $3/7 = 12/x$, then $3/12 =$ ____.

(k) If $3/8 = a/b$, then $\dfrac{3 + 8}{8} = \dfrac{}{b}$ and $\dfrac{3 - 8}{8} = \dfrac{}{b}$.

(l) If $x/5 = 12/60$, then $\dfrac{x + 5}{5} = \dfrac{}{60}$ and $\dfrac{x - 5}{5} = \dfrac{}{60}$.

(m) If $x/3 = 1/11$, then $\dfrac{x + 3}{3} = \dfrac{}{11}$ and $\dfrac{x - 3}{3} = \dfrac{}{11}$.

(n) If $4/9 = x/2$, then $13/9 = \dfrac{}{2}$.

(o) If $3/7 = 12/x$, then $10/7 = \dfrac{}{x}$.

(p) If $\dfrac{a + b}{b} = 13/10$, then $a/b = $ _____.

(q) If $\dfrac{a - c}{c} = 9/10$, then $a/c = $ _____.

3. Given several sequences of numbers, an effective way of telling whether any two or more sequences are proportional is to divide each sequence by the first number of that sequence; for example, given the sequences

$$2, \quad 5, \quad 11$$
$$6, \quad 15, \quad 33$$
$$\tfrac{7}{2}, \quad \tfrac{35}{4}, \quad \tfrac{77}{4},$$

we divide the numbers in the three sequences by 2, 6, and $\tfrac{7}{2}$ respectively:

$$1, \quad \tfrac{5}{2}, \quad \tfrac{11}{2}$$
$$1, \quad \tfrac{15}{6} = \tfrac{5}{2}, \quad \tfrac{33}{6} = \tfrac{11}{2}$$
$$1, \quad \tfrac{5}{2}, \quad \tfrac{11}{2}$$

The fact that this procedure has resulted in identical sequences means that any two of the sequences are proportional, that is,

$$\tfrac{2}{6} = \tfrac{5}{15} = \tfrac{11}{33}$$

$$\frac{2}{\tfrac{7}{2}} = \frac{5}{\tfrac{35}{4}} = \frac{11}{\tfrac{77}{4}}$$

or

$$\frac{6}{\tfrac{7}{2}} = \frac{15}{\tfrac{35}{4}} = \frac{33}{\tfrac{77}{4}}$$

In the following list, identify the sequences that are proportional.

(a) 2, 3, 7   (b) 5, 7, 11
(c) $\tfrac{2}{3}$, 1, $\tfrac{7}{3}$ (d) 14, 17, 19

(e) $1, \frac{7}{5}, \frac{11}{5}$ (f) $\frac{15}{2}, \frac{21}{2}, \frac{33}{2}$
(g) $7, 8\frac{1}{2}, 9\frac{1}{2}$ (h) $4, \frac{34}{7}, \frac{38}{7}$
(i) $6, 9, 13$ (j) $7\frac{1}{2}, 10\frac{1}{2}, 16\frac{1}{2}$

4. The reader may recall that the arithmetic mean (or average) of two positive numbers a and c is $\frac{a+c}{2}$; by definition (in this section) the geometric mean of a and c is $\sqrt{ac}$. Find the arithmetic mean and the geometric mean of the following pairs of numbers.

(a) 2, 8   (b) 5, 45   (c) 6, 24   (d) 8, 10
(e) 7, 3   (f) 3, 11   (g) 4, 3   (h) 1, 1

13.3 SIMILAR TRIANGLES

We are now ready to define *similar triangles*. Two triangles are said to be **similar triangles** iff a correspondence between vertices exists such that the corresponding angles are congruent and the lengths of the corresponding sides are proportional. For example, given $\triangle ABC$ and $\triangle A'B'C'$,

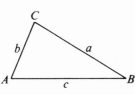 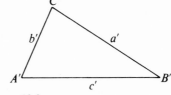

Figure 13.2

where

$$m(\overline{AB}) = c, \quad m(\overline{AC}) = b, \quad m(\overline{BC}) = a,$$
$$m(\overline{A'B'}) = c', \quad m(\overline{A'C'}) = b', \quad m(\overline{B'C'}) = a'$$

$\triangle ABC$ is similar to $\triangle A'B'C'$, symbolized as

$$\triangle ABC \sim \triangle A'B'C'$$

iff

$$ABC \longleftrightarrow A'B'C'$$

and

$$\angle A \cong \angle A', \quad \angle B = \angle B', \quad \angle C = \angle C'$$

and

$$a/a' = b/b' = c/c'$$

Notice that this definition places two requirements on the figures. They are similar if and only if (1) their corresponding angles are congruent and (2) the lengths of their corresponding sides are proportional. These requirements become particularly important when the concept of similarity is extended to polygons having more than three sides. For example, consider the square and rhombus in Figure 13.3. If the lengths of the sides of the square and rhombus are as indicated in the drawings, then (under the correspondence $ABCD \leftrightarrow A'B'C'D'$) the lengths of their corresponding sides are proportional. Yet they do not have the "same shape," that is, they are not similar. The one requirement that the lengths of their corresponding sides be proportional is not enough to establish similarity for quadrilaterals (four-sided polygons).

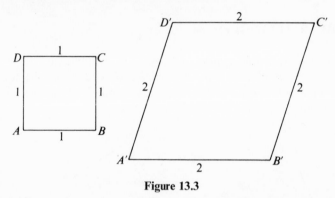

Figure 13.3

We can illustrate, also, that similarity of quadrilaterals cannot be established by the lone requirement that corresponding angles be congruent. Consider the two rectangles in Figure 13.4. Since every angle in each rectangle is a right angle, then (under the correspondence $ABCD \leftrightarrow A'B'C'D'$) the corresponding angles are congruent. It is quite apparent, however, that the rectangles are not similar; their shapes are certainly not the same.

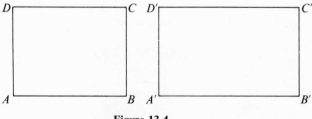

Figure 13.4

We shall see that, for the case of correspondences between triangles, either one of the requirements for similarity is sufficient. Preliminary to showing this, we offer a theorem that is often called the basic proportionality theorem.

Theorem 13.1 (the basic proportionality theorem): If a line that is parallel to one side of a triangle intersects the other two sides in two distinct points, then this line divides these two sides into segments whose lengths are proportional to the lengths of these sides.

A proof of this theorem follows:

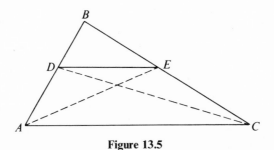

Figure 13.5

Given: $\triangle ABC$ with D and E in $\overline{AB}$ and $\overline{BC}$ respectively such that $\overleftrightarrow{DE} \parallel \overleftrightarrow{AC}$.

Prove: $\dfrac{m(\overline{AB})}{m(\overline{AD})} = \dfrac{m(\overline{BC})}{m(\overline{CE})}$ and $\dfrac{m(\overline{AB})}{m(\overline{BD})} = \dfrac{m(\overline{BC})}{m(\overline{BE})}$.

PROOF

Statements	*Reasons*

Consider $\triangle ADE$ and $\triangle BDE$ with $\overline{AD}$ and $\overline{BD}$, respectively, as the bases.

1. Points D and E are in sides $\overline{AB}$ and $\overline{BC}$ of $\triangle ABC$, respectively; $\overleftrightarrow{DE} \parallel \overleftrightarrow{AC}$.	1. Given
2. $\dfrac{A(\triangle ADE)}{A(\triangle BDE)} = \dfrac{m(\overline{AD})}{m(\overline{BD})}$	2. Theorem 12.6.

Consider $\triangle BDE$ and $\triangle CDE$ with $\overline{BE}$ and $\overline{CE}$, respectively, as bases. These triangles have a common altitude from D to $\overline{BC}$.

3. $\dfrac{A(\triangle CDE)}{A(\triangle BDE)} = \dfrac{m(\overline{CE})}{m(\overline{BE})}$	3. Theorem 12.6.

Consider $\triangle ADE$ and $\triangle CDE$ with $\overline{DE}$ as common base.

4. The distance from A to $\overleftrightarrow{DE}$ equals the distance from C to $\overleftrightarrow{DE}$.	4. Parallel lines are everywhere equidistant.

Statements	*Reasons*
5. The altitudes of $\triangle ADE$ and $\triangle CDE$ from A and C, respectively, to $\overleftrightarrow{DE}$ are congruent.	5. Definition of altitude and Statement 4.
6. $A(\triangle ADE) = A(\triangle CDE)$.	6. Theorem 12.7.
7. $\dfrac{m(\overline{AD})}{m(\overline{BD})} = \dfrac{m(\overline{CE})}{m(\overline{BE})}$	7. Statements 2, 3, and 6.
8. $\dfrac{m(\overline{AD}) + m(\overline{BD})}{m(\overline{BD})} = \dfrac{m(\overline{CE}) + m(\overline{BE})}{m(\overline{BE})}$	8. Algebraic property 4, Section 13.2.
9. But $m(\overline{AD}) + m(\overline{BD}) = m(\overline{AB})$ and $m(\overline{CE}) + m(\overline{BE}) = m(\overline{BC})$.	9. Assumption 11.2. The addition postulate.
10. Thus $\dfrac{m(\overline{AB})}{m(\overline{BD})} = \dfrac{m(\overline{BC})}{m(\overline{BE})}$	10. Statements 8 and 9.
11. If $\dfrac{m(\overline{AD})}{m(\overline{BD})} = \dfrac{m(\overline{CE})}{m(\overline{BE})}$, then $\dfrac{m(\overline{BD})}{m(\overline{AD})} = \dfrac{m(\overline{BE})}{m(\overline{CE})}$	11. Algebraic property 3, Section 13.2.
12. $\dfrac{m(\overline{BD}) + m(\overline{AD})}{m(\overline{AD})} = \dfrac{m(\overline{BE}) + m(\overline{CE})}{m(\overline{CE})}$	12. Algebraic property 4, Section 13.2.
13. $\dfrac{m(\overline{AB})}{m(\overline{AD})} = \dfrac{m(\overline{BC})}{m(\overline{CE})}$	13. Statements 9 and 12.

The converse of Theorem 13.1 is also true:

Theorem 13.2: If a line intersects two sides of a triangle and divides these sides into segments whose lengths are proportional to the lengths of these sides, then the line is parallel to the third side of the triangle.

A proof of this theorem follows.

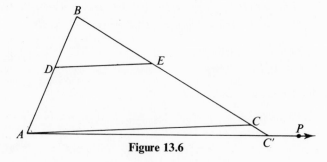

Figure 13.6

Given: $\triangle ABC$ with D and E in $\overline{AB}$ and $\overline{BC}$ respectively so that

$$\frac{m(\overline{AB})}{m(\overline{BD})} = \frac{m(\overline{BC})}{m(\overline{BE})}.$$

Prove: $\overleftrightarrow{DE} \parallel \overline{AC}.$

PROOF

Statements	*Reasons*
1. Points D and E are in sides $\overline{AB}$ and $\overline{BC}$ of $\triangle ABC$, respectively; $\dfrac{m(\overline{AB})}{m(\overline{BD})} = \dfrac{m(\overline{BC})}{m(\overline{BE})}.$	1. Given.
2. There exists a line $\overleftrightarrow{AP}$ through A that is parallel to $\overleftrightarrow{DE}$.	2. Assumption 10.13. The parallel postulate.
3. $\overleftrightarrow{AP}$ intersects $\overleftrightarrow{BC}$ in a point that we shall call C'.	3. Theorem 10.7.
4. $\dfrac{m(\overline{AB})}{m(\overline{BD})} = \dfrac{m(\overline{BC'})}{m(\overline{BE})}.$	4. Theorem 13.1.
5. $m(\overline{BC'}) = m(\overline{BE}) \cdot \dfrac{m(\overline{AB})}{m(\overline{BD})}$	5. Statement 4 and the multiplication property of equality of real numbers.
6. But $m(\overline{BC}) = m(\overline{BE}) \cdot \dfrac{m(\overline{AB})}{m(\overline{BD})}$	6. Given, and the multiplication property of equality of real numbers.
7. $m(\overline{BC}) = m(\overline{BC'})$	7. Transitive property of equality of real numbers.
8. $C = C'$.	8. The point location postulate.
9. Therefore $\overline{AC} = \overline{AC'}$ and $\overline{AC}$ lies in $\overleftrightarrow{AP}$.	9. Statements 2 and 7.
10. $\overleftrightarrow{DE} \parallel \overline{AC}.$	10. Statements 2 and 9.

EXERCISES 13.3

1. Prove this theorem: The triangle whose vertices are the midpoints of a given triangle is similar to the given triangle. *Hint:* Use the theorem in Exercise 18, Exercises 11.10, and the definition of similar triangles.

2. Prove this theorem: If D and E are the midpoints of sides $\overline{AC}$ and $\overline{BC}$, respectively, of $\triangle ABC$, then $\triangle DEC \sim \triangle ABC$. *Hint:* Use the same theorem and definition as suggested for Exercise 1.

3. Refer to the sequences of numbers given in Exercises 13.2, Exercise 3. Which pairs of these sequences could represent the lengths of the sides of pairs of similar triangles?

4. In this drawing, $\angle DEA \cong \angle CBE$.

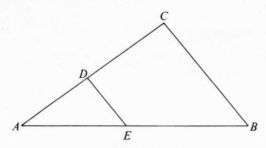

Figure 13.7

(a) If $m(\overline{AD}) = 4$, $m(\overline{AC}) = 10$ and $m(\overline{AE}) = 5$, find $m(\overline{AB})$.

(b) If $m(\overline{AE}) = 3$, $m(\overline{BE}) = 2$ and $m(\overline{AC}) = 4$, find $m(\overline{AD})$ and $m(\overline{CD})$.

(c) If $m(\overline{AE}) = m(\overline{BE})$ and $m(\overline{DE}) = 5$, find $m(\overline{BC})$.

(d) If $m(\angle C) = 90$, $m(\angle B) = 30$, $m(\overline{AC}) = 12$ and $m(\overline{AE}) = 8$, find $m(\overline{AB})$, $m(\overline{AD})$, $m(\overline{BE})$ and $m(\overline{CD})$.

5. Refer to the accompanying drawing and tell which of the following sets of given measurements make $\overline{DE} \parallel \overline{AB}$.

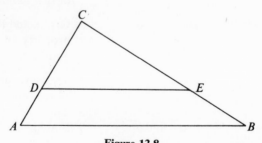

Figure 13.8

(a) $m(\overline{AC}) = 16$, $m(\overline{CD}) = 4$, $m(\overline{BC}) = 20$, $m(\overline{CE}) = 4$.

(b) $m(\overline{AD}) = 4$, $m(\overline{CD}) = 3$, $m(\overline{BE}) = 6$, $m(\overline{CE}) = 4$.

(c) $m(\overline{BC}) = 15$, $m(\overline{BE}) = 10$, $m(\overline{AC}) = 9$, $m(\overline{CD}) = 3$.

(d) $m(\overline{AC}) = 18$, $m(\overline{CD}) = 6$, $m(\overline{BE}) = 9$, $m(\overline{CE}) = 4\frac{1}{2}$.

(e) $m(\overline{CD}) = 3$, $m(\overline{AD}) = 9$, $m(\overline{BC}) = 16$, $m(\overline{BE}) = 12$.

(f) $m(\overline{AC}) = 20$, $m(\overline{CD}) = 6$, $m(\overline{BC}) = 30$, $m(\overline{BE}) = 21$.

(g) $m(\overline{CD}) = 5$, $m(\overline{AD}) = 7$, $m(\overline{CE}) = 8$, $m(\overline{BE}) = 10$.

6. Complete the proof of the following theorem: If three or more coplanar parallel lines are cut by two transversals, the lengths of the intercepted segments on the two transversals are proportional.

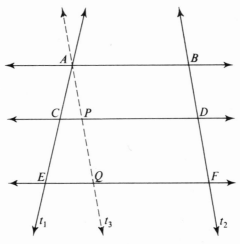

Figure 13.9

Given: Coplanar parallel lines $\overleftrightarrow{AB}$, $\overleftrightarrow{CD}$, and $\overleftrightarrow{EF}$ with transversals t_1 and t_2. The *intercepted segments* on t_1 are $\overline{AC}$ and $\overline{CE}$; the *intercepted segments* on t_2 are $\overline{BD}$ and $\overline{DF}$.

Prove: $\dfrac{m(\overline{AC})}{m(\overline{CE})} = \dfrac{m(\overline{BD})}{m(\overline{DF})}$

PROOF

Statements	*Reasons*
1. Coplanar lines $\overleftrightarrow{AB}$, $\overleftrightarrow{CD}$, and $\overleftrightarrow{EF}$ are parallel and are cut by transversals t_1 and t_2.	1. Why?
2. There exists a line t_3, through A, that is parallel to t_2.	2. Why?
3. $\overleftrightarrow{CD}$ and $\overleftrightarrow{EF}$ intersect t_3 in points that we call P and Q, respectively.	3. Why?
4. $APDB$ and $PQFD$ are parallelograms.	4. Why?
5. $m(\overline{AP}) = m(\overline{BD})$ and $m(\overline{PQ}) = m(\overline{DF})$.	5. Why?
6. $\dfrac{m(\overline{AE})}{m(\overline{CE})} = \dfrac{m(\overline{AQ})}{m(\overline{PQ})}$.	6. Why?
7. $\dfrac{m(\overline{AC}) + m(\overline{CE})}{m(\overline{CE})} = \dfrac{m(\overline{AP}) + m(\overline{PQ})}{m(\overline{PQ})}$.	7. Why?

Statements	Reasons

8. $\dfrac{m(\overline{AC})}{m(\overline{CE})} = \dfrac{m(\overline{AP})}{m(\overline{PQ})}$.   8. Why?

9. $\dfrac{m(\overline{AC})}{m(\overline{CE})} = \dfrac{m(\overline{BD})}{m(\overline{DF})}$.   9. Why?

13.4 THE BASIC SIMILARITY THEOREMS

Recall that the definition of similar triangles specified two conditions for similarity: (1) that the corresponding angles be congruent and (2) that the corresponding sides be proportional. We have previously stated that either one of these conditions is sufficient to establish similarity for triangles. We present a theorem involving the first condition.

> **Theorem 13.3** (the AAA similarity theorem):   Under a given correspondence between the vertices of two triangles, if their corresponding angles are congruent, the triangles are similar.

A proof of this theorem follows.

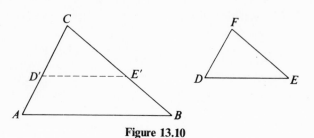

Figure 13.10

Given: $\triangle ABC$ and $\triangle DEF$. Under the correspondence $ABC \leftrightarrow DEF$, $\angle A \cong \angle D$, $\angle B \cong \angle E$, and $\angle C \cong \angle F$.

Prove: $\triangle ABC \sim \triangle DEF$.

PROOF

Statements	Reasons

1. In $\triangle ABC$ and $\triangle DEF$ where   1. Given.
   $ABC \leftrightarrow DEF$, $\angle A \cong \angle D$,
   $\angle B \cong \angle E$, and $\angle C \cong \angle F$.

Statements	*Reasons*

Consider two corresponding sides of $\triangle ABC$ and $\triangle DEF$; for example, $\overline{AC}$ and $\overline{DF}$:

2. If $\overline{AC} \cong \overline{DF}$, then $\triangle ABC \cong \triangle DEF$.	2. The ASA theorem (Theorem 11.21).
3. If $\triangle ABC \cong \triangle DEF$, then $\triangle ABC \sim \triangle DEF$.	3. Definition of congruent triangles and definition of similar triangles.
4. If $\overline{AC} \ncong \overline{DF}$, then $m(\overline{AC}) > m(\overline{DF})$ or $m(\overline{AC}) < m(\overline{DF})$.	4. The trichotomy property of real numbers.

Suppose $m(\overline{AC}) > m(\overline{DF})$:

5. A point D' exists on $\overline{AC}$ such that $\overline{CD'} \cong \overline{DF}$.	5. The point location theorem (Theorem 11.1).
6. Construct $\overline{D'E'}$ such that E' is on $\overline{BC}$ and $\overline{D'E'} \parallel \overline{AB}$.	6. The parallel postulate (Assumption 10.13).
7. Then $\angle CD'E' \cong \angle A$.	Exercises 11.10.
8. $\angle A \cong \angle D$.	8. Given.
9. $\angle CD'E' \cong \angle D$.	9. Transitive property of congruence.
10. $\triangle CD'E' \cong \triangle FDE$.	10. The ASA theorem (Theorem 11.21).
11. Then $\overline{E'C'} \cong \overline{EF}$.	11. Definition of congruent triangles.
12. $\dfrac{m(\overline{D'C})}{m(\overline{AC})} = \dfrac{m(\overline{E'C})}{m(\overline{BC})}$.	12. Step 6 and the basic proportionality theorem (Theorem 13.1).
13. $\dfrac{m(\overline{DF})}{m(\overline{AC})} = \dfrac{m(\overline{EF})}{m(\overline{BC})}$	13. Steps 5, 11, and 12.

In a similar manner, it can be proved that $\dfrac{m(\overline{DF})}{m(\overline{AC})} = \dfrac{m(\overline{DE})}{m(\overline{AB})}$.

14. $\triangle ABC \sim \triangle DEF$.	14. Definition of similar triangles.

If, following Step 4, we were to assume $m(\overline{AC}) < m(\overline{DF})$, Theorem 13.3 can be proved in a similar manner.

We have proved in Exercises 11.10, Exercise 5, that if two angles of one triangle are congruent to the corresponding two angles of a second triangle, then the third angle of the first triangle is congruent to the third angle of the second triangle. A corollary to Theorem 13.3 follows directly from this theorem:

> **Corollary 1** (the AA corollary):   Given a correspondence between two triangles, if two pairs of corresponding angles are congruent, then the triangles are similar.

A second corollary follows quite directly from Theorem 13.3:

> **Corollary 2:**   If a line that is parallel to one side of a triangle intersects the other two sides in distinct points, then a second triangle is formed that is similar to the first. For example, given $\triangle ABC$ as in Figure 13.11, let $\overleftrightarrow{DE}$ be a line that is parallel to $\overline{AB}$. If $\overleftrightarrow{DE}$ intersects $\overline{AC}$ and $\overline{BC}$ in points D and E respectively, then $\triangle ABC \sim \triangle CDE$.

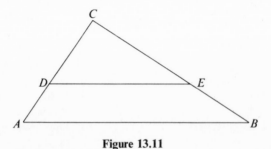

Figure 13.11

There are two additional basic similarity theorems. We state them here; their proofs will be considered in Exercises 8 and 9.

> **Theorem 13.4** (the SAS similarity theorem):   Given a correspondence between the vertices of two triangles, if the length of two pairs of corresponding sides are proportional and the included angles are congruent, then the triangles are similar.

> **Theorem 13.5** (the SSS similarity theorem):   Given a correspondence between the vertices of two triangles, if their corresponding sides are proportional, then the triangles are similar.

EXERCISES 13.4

1. Which of these similarity theorems have related congruence theorems: SAS, AAA, SSS, AA?

2. Consider the triangles in Figure 13.12. State whether or not the triangles of each pair are similar. If they are similar, identify the theorem by which you can justify the similarity.

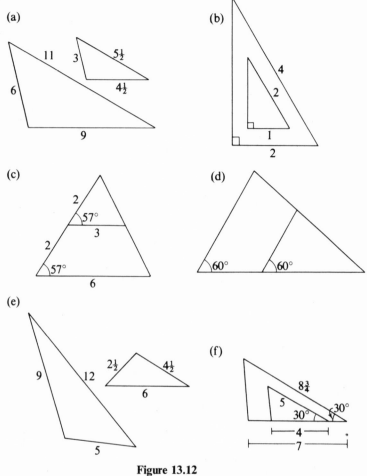

Figure 13.12

3. Given $\triangle ABC$ as shown with $\overline{AC} \perp \overline{BC}$ and $\overline{CD} \perp \overline{AB}$ (Fig. 13.13),

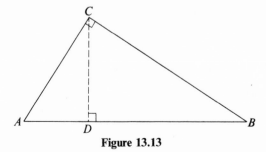

Figure 13.13

(a) Name all pairs of complementary angles in the figure.
(b) Find an angle that is congruent to $\angle ACD$.
(c) Find an angle that is congruent to $\angle BCD$.

    (d) Find two angles that are congruent to $\angle ACB$.

    (e) Name as many pairs of similar triangles as you can find.

    (f) Write statements of proportionality for the sides of the similar triangles.

    (g) In the statements of part (f), find those statements that indicate that the measure of one side is the mean proportional between the measures of two other sides.

4. *Prove:* The measure of the altitude to the hypotenuse of a right triangle is the mean proportional of the measures of the segments into which the hypotenuse is divided (by the altitude).

5. In the illustration (Fig. 13.14),

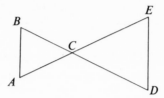

Figure 13.14

    (a) If $m(\overline{AC}) = 4$, $m(\overline{CE}) = 6$, $\overline{BC} \cong \overline{AC}$ and $\overline{CE} \cong \overline{CD}$, is $\triangle ABC \sim \triangle CDE$? Explain your answer.

    (b) If $\angle A = \angle E$, is $\triangle ABC \sim \triangle CDE$? Explain your answer.

    (c) If $\overline{AC} \cong \overline{BC}$ and $\overline{CD} \cong \overline{CE}$, is $\triangle ABC \sim \triangle CDE$? Explain your answer.

    (d) If $m(\overline{AB}) = 2$, $m(\overline{DE}) = 3$, $m(\overline{AC}) = 4$ and $m(\overline{CE}) = 6$, is $\triangle ABC \sim \triangle CDE$? Explain your answer.

6. Prove Corollary 1 (the AA corollary) of Theorem 13.3: Given a correspondence between two triangles, if two pairs of corresponding angles are congruent, then the triangles are similar.

7. Prove Corollary 2 of Theorem 13.3: If a line that is parallel to one side of a triangle intersects the other two sides in distinct points, then a second triangle is formed that is similar to the first.

8. Complete this proof of Theorem 13.4 (the SAS similarity theorem): Given a correspondence between the vertices of two triangles, if the lengths of two pairs of

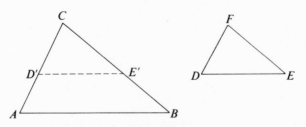

Figure 13.15

corresponding sides are proportional and the included angles are congruent, then the triangles are similar.

Given: $\triangle ABC$ and $\triangle DEF$ where $ABC \leftrightarrow DEF$; $\dfrac{m(\overline{AC})}{m(\overline{DF})} = \dfrac{m(\overline{BC})}{m(\overline{EF})}$ and $\angle C \cong \angle F$.
Assume $m(\overline{AC}) > m(\overline{DF})$, $m(\overline{BC}) > m(\overline{EF})$.
Prove: $\triangle ABC \sim \triangle DEF$.

PROOF

Statements	*Reasons*
1. In $\triangle ABC$ and $\triangle DEF$, $\dfrac{m(\overline{AC})}{m(\overline{DF})} = \dfrac{m(\overline{BC})}{m(\overline{EF})}$; $\angle C \cong \angle F$. We assume $m(\overline{AC}) > m(\overline{DF})$, $m(\overline{BC}) > m(\overline{EF})$.	1. Why?
2. There exists on $\overline{AC}$ a point D' such that $\overline{CD'} \cong \overline{DF}$; there exists on $\overline{BC}$ a point E' such that $\overline{CE'} \cong \overline{EF}$.	2. Why?
3. $\triangle D'E'C \cong \triangle DEF$.	3. Why?
4. $\angle CD'E' \cong \angle D$.	4. Why?
5. $\dfrac{m(\overline{AC})}{m(\overline{CD'})} = \dfrac{m(\overline{BC})}{m(\overline{CE'})}$.	5. Why?
6. Then $\overline{D'E'} \parallel \overline{AB}$.	6. Why?
7. $\angle CD'E' \cong \angle A$.	7. Why?
8. $\angle A \cong \angle D$.	8. Why?
9. $\triangle ABC \sim \triangle DEF$.	9. Why?

9. Complete this proof of Theorem 13.5. (The SSS similarity theorem): Given a correspondence between two triangles, if their corresponding sides are proportional, then the triangles are similar.

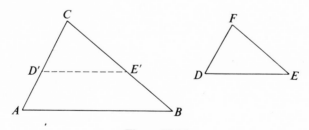

Figure 13.16

Given: $\triangle ABC$ and $\triangle DEF$ where $ABC \leftrightarrow DEF$; $\dfrac{m(\overline{AB})}{m(\overline{DE})} = \dfrac{m(\overline{AC})}{m(\overline{DF})} = \dfrac{m(\overline{BC})}{m(\overline{EF})}$.

Assume $m(\overline{AC}) > m(\overline{DF})$, $m(\overline{BC}) > m(\overline{EF})$, and $m(\overline{AB}) > m(\overline{DE})$.
Prove: $\triangle ABC \sim \triangle DEF$.

PROOF

| Statements | Reasons |

1. In $\triangle ABC$ and $\triangle DEF$,   1. Why?

$$\frac{m(\overline{AB})}{m(\overline{DE})} = \frac{m(\overline{AC})}{m(\overline{DF})} = \frac{m(\overline{BC})}{m(\overline{EF})}.$$

We assume $m(\overline{AC}) > m(\overline{DF})$, $m(\overline{BC}) > m(\overline{EF})$, and $m(\overline{AB}) > m(\overline{DE})$.

2. There exists a point D' in $\overline{AC}$ such that $\overline{CD'} \cong \overline{DF}$ and there exists a point E' in $\overline{BC}$ such that $\overline{CE'} \cong \overline{EF}$.   2. Why?

3. $\dfrac{m(\overline{AC})}{m(\overline{CD'})} = \dfrac{m(\overline{BC})}{m(\overline{CE'})}.$ 3. Why?

4. $\overline{D'E'} \parallel \overline{AB}.$ 4. Why?

5. $\angle CD'E' \cong \angle A.$ 5. Why?

6. $\triangle ABC \sim \triangle D'E'C$ 6. Why?

7. $\dfrac{m(\overline{D'E'})}{m(\overline{AB})} = \dfrac{m(\overline{CD'})}{m(\overline{AC})}.$ 7. Why?

8. $\dfrac{m(\overline{CD'})}{m(\overline{AC})} = \dfrac{m(\overline{DF})}{m(\overline{AC})}.$ 8. Why?

9. $\dfrac{m(\overline{D'E'})}{m(\overline{AB})} = \dfrac{m(\overline{DF})}{m(\overline{AC})}$ 9. Transitive property of equality of real numbers.

10. $m(\overline{D'E'}) = m(\overline{AB}) \cdot \dfrac{m(\overline{DF})}{m(\overline{AC})}.$ 10. Why?

11. $\dfrac{m(\overline{DE})}{m(\overline{AB})} = \dfrac{m(\overline{DF})}{m(\overline{AC})}.$ 11. Why?

12. $m(\overline{DE}) = m(\overline{AB}) \cdot \dfrac{m(\overline{DF})}{m(\overline{AC})}.$ 12. Why?

13. $m(\overline{D'E'}) = m(\overline{DE}).$ 13. Why?

14. $\triangle CD'E' \cong \triangle DEF.$ 14. Why?

15. $\angle C \cong \angle F.$ 15. Why?

16. $\triangle ABC \sim \triangle DEF.$ 16. Why?

13.5   TRIGONOMETRIC RATIOS

Trigonometry, as the Greek derivation of the word indicates, involves measurement of triangles. We offer a brief introduction to the topic of trigonometric functions of an acute angle of a right triangle. We begin with a theorem about similar right triangles.

Theorem 13.6: If an acute angle of one right triangle is congruent to an acute angle of a second right triangle, then the triangles are similar.

This theorem follows directly from the AA corollary to the AAA similarity theorem and from the fact that all right angles are congruent. Before proceeding further, we need to define a new term that will be used in our discussion. For any $\triangle ABC$, recall that $\overline{BC}$ is the side *opposite* $\angle A$. Sides $\overline{AC}$ and $\overline{AB}$ are said to be **adjacent** to $\angle A$. In general, if a side of a triangle is not opposite a given angle, then it is **adjacent** to the angle.

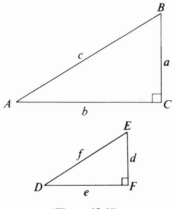

Figure 13.17

Let us examine two right triangles that are similar by Theorem 13.6: $\triangle ABC$ and $\triangle DEF$ where $m(\angle C) = m(\angle F) = 90$, and $\angle A \cong \angle D$. The lengths of the sides of these triangles are as indicated in the illustration (Fig. 13.17): $m(\overline{AB}) = c$, $m(\overline{AC}) = b$, etc. Since $\triangle ABC \sim \triangle DEF$, it follows that

$$\frac{a}{d} = \frac{c}{f}, \quad \frac{b}{e} = \frac{c}{f}, \quad \frac{a}{d} = \frac{b}{e}$$

We use one of the algebraic properties of a proportion introduced in Section 13.2 to rewrite these equations as:

$$\frac{a}{c} = \frac{d}{f}, \quad \frac{b}{c} = \frac{e}{f}, \quad \frac{a}{b} = \frac{d}{e}.$$

The facts that $\triangle ABC$ and $\triangle DEF$ are right triangles and that $\angle A \cong \angle D$ are sufficient conditions for the equality of these ratios. Therefore, we consider each of these ratios to be directly related to the measure of $\angle A$ (or $\angle D$). We call them functions of $\angle A$ (or $\angle D$) and we give them special names.

In general, the measure of an acute angle of a right triangle uniquely determines functions that may be defined as follows:

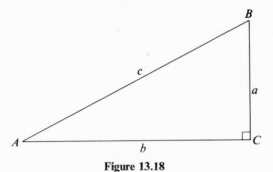

Figure 13.18

If angle A is an acute angle in a right triangle ABC, as pictured in Figure 13.18, with lengths of sides as indicated in the drawing,

1. The **sine of** $\angle A$ is the ratio of the length of the side opposite $\angle A$ to the length of the hypotenuse. We abbreviate the statement

$$\sin A = \frac{\text{opp. side}}{\text{hypotenuse}} = \frac{a}{c}$$

2. The **cosine of** $\angle A$ is the ratio of the length of the side adjacent to $\angle A$ to the length of the hypotenuse. Abbreviated, this is,

$$\cos A = \frac{\text{adj. side}}{\text{hypotenuse}} = \frac{b}{c}$$

3. The **tangent of** $\angle A$ is the ratio of the length of the side opposite $\angle A$ to the length of the side adjacent to $\angle A$. Abbreviated, this is,

$$\tan A = \frac{\text{opp. side}}{\text{adj. side}} = \frac{a}{b}$$

There are other trigonometric functions that we shall not consider.

For most acute angles, the functions we have defined are transcendental numbers and approximations of them are computed by methods far beyond the scope of this book. We have included, in the Appendix, a table of trigonometric functions for angles with integral measures from 0 to 90. These tabular values, with few exceptions, are approximations. There are, however, a few angles whose sine, cosine, and tangent we can compute exactly with information available to us. One of these special angles is an angle whose measure is 30. Consider a right triangle where $m(\angle A) = 30$, as shown in Figure 13.19.

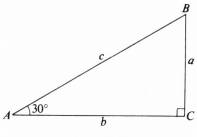

Figure 13.19

The measures of the sides are as indicated in the drawing. We know that $a = \frac{1}{2}c$, or $c = 2a$ in this triangle. Therefore we can choose values $a = 1$ and $c = 2$ (or any other values where $c = 2a$) for the purpose of finding the trigonometric functions of $30°$. By the Pythagorean theorem,

$$a^2 + b^2 = c^2$$
$$1 + b^2 = 2^2$$
$$1 + b^2 = 4$$
$$b^2 = 4 - 1$$
$$b^2 = 3$$
$$b = \sqrt{3}$$

Thus,

$$\sin 30° = 1/2 \quad \cos 30° = \sqrt{3}/2 \quad \tan 30° = 1/\sqrt{3} = \sqrt{3}/3$$

A second special angle whose trigonometric functions can be computed exactly is an angle whose measure is 45. Let $\triangle PQR$ be a right triangle, as pictured in Figure 13.20, with the measures of the sides as indicated. In a right triangle, if the measure of one acute angle is 45, then the measure of the other acute angle is 45; thus the triangle is isosceles. Hence $\overline{PR} \cong \overline{QR}$. To find the sine, cosine, and tangent of $\angle P = 45°$, we may choose any values for p and q that are equal; we shall choose $p = q = 1$. By the Pythagorean theorem,

$$p^2 + q^2 = r^2$$
$$1^2 + 1^2 = r^2$$
$$r^2 = 2$$
$$r = \sqrt{2}$$

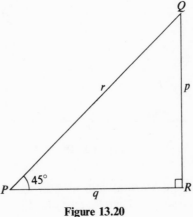

Figure 13.20

Then

$$\sin 45° = 1/\sqrt{2} = \sqrt{2}/2, \quad \cos 45° = 1/\sqrt{2} = \sqrt{2}/2, \quad \tan 45° = 1/1 = 1$$

We can compute the sine, cosine, and tangent of an angle whose measure is 60, also. This will be left to the student as an exercise. When trigonometric functions of angles other than 30°, 45°, or 60° are needed, we may use the table of trigonometric functions in the Appendix. For example, suppose we need to know sin 35°. The table has two columns with the heading "Angle." The one on the far left of the page has angles measuring 0 to 45; the one near the center of the page has angles measuring 46 to 90. We have reproduced a part of that table here.

Angle	Sine	Cosine	Tangent
⋮	⋮	⋮	⋮
25	0.423	0.906	0.466
26	0.438	0.899	0.488
27	0.454	0.891	0.510
28	0.469	0.883	0.532
29	0.485	0.875	0.554
30	0.500	0.866	0.577
31	0.515	0.857	0.601
32	0.530	0.848	0.625
33	0.545	0.839	0.649
34	0.559	0.829	0.675
35	0.574	0.819	0.700
36	0.588	0.809	0.727
37	0.602	0.799	0.754
38	0.616	0.788	0.781
39	0.629	0.777	0.810
40	0.643	0.766	0.839
⋮	⋮	⋮	⋮

We look under the left-hand column for 35. Sin 35° appears opposite the entry 35, in the first column to the right whose heading is "Sine"; as our eye moves horizontally from left to right, the entry in the next column will give us cos 35°, and the entry in the fourth column will give us tan 35°, as the headings indicate. We find that sin 35° = 0.574, cos 35° = 0.819, and tan 35° = 0.700.

Given a value for a function of an angle, we can also use the table to find the angle that has approximately the given value for its function. For example, if we know that sin A = 0.485, we look for the value 0.485 under the heading "Sine," and move to the left to identify the angle. We find that A = 29°, since sin 29° = 0.485.

We can use the table of trigonometric functions to "solve right triangles." We shall demonstrate this use now. Given the right angle of a triangle and the measures of any other two parts (one of which must be a side), we can find the approximate measures of the remaining three parts. Finding these measures is what we call "solving the triangle." In our solutions of triangles, we shall consider only integral lengths of sides and integral degree measure for angles. Let us solve right △ABC where $m(\angle C)$ = 90, a = 4, and c = 5 (Fig. 13.21).

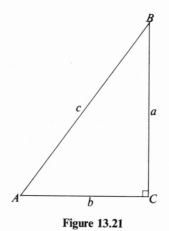

Figure 13.21

Our problem is to find approximations for $m(\angle A)$, $m(\angle B)$, and $m(\overline{AC})$ (or b). We can use the given values for a and c to find sin A.

$$\sin A = \tfrac{4}{5} = 0.800$$

Now we use the table to find the angle whose sine is nearest to 0.800. We find

$$\sin 53° = 0.799$$

Since this is the nearest tabular value to 0.800, we conclude that

$$m(\angle A) = 53 \quad \text{(approximately)}$$

Since the acute angles of a right triangle are complementary,

$$m(\angle B) = 90 - 53$$

$$= 37 \quad \text{(approximately)}$$

To find $m(\overline{AC})$, that is, the value of b, we could use the Pythagorean theorem; but we choose to find it by using the cosine of $\angle A$:

$$\frac{b}{c} = \cos A$$

$$\frac{b}{5} = \cos 53°$$

Here we use the table to find an approximation for $\cos 53°$; we find $\cos 53° = 0.602$. Thus,

$$\frac{b}{5} = 0.602$$

$$b = 5(0.602)$$

$$= 3.010 \quad \text{(approximately)}$$

We round off the value of b to $b = 3$. This concludes the solution of this triangle.

EXERCISES 13.5

1. Find, without using a table of trigonometric functions, the sine, cosine, and tangent of an angle whose measure is 60. *Hint:*   Recall that, in a right triangle where $m(\angle A) = 60$ and $m(\angle C) = 90$, $m(\angle B) = 30$.

2. In this exercise we shall assure ourselves that, in the computation of the trigonometric ratios of 30°, the choice of values for the lengths of the opposite side and the hypotenuse may be any numbers whose ratio is $\frac{1}{2}$. Let us use lengths a and $2a$ respectively. See Figure 13.22.

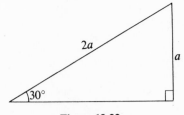

Figure 13.22

Compute the value of the adjacent side and then find sin 30°, cos 30°, and tan 30°.

3. Use the given right triangle (Fig. 13.23) to find sin $x°$, cos $x°$, and tan $x°$.

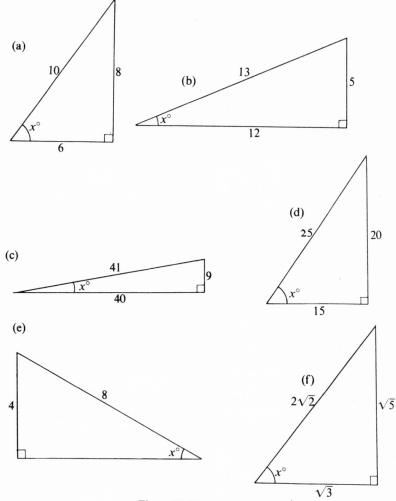

Figure 13.23

4. Consider the given $\triangle ABC$ (Fig. 13.24), construct related right triangles and find

(a) sin A (b) sin B (c) cos A
(d) cos B (e) tan A (f) tan B

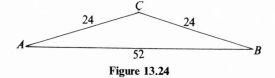

Figure 13.24

5. Use the table of trigonometric functions in the Appendix to find

(a) $\sin 72°$ (b) $\tan 13°$ (c) $\cos 65°$

(d) $\tan 71°$ (e) $\sin 87°$ (f) $\cos 43°$

6. Use the table of trigonometric functions in the Appendix to find $m(\angle X)$ if

(a) $\tan X = 0.194$ (b) $\cos X = 0.438$ (c) $\sin X = 0.656$

(d) $\sin X = 0.999$ (e) $\tan X = 7.115$ (f) $\cos X = 0.809$

7. Given $\triangle ABC$ with $m(\angle C) = 90$, $m(\angle A) = 32$, and $m(\overline{AB}) = 9$,

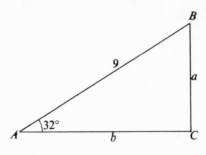

Figure 13.25

find a, b, and $m(\angle B)$.

8. A 15-foot ladder is resting against a building so that the degree measure of the angle it makes with the ground is 75.

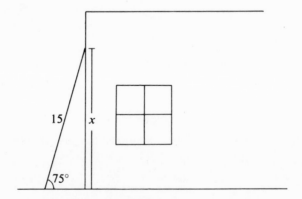

Figure 13.26

Find the distance from the base of the building to the place where the ladder rests against the wall.

9. Find the area of an isosceles triangle if the vertex angle is 38° and the length of the altitude from the vertex angle is 8.

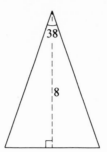

Figure 13.27

Appendix

N	N^2	$\sqrt{N}$	N	N^2	$\sqrt{N}$
1	1	1	51	2,601	7.141
2	4	1.414	52	2,704	7.211
3	9	1.732	53	2,809	7.280
4	16	2	54	2,916	7.348
5	25	2.236	55	3,025	7.416
6	36	2.449	56	3,136	7.483
7	49	2.646	57	3,249	7.550
8	64	2.828	58	3,364	7.616
9	81	3	59	3,481	7.681
10	100	3.162	60	3,600	7.746
11	121	3.317	61	3,721	7.810
12	144	3.464	62	3,844	7.874
13	169	3.606	63	3,969	7.937
14	196	3.742	64	4,096	8
15	225	3.873	65	4,225	8.062
16	256	4	66	4,356	8.124
17	289	4.123	67	4,489	8.185
18	324	4.243	68	4,624	8.246
19	361	4.359	69	4,761	8.307
20	400	4.472	70	4,900	8.367
21	441	4.583	71	5,041	8.426
22	484	4.690	72	5,184	8.485
23	529	4.796	73	5,329	8.544
24	576	4.899	74	5,476	8.602
25	625	5	75	5,625	8.660
26	676	5.099	76	5,776	8.718
27	729	5.196	77	5,929	8.775
28	784	5.292	78	6,084	8.832
29	841	5.385	79	6,241	8.888
30	900	5.477	80	6,400	8.944
31	961	5.568	81	6,561	9
32	1,024	5.657	82	6,724	9.055
33	1,089	5.745	83	6,889	9.110
34	1,156	5.831	84	7,056	9.165
35	1,225	5.916	85	7,225	9.220
36	1,296	6	86	7,396	9.274
37	1,369	6.083	87	7,569	9.327
38	1,444	6.164	88	7,744	9.381
39	1,521	6.245	89	7,921	9.434
40	1,600	6.325	90	8,100	9.487
41	1,681	6.403	91	8,281	9.539
42	1,764	6.481	92	8,464	9.592
43	1,849	6.557	93	8,649	9.644
44	1,936	6.633	94	8,836	9.695
45	2,025	6.708	95	9,025	9.747
46	2,116	6.782	96	9,216	9.798
47	2,209	6.856	97	9,409	9.849
48	2,304	6.928	98	9,604	9.899
49	2,401	7	99	9,801	9.950
50	2,500	7.071	100	10,000	10

TABLE II: TABLE OF TRIGONOMETRIC FUNCTIONS

Angle	Sine	Cosine	Tangent	Angle	Sine	Cosine	Tangent
0	0.000	1.000	0.000				
1	0.017	1.000	0.017	46	0.719	0.695	1.036
2	0.035	0.999	0.035	47	0.731	0.682	1.072
3	0.052	0.999	0.052	48	0.743	0.669	1.111
4	0.070	0.998	0.070	49	0.755	0.656	1.150
5	0.087	0.996	0.088	50	0.766	0.643	1.192
6	0.105	0.995	0.105	51	0.777	0.629	1.235
7	0.122	0.993	0.123	52	0.788	0.616	1.280
8	0.139	0.990	0.141	53	0.799	0.602	1.327
9	0.156	0.988	0.158	54	0.809	0.588	1.376
10	0.174	0.985	0.176	55	0.819	0.574	1.428
11	0.191	0.982	0.194	56	0.829	0.559	1.483
12	0.208	0.978	0.213	57	0.839	0.545	1.540
13	0.225	0.974	0.231	58	0.848	0.530	1.600
14	0.242	0.970	0.249	59	0.857	0.515	1.664
15	0.259	0.966	0.268	60	0.866	0.500	1.732
16	0.276	0.961	0.287	61	0.875	0.485	1.804
17	0.292	0.956	0.306	62	0.883	0.469	1.881
18	0.309	0.951	0.325	63	0.891	0.454	1.963
19	0.326	0.946	0.344	64	0.899	0.438	2.050
20	0.342	0.940	0.364	65	0.906	0.423	2.145
21	0.358	0.934	0.384	66	0.914	0.407	2.246
22	0.375	0.927	0.404	67	0.921	0.391	2.356
23	0.391	0.921	0.424	68	0.927	0.375	2.475
24	0.407	0.914	0.445	69	0.934	0.358	2.605
25	0.423	0.906	0.466	70	0.940	0.342	2.747
26	0.438	0.899	0.488	71	0.946	0.326	2.904
27	0.454	0.891	0.510	72	0.951	0.309	3.078
28	0.469	0.883	0.532	73	0.956	0.292	3.271
29	0.485	0.875	0.554	74	0.961	0.276	3.487
30	0.500	0.866	0.577	75	0.966	0.259	3.732
31	0.515	0.857	0.601	76	0.970	0.242	4.011
32	0.530	0.848	0.625	77	0.974	0.225	4.331
33	0.545	0.839	0.649	78	0.978	0.208	4.705
34	0.559	0.829	9.675	79	0.982	0.191	5.145
35	0.574	0.819	0.700	80	0.985	0.174	5.671
36	0.588	0.809	0.727	81	0.988	0.156	6.314
37	0.602	0.799	0.754	82	0.990	0.139	7.115
38	0.616	0.788	0.781	83	0.993	0.122	8.144
39	0.629	0.777	0.810	84	0.995	0.105	9.514
40	0.643	0.766	0.839	85	0.996	0.087	11.43
41	0.656	0.755	0.869	86	0.998	0.070	14.30
42	0.669	0.743	0.900	87	0.999	0.052	19.08
43	0.682	0.731	0.933	88	0.999	0.035	28.64
44	0.695	0.719	0.966	89	1.000	0.017	57.29
45	0.707	0.707	1.000	90	1.000	0.000	

Answers for Selected Odd-Numbered Exercises

CHAPTER 1

Exercises 1.1

1. (a) $V = \{a, e, i, o, u\}$
  (b) $C = \{b, c, d, f, g, h, j, k, l, m, n, p, q, r, s, t, v, w, x, y, z\}$
  (c) $X = \{$Erie, Huron, Ontario, Michigan, Superior$\}$
  (d) $S_1 = \{5, 7, 9, 11, 13, 15\}$
  (e) $S_2 = \{5, 10, 15\}$
  (f) $Y = \{\quad\}$
  (g) $X = \{2, 4, 6, \dots\}$
  (h) $A = \{6, 12, 18, \dots\}$

3. Examples of acceptable answers are:

  (a) $\{\quad\}$, the set of all astronauts who have landed on Venus.
    $\{\quad\}$, the set of all college students over 10 feet tall.
  (b) $\{$red, yellow, blue$\}$, the set of primary colors.
    $\{m, a, t, h, e, i, c, s\}$, the set of letters (English Alphabet) used to spell "mathematics."
  (c) $\{2, 4, 6, \dots\}$, the set of all even counting numbers.
    $\{5, 10, 15, \dots\}$, the set of all counting numbers that are multiples of five.

311

Exercises 1.3

1. (a) is equivalent to
   (b) is an element of
   (c) is a subset of; is a proper subset of
   (d) is a subset of; is equal to; is equivalent to
   (e) is a subset of; is a proper subset of
   (f) is a subset of; is equal to
   (g) none of the given relations apply
   (h) is a subset of; is a proper subset of

3. (a) yes. (b) yes. (c) yes. (d) yes.

5. (a) yes.

7. (a)   (b)   (c)

(d)   (e)   (f)

 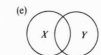 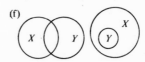

9. (a)

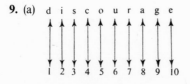

(b)

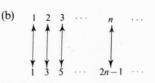

(If we multiply each of the counting numbers by 2 and subtract 1 from the result, we get a corresponding odd number; thus n corresponds to $2n - 1$.)

(c)   (d)

(e)   (f)

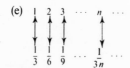 

Exercises 1.4

1. Examples of correct answers are (many choices are possible):

   (a) The set of all makes of automobiles, or the set of all American made cars.
   (b) The set of all items of classroom equipment in our room.

(c) The set of all tools used to carry out jobs, or $\{S_1, S_2, S_3, S_4\}$.

(d) The set of all conditions and professions among human beings.

(e) The set of all even numbers, or the set of all even numbers from 0 to 24 inclusive.

(f) The set of all consonants in the English alphabet, or the set of all letters in the English alphabet.

3. $A \cup C = \{a, b, c, d, e, f, g, h, n, o, r\}$
$A \cap C = \{\ \}$ or $\varnothing$

5. $\{\ \}, \{a\}, \{b\}, \{c\}, \{d\}, \{a, b\}, \{a, c\}, \{a, d\}, \{b, c\}, \{b, d\}, \{c, d\}, \{a, b, c\}, \{a, b, d\}, \{a, c, d\},$
$\{b, c, d\}, \{a, b, c, d\}$

7. (a) false.   (b) false.   (c) true.   (d) true.   (e) true.
    (f) false.   (g) true.   (h) true.   (i) false.   (j) true.
    (k) true.   (l) false.

9. (a) $\{\ \}$ or $\varnothing$.   (b) $\{\ \}$ or $\varnothing$.   (c) $\{1, 2, 3\}$ or U.   (d) $\{1, 2, 3\}$ or U.
    (e) $\{\ \}$ or $\varnothing$.   (f) $\{1, 2\}$ or A.   (g) $\{\ \}$ or $\varnothing$.   (h) $\{1, 2\}$ or A.
    (i) $\{(1, (1, 2)), (1, (1, 3)), (1, (3, 2)), (1, (3, 3)), (2, (1, 2)), (2, (1, 3)), (2, (3, 2)), (2, (3, 3))\}$.
    (j) $\{((1, 1), 2), ((1, 1), 3), ((1, 3), 2), ((1, 3), 3), ((2, 1), 2), ((2, 1), 3), ((2, 3), 2), ((2, 3), 3)\}$.
    (k) $\{1, 2\}$ or A.   (l) $\{1, 2\}$ or A.   (m) $\{1, 2, 3\}$ or U.   (n) $\{1, 2, 3\}$ or U.

CHAPTER 2

Exercises 2.2

1. (a) $\{(a, 1), (a, 2), (b, 1), (b, 2), (c, 1), (c, 2)\}$

(b) $\{(1, a), (1, b), (1, c), (2, a), (2, b), (2, c)\}$

(c) No.

(d) Yes. There are two cases. When both sets have identical elements or when one or both sets are empty.

3. (a) $(P \cap Q) \cap R = (\{1, 2, 3, 4, 5\} \cap \{1, 5, 9\}) \cap \{2, 4, 6, 8, 10\}$
$= \{1, 5\} \cap \{2, 4, 6, 8, 10\} = \{\ \}$
$P \cap (Q \cap R) = \{1, 2, 3, 4, 5\} \cap (\{1, 5, 9\} \cap \{2, 4, 6, 8, 10\})$
$= \{1, 2, 3, 4, 5\} \cap \{\ \} = \{\ \}$

(b) $(P \cup Q) \cup R = (\{1, 2, 3, 4, 5\} \cup \{1, 5, 9\}) \cup \{2, 4, 6, 8, 10\}$
$= \{1, 2, 3, 4, 5, 9\} \cup \{2, 4, 6, 8, 10\}$
$= \{1, 2, 3, 4, 5, 6, 8, 9, 10\}$
$P \cup (Q \cup R) = \{1, 2, 3, 4, 5\} \cup (\{1, 5, 9\} \cup \{2, 4, 6, 8, 10\})$
$= \{1, 2, 3, 4, 5\} \cup \{1, 2, 4, 5, 6, 8, 9, 10\}$
$= \{1, 2, 3, 4, 5, 6, 8, 9, 10\}$

(c) $P \cup (Q \cap R) = \{1, 2, 3, 4, 5\} \cup (\{1, 5, 9\} \cap \{2, 4, 6, 8, 10\})$
$= \{1, 2, 3, 4, 5\} \cup \{\ \} = \{1, 2, 3, 4, 5\}$
$(P \cup Q) \cap (P \cup R) = (\{1, 2, 3, 4, 5\} \cup \{1, 5, 9\}) \cap (\{1, 2, 3, 4, 5\} \cup \{2, 4, 6, 8, 10\})$
$= \{1, 2, 3, 4, 5, 9\} \cap \{1, 2, 3, 4, 5, 6, 8, 10\}$
$= \{1, 2, 3, 4, 5\}$

(d) $P \cap (Q \cup R) = \{1, 2, 3, 4, 5\} \cap (\{1, 5, 9\} \cup \{2, 4, 6, 8, 10\})$
$= \{1, 2, 3, 4, 5\} \cap \{1, 2, 4, 5, 6, 8, 9, 10\}$
$= \{1, 2, 4, 5\}$
$(P \cap Q) \cup (P \cap R) = (\{1, 2, 3, 4, 5\} \cap \{1, 5, 9\}) \cup (\{1, 2, 3, 4, 5\} \cap \{2, 4, 6, 8, 10\})$
$= \{1, 5\} \cup \{2, 4\}$
$= \{1, 2, 4, 5\}$

(e) $(P \cup Q)' = \{1, 2, 3, 4, 5, 9\}' = \{6, 7, 8, 10\}$
$P' \cap Q' = \{6, 7, 8, 9, 10\} \cap \{2, 3, 4, 6, 7, 8, 10\}$
$= \{6, 7, 8, 10\}$

(f) $(P \cap Q)' = \{1, 5\}' = \{2, 3, 4, 6, 7, 8, 9, 10\}$
$P' \cup Q' = \{6, 7, 8, 9, 10\} \cup \{2, 3, 4, 6, 7, 8, 10\}$
$= \{2, 3, 4, 6, 7, 8, 9, 10\}$

5. (a)

Euler diagram:

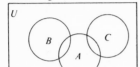

Venn diagram:

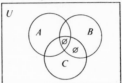

(b)

Euler diagram:

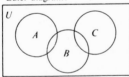

Venn diagram:

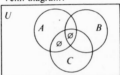

(c)

Euler diagram:

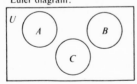

Venn diagram:

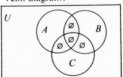

7. 12

9. (a) 13   (b) 224   (c) 59   (d) 13   (e) 73

Exercises 2.3

1. Theorem IV: The complement of the universe is the empty set.

PROOF   By definition of complement, the complement of the universe contains just those elements in the universe that are not contained in the universe. By definitions of the universe and the empty set, the complement of the universe is the empty set.

Theorem V: The union of any set A with the universe U is the universe.

PROOF $A \cup U = U \cup A$ because union is commutative. By definition of union, $A \cup U$ contains each element of A, each element of U and no other elements. Since U, by definition, contains all elements, $A \cup U = U \cup A = U$.

Theorem VI: The intersection of any set A with the universe U is the set A.

PROOF $A \cap U = U \cap A$ because intersection is commutative. By definition of intersection, $A \cap U$ contains each element of A that is also an element of U and no other elements. Since, by definition, U contains all elements, $A \cap U = U \cap A = A$.

Theorem VII: $A \cap A = A$.

PROOF   By definition of intersection, $A \cap A$ is the set containing just those elements that are common to A and A. Thus $A \cap A = A$.

Theorem VIII: $A \cup A = A$.

PROOF   By definition of union, $A \cup A$ is the set containing each element of A, each element of A and no other elements. Thus $A \cup A = A$.

3. (a) Set, element or member, belongs to, contains, is contained in.
   (b) Well-defined set, empty set, subset, proper subset, equal sets, equivalent sets, disjoint sets, union, intersection, complement, ordered pair, cartesian product, one-to-one correspondence.
   (c) Commutative property of intersection, commutative property of union, associative property of union, associative property of intersection, closure property of intersection, closure property of union, distributive property of union over intersection, distributive property of intersection over union, DeMorgan's law for the complement of a union, DeMorgan's law for the complement of an intersection.
   (d) Special properties of the empty set: $A \cap \varnothing = \varnothing \cap A = \varnothing$; $A \cup \varnothing = \varnothing \cup A = A$ (identity element of union); $\varnothing' = U$.
       Special properties of the universe: $A \cap U = U \cap A = A$; $A \cup U = U \cup A = U$; $U' = \varnothing$.
       Idempotent properties: $A \cap A = A$; $A \cup A = A$.
       Any other theorems presented in text or exercises.

CHAPTER 3

Exercises 3.6

1. (a) Venn diagrams for *No x's are y's:*

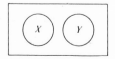   Where all not-y's are x's.

Negation: Some x's are y's. Venn diagrams for negation:

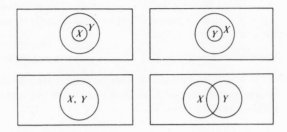

(b) Venn diagrams for *Some x's are not-y's*:

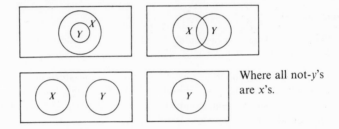

Where all not-y's are x's.

Negation: All x's are y's. Venn diagrams for negation:

(c) Venn diagrams for *All x's are y's* appear in part (b).
Negation: Some x's are not-y's. Venn diagrams for negation appear in part (b).

3. (a) p: The student can pass a test on the U.S. Constitution;
  q: He is required to take a course in it.
  $p \rightarrow (\sim q)$
(b) p: Each student is required to take a course in U.S. Constitution; q: He must pass a test on it.
  $p \wedge q$
(c) p: A student takes a course in U.S. Constitution; q: He passes a test on it; r: He has fulfilled one requirement for graduation.
  $(p \vee q) \rightarrow r$
(d) p: John graduated last June; q: He took a course in U.S. Constitution; r: He passed a test on it.
  $p \rightarrow (q \wedge r)$
(e) p: One must take a course in U.S. Constitution; q: One must pass a test on it; r: One must complete a major; s: One is to graduate.
  $s \rightarrow [(p \vee q) \wedge r]$

(f) p: He passes the test on U.S. Constitution; q: He takes a course in it; r: He has completed a major; s: He will graduate.

$[\sim(p \vee q) \wedge r] \rightarrow (\sim s)$

Exercises 3.7

p	q	1. $p \vee q$	3. $p \leftrightarrow q$
T	T	T	T
T	F	T	F
F	T	T	F
F	F	F	T

Exercises 3.8

p	q	1. $(\sim p) \wedge q$	3. $(\sim p) \wedge (\sim q)$	5. $p \vee (\sim q)$
T	T	F	F	T
T	F	F	F	T
F	T	T	F	F
F	F	F	T	T

p	q	7. $(\sim p) \rightarrow q$	9. $(\sim p) \rightarrow (\sim q)$	11. $p \leftrightarrow (\sim q)$
T	T	T	T	F
T	F	T	T	T
F	T	T	F	T
F	F	F	T	F

Exercises 3.9

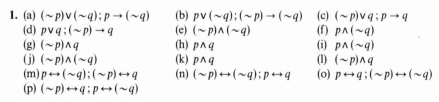

1. (a) $(\sim p) \vee (\sim q); p \rightarrow (\sim q)$ (b) $p \vee (\sim q); (\sim p) \rightarrow (\sim q)$ (c) $(\sim p) \vee q; p \rightarrow q$
   (d) $p \vee q; (\sim p) \rightarrow q$ (e) $(\sim p) \wedge (\sim q)$ (f) $p \wedge (\sim q)$
   (g) $(\sim p) \wedge q$ (h) $p \wedge q$ (i) $p \wedge (\sim q)$
   (j) $(\sim p) \wedge (\sim q)$ (k) $p \wedge q$ (l) $(\sim p) \wedge q$
   (m) $p \leftrightarrow (\sim q); (\sim p) \leftrightarrow q$ (n) $(\sim p) \leftrightarrow (\sim q); p \leftrightarrow q$ (o) $p \leftrightarrow q; (\sim p) \leftrightarrow (\sim q)$
   (p) $(\sim p) \leftrightarrow q; p \leftrightarrow (\sim q)$

3. (a) $p \rightarrow q$ (b) $p \wedge q$ (c) $p \vee (\sim q)$
   (d) $p \leftrightarrow q$ (e) $q \rightarrow p$ (f) $(\sim p) \rightarrow (\sim q)$
   (g) $(\sim q) \rightarrow (\sim p)$ (h) $[p \vee (\sim q)] \leftrightarrow [(\sim p) \rightarrow (\sim q)]$

5. (a) Statement in part (e).   (b) Statement in part (f).   (c) Statement in part (g).

7. (a) All the boys at the Institute have other homes.
   (b) Some of the boys are considered problem children.
   (c) Some of the boys are not taking advantage of the educational program.
   (d) All the boys attending classes live at the Institute.

(e) Pascal did not meet Desargues in Paris, or he did not become interested in Desargues' work in projective geometry.

(f) Some of Desargues' colleagues did not call him crazy, or some did not forget about projective geometry.

(g) A figure is a parabola iff it is not the locus of points equidistant from a fixed point and a fixed line. Or: A figure is not a parabola iff it is the locus of points equidistant from a fixed point and a fixed line.

Exercises 3.10

1.

		(a) $(p \wedge q) \leftrightarrow (q \wedge p)$			(b) $(p \vee q) \leftrightarrow (q \vee p)$			(c) $(p \rightarrow q) \leftrightarrow (q \rightarrow p)$		
p	q									
T	T	T	T	T	T	T	T	T	T	T
T	F	F	T	F	T	T	T	F	F	T
F	T	F	T	F	T	T	T	T	F	F
F	F	F	T	F	F	T	F	T	T	T

		(d) $(p \leftrightarrow q) \leftrightarrow (q \leftrightarrow p)$		
p	q			
T	T	T	T	T
T	F	F	T	F
F	T	F	T	F
F	F	T	T	T

			(e) $[(p \wedge q) \wedge r] \leftrightarrow [p \wedge (q \wedge r)]$						(f) $[(p \vee q) \vee r] \leftrightarrow [p \vee (q \vee r)]$					
p	q	r												
T	T	T	T	T	T	T	T	T	T	T	T	T	T	T
T	T	F	T	F	F	T	T	F	F	T	T	F	T	T	T
T	F	T	F	F	T	T	T	F	F	T	T	T	T	T	T
T	F	F	F	F	F	T	T	F	F	T	T	F	T	T	F
F	T	T	F	F	T	T	F	F	T	T	T	T	T	F	T
F	T	F	F	F	F	T	F	F	F	T	T	F	T	F	T
F	F	T	F	F	T	T	F	F	F	F	T	T	T	F	T
F	F	F	F	F	F	T	F	F	F	F	T	F	T	F	F

			(g) $[(p \rightarrow q) \rightarrow r] \leftrightarrow [p \rightarrow (q \rightarrow r)]$						(h) $[p \wedge (q \vee r)] \leftrightarrow [(p \wedge q) \vee (p \wedge r)]$					
p	q	r												
T	T	T	T	T	T	T	T	T	T	T	T	T	T	T
T	T	F	T	F	F	T	T	F	F	T	T	T	T	T	F
T	F	T	F	T	T	T	T	T	T	T	T	T	F	T	T
T	F	F	F	T	F	T	T	T	T	T	F	F	T	F	F
F	T	T	T	T	T	T	F	T	T	F	F	T	T	F	F
F	T	F	T	F	F	F	F	T	F	F	F	T	T	F	F
F	F	T	T	T	T	T	F	T	T	F	F	T	T	F	F
F	F	F	T	F	F	F	F	T	T	F	F	F	T	F	F

(i)

p	q	r	$[p \lor (q \land r)] \leftrightarrow [(p \lor q) \land (p \lor r)]$

p	q	r	p	$\lor$	$(q$	$\land$	$r)$	$\leftrightarrow$	$(p$	$\lor$	$q)$	$\land$	$(p$	$\lor$	$r)$
T	T	T	T	T	T	T	T	T	T	T		T	T	T	T
T	T	F	T	T	T	F	T	T	T	T		T	T	T	T
T	F	T	T	T	T	F	T	T	T	T		T	T	T	T
T	F	F	T	T	T	F	T	T	T	T		T	T	T	T
F	T	T	F	T	T	T	T	T	T	T		T	T	T	T
F	T	F	F	F	F	F	T	T	T	F		F	T	F	F
F	F	T	F	F	F	F	T	F	F	F		F	T	T	T
F	F	F	F	F	F	F	T	F	F	F		F	F	F	F

(j)			(k)		

p	t	$(p \land t) \leftrightarrow p$	p	f	$(p \lor f) \leftrightarrow p$
T	T	T T T	T	F	T T T
F	T	F T F	F	F	F T F

Note: All properties are true except those in parts (c) and (g).

3. (a) proposition   (b) conjunction
     true   disjunction
     false   implication
     is   biconditional
     are   negation
                               self-contradiction
                               simple proposition
                               compound proposition

(c) A conjunction is true iff both simple propositions involved are true.
     A disjunction is false iff both simple propositions involved are false.
     An implication is false iff a true proposition is followed by a false one.
     A biconditional is true iff both propositions are true or both are false.

(d) The commutative property of conjunction for propositions.
     The commutative property of disjunction for propositions.
     The commutative property of a biconditional for propositions.
     The associative property of conjunction for propositions.
     The associative property of disjunction for propositions.
     The associative property of biconditionals for propositions.
     The distributive property of conjunction over disjunction for propositions.
     The distributive property of disjunction over conjunction for propositions.
     There is an identity element t for conjunction.
     There is an identity element f for disjunction.
     (And any other statements that have been proved by using a truth table.)

Exercises 3.11

1. (a) f (b) t

3.

p	q	(a) (p→q)→[(p∨q)↔q]					(b) (p∧q)→p		
T	T	T	T	T	T	T	T	T	T
T	F	F	T	T	T	T	F	T	T
F	T	T	T	T	T	T	T	T	F
F	F	T	T	F	T	F	F	T	F

(c) p→(p∨q)			(d) [p∨(p∧q)]↔p				
T	T	T	T	T	T	T	T
T	T	T	T	T	F	T	T
F	T	T	F	F	F	T	F
F	T	F	F	F	F	T	F

(e)   (f)

p	q	r	[p∧(q∧r)]↔[(p∧q)∧(p∧r)]							[p∨(q∨r)]↔[(p∨q)∨(p∨r)]						
T	T	T	T	T	T	T	T	T	T	T	T	T	T	T	T	T
T	T	F	T	F	F	T	T	F	F	T	T	T	T	T	T	T
T	F	T	T	F	F	T	F	F	T	T	T	T	T	T	T	T
T	F	F	T	F	F	T	F	F	F	T	T	F	T	T	T	T
F	T	T	F	F	T	T	F	F	F	F	T	T	T	T	T	T
F	T	F	F	F	F	T	F	F	F	F	T	T	T	T	T	F
F	F	T	F	F	F	T	F	F	F	F	T	T	T	F	T	T
F	F	F	F	F	F	T	F	F	F	F	F	F	T	F	F	F

Exercises 3.13

1. (a) $(p∧q)∨[p∧(∼q)]∨[(∼p)∧q]∨[(∼p)∧(∼q)]$

(b) $(p∧q)∨[(∼p)∧q]$

(c) $(p∧q)∨[p∧(∼q)]$

(d) $(p∧q)∨[(∼p)∧(∼q)]$

(e) $[p∧(∼q)]∨[(∼p)∧q]∨[(∼p)∧(∼q)]$

(f) $[(∼p)∧q]$

(g) $[p∧(∼q)]∨[(∼p)∧(∼q)]$

(h) $[(∼p)∧(∼q)]$

3. (a) $(p∧q∧r)∨[p∧(∼q)∧(∼r)]∨[(∼p)∧q∧r]∨[(∼p)∧(∼q)∧r]$

(b) $(p∧q∧r)∨[(∼p)∧q∧r]∨[(∼p)∧(∼q)∧r]$

(c) $(p∧q∧r)∨[p∧q∧(∼r)]∨[(∼p)∧q∧r]$

(d) $[p∧(∼q)∧r]∨[p∧(∼q)∧(∼r)]∨[(∼p)∧(∼q)∧(∼r)]$

(e) $[p∧(∼q)∧r]∨[(∼p)∧(∼q)∧(∼r)]$

(f) $[p∧q∧(∼r)]∨[p∧(∼q)∧r]∨[(∼p)∧(∼q)∧r]∨[(∼p)∧(∼q)∧(∼r)]$

5. (a) Network will be completed under all possible combinations of P and Q open and/or closed.

(b) Completed when P and Q are both closed, when P is open and Q is closed, or when P and Q are both open.

(c) Network *not* completed only when P is closed, Q is open and R is open, or when P is open, Q is closed and R is open, or when P, Q and R are all open.

(d) Network *not* completed only when P, Q and R are all open.

 7.

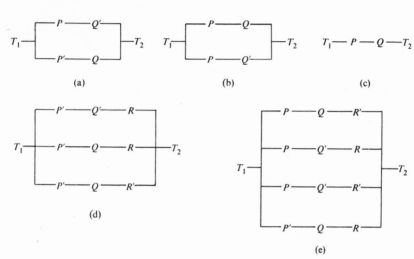

CHAPTER 4

Exercises 4.2

1. (a) 16   (b) 32   (c) 64   (d) 1   (e) 81   (f) 64
   (g) 1   (h) 125   (i) 625   (j) 3125   (k) 15625   (l) 1
   (m) 64   (n) 512   (o) 1   (p) 144   (q) 1728   (r) 20736
   (s) 248832   (t) 1

3. 7; 90; 700

5. (a) It represents 2 tens; hence we "carry" it to the tens column.
   (b) $5 \times 24 = 5 \cdot (20 + 4) = 5 \cdot 20 + 5 \cdot 4 = 100 + 20 = 120$

Exercises 4.3

1. (a) 21   (b) 39   (c) 82   (d) 295   (e) 388   (f) 624
   (g) 696   (h) 2234   (i) 14433

3. (a) 303   (b) 1130   (c) 2344   (d) 10224   (e) 11000   (f) 13000
   (g) 40334   (h) 124401   (i) 130002   (j) 223044   (k) 344314   (l) 1334104

Exercises 4.4

1. (a) 19   (b) 42   (c) 110   (d) 122   (e) 568   (f) 1559
   (g) 12666   (h) 52976   (i) 217141   (j) 602843

Exercises 4.5

1. 1, 10, 11, 100, 101, 110, 111, 1000, 1001, 1010, 1011, 1100, 1101, 1110, 1111

3. (a) 110   (b) 1100   (c) 100000   (d) 110001   (e) 1001011   (f) 1011111
    (g) 10011101   (h) 11100001   (i) 101000101 (j) 110100000

Exercises 4.7

1.

+	0_5	1_5	2_5	3_5	4_5
0_5	0_5	1_5	2_5	3_5	4_5
1_5	1_5	2_5	3_5	4_5	10_5
2_5	2_5	3_5	4_5	10_5	11_5
3_5	3_5	4_5	10_5	11_5	12_5
4_5	4_5	10_5	11_5	12_5	13_5

×	0_5	1_5	2_5	3_5	4_5
0_5	0_5	0_5	0_5	0_5	0_5
1_5	0_5	1_5	2_5	3_5	4_5
2_5	0_5	2_5	4_5	11_5	13_5
3_5	0_5	3_5	11_5	14_5	22_5
4_5	0_5	4_5	13_5	22_5	31_5

3. (a) 4424   (b) 1113   (c) 14034   (d) 2203   (e) 589e   (f) e037
    (g) 1790   (h) 2711   (i) 19d4e   (j) 1000   (k) 11111   (l) 10010
    (m) 100000

5. (a) 13   (b) 35   (c) 55   (d) 131   (e) 617   (f) 32456
    (g) 111,110   (h) 101,100,011   (i) 111,100,000,010
    (j) 110,101,001,011,101   (k) 11,100,010,111,110,010
    (l) 1,010,011,100,101,110,111

Exercises 4.8

1. 01000 = 8   **3.** 01001 = 9   **5.** 01110 = 14   **7.** 01101 = 13
9. 01101 = 13   **11.** 11110 = ⁻1   **13.** 10011 = ⁻12   **15.** 00100 = 4
17. 00010 = 2   **19.** 00100 = 4   **21.** 01001 = 9   **23.** 11000 = ⁻7
25. 01011 = 11   **27.** 01001 = 9   **29.** 00011 = 3

CHAPTER 5

Exercises 5.5

1. Yes; No; No; No.

3. (a) $27 = 27$; commutative property of addition.
    (b) $18 + 6 \overset{?}{=} 13 + 11, 24 = 24$; associative property of addition.
    (c) $18 + 6 \overset{?}{=} 6 + 18, 24 = 24$; commutative property of addition.
    (d) $18 + 0 \overset{?}{=} 18, 18 = 18$; identity element for addition.

Exercises 5.7

1. (a) $48 = 48$; commutative property of multiplication.
    (b) $24 \cdot 5 \overset{?}{=} 4 \cdot 30, 120 = 120$; associative property of multiplication.
    (c) $24 \cdot 1 \overset{?}{=} 24, 24 = 24$; identity element for multiplication.
    (d) $19 \cdot 4 \overset{?}{=} 4 \cdot 19, 76 = 76$; commutative property of multiplication.

3. If $n(A) = p$ and $n(B) = 1$, $n(A) \cdot n(B) = p \cdot 1$; $p \cdot 1 = 1 \cdot p$ by the commutative property of multiplication; $1 \cdot p = p$ by the definition of multiplication of whole numbers.

5.

(a)

×	a	b	c	d
1	(1, a)	(1, b)	(1, c)	(1, d)
2	(2, a)	(2, b)	(2, c)	(2, d)
3	(3, a)	(3, b)	(3, c)	(3, d)

(c)

×	r	s
a	(a, r)	(a, s)
b	(b, r)	(b, s)
c	(c, r)	(c, s)
d	(d, r)	(d, s)

(b)

×	r	s
(1, a)	((1, a), r)	((1, a), s)
(1, b)	((1, b), r)	((1, b), s)
(1, c)	((1, c), r)	((1, c), s)
(1, d)	((1, d), r)	((1, d), s)
(2, a)	((2, a), r)	((2, a), s)
(2, b)	((2, b), r)	((2, b), s)
(2, c)	((2, c), r)	((2, c), s)
(2, d)	((2, d), r)	((2, d), s)
(3, a)	((3, a), r)	((3, a), s)
(3, b)	((3, b), r)	((3, b), s)
(3, c)	((3, c), r)	((3, c), s)
(3, d)	((3, d), r)	((3, d), s)

(d)

×	(a, r)	(a, s)	(b, r)	(b, s)	(c, r)	(c, s)	(d, r)	(d, s)
1	(1, (a, r))	(1, (a, s))	(1, (b, r))	(1, (b, s))	(1, (c, r))	(1, (c, s))	(1, (d, r))	(1, (d, s))
2	(2, (a, r))	(2, (a, s))	(2, (b, r))	(2, (b, s))	(2, (c, r))	(2, (c, s))	(2, (d, r))	(2, (d, s))
3	(3, (a, r))	(3, (a, s))	(3, (b, r))	(3, (b, s))	(3, (c, r))	(3, (c, s))	(3, (d, r))	(3, (d, s))

7. Let $A = \{a, b, c, d, e\}$, $B = \{x, y\}$, $C = \{1, 2, 3\}$.

$A \times B$:

×	x	y
a	(a, x)	(a, y)
b	(b, x)	(b, y)
c	(c, x)	(c, y)
d	(d, x)	(d, y)
e	(e, x)	(e, y)

$(A \times B) \times C$:

×	1	2	3
(a, x)	((a, x), 1)	((a, x), 2)	((a, x), 3)
(a, y)	((a, y), 1)	((a, y), 2)	((a, y), 3)
(b, x)	((b, x), 1)	((b, x), 2)	((b, x), 3)
(b, y)	((b, y), 1)	((b, y), 2)	((b, y), 3)
(c, x)	((c, x), 1)	((c, x), 2)	((c, x), 3)
(c, y)	((c, y), 1)	((c, y), 2)	((c, y), 3)
(d, x)	((d, x), 1)	((d, x), 2)	((d, x), 3)
(d, y)	((d, y), 1)	((d, y), 2)	((d, y), 3)
(e, x)	((e, x), 1)	((e, x), 2)	((e, x), 3)
(e, y)	((e, y), 1)	((e, y), 2)	((e, y), 3)

$B \times C$:

×	1	2	3
x	(x, 1)	(x, 2)	(x, 3)
y	(y, 1)	(y, 2)	(y, 3)

$A \times (B \times C)$:

×	(x, 1)	(x, 2)	(x, 3)	(y, 1)	(y, 2)	(y, 3)
a	(a, (x, 1))	(a, (x, 2))	(a, (x, 3))	(a, (y, 1))	(a, (y, 2))	(a, (y, 3))
b	(b, (x, 1))	(b, (x, 2))	(b, (x, 3))	(b, (y, 1))	(b, (y, 2))	(b, (y, 3))
c	(c, (x, 1))	(c, (x, 2))	(c, (x, 3))	(c, (y, 1))	(c, (y, 2))	(c, (y, 3))
d	(d, (x, 1))	(d, (x, 2))	(d, (x, 3))	(d, (y, 1))	(d, (y, 2))	(d, (y, 3))
e	(e, (x, 1))	(e, (x, 2))	(e, (x, 3))	(e, (y, 1))	(e, (y, 2))	(e, (y, 3))

Note: $n(A) = 5, n(B) = 2, n(C) = 3$; $n[(A \times B) \times C] = [n(A) \cdot n(B)] \cdot n(C) = (5 \cdot 2) \cdot 3$ and $n[A \times (B \times C)] = n(A) \cdot [n(B) \cdot n(C)] = 5 \cdot (2 \cdot 3)$. From the tables: $n[(A \times B) \times C] = 30$ and $n[A \times (B \times C)] = 30$. Therefore, $(5 \cdot 2) \cdot 3 = 5 \cdot (2 \cdot 3)$.

Exercises 5.8

1. (a) $6 \cdot 8 \overset{?}{=} 30 + 18, 48 = 48$
   (b) $64 + 112 \overset{?}{=} 16 \cdot 11, 176 = 176$
   (c) $22 \cdot 7 \overset{?}{=} 112 + 42, 154 = 154$
   (d) $6 \cdot 21 \overset{?}{=} 30 + 18 + 54 + 24, 126 = 126$

3. $A \cup (B \times C) = \{a, b\} \cup \{(c, f), (c, g), (c, h), (d, f), (d, g), (d, h), (e, f), (e, g), (e, h)\}$
   $= \{a, b, (c, f), (c, g), (c, h), (d, f), (d, g), (d, h), (e, f), (e, g), (e, h)\}$
   $(A \cup B) \times (A \cup C) = \{a, b, c, d, e\} \times \{a, b, f, g, h\}$
   $= \{(a, a), (a, b), (a, f), (a, g), (a, h), (b, a), (b, b), (b, f), (b, g), (b, h),$
   $(c, a), (c, b), (c, f), (c, g), (c, h), (d, a), (d, b), (d, f), (d, g), (d, h),$
   $(e, a), (e, b), (e, f), (e, g), (e, h)\}$
   Thus $n[A \cup (B \times C)] \neq n[(A \cup B) \times (A \cup C)]$. Therefore, $n(A) + [n(B) \cdot n(C)] \neq [n(A) \cdot n(B)] + [n(A) \cdot n(C)]$.

Exercises 5.9

1. For any two whole numbers p and q, the difference $p - q$ is not always a whole number; for example, $3 - 5$ is not a whole number. No.

3. No; for example, $5 - 3 \neq 3 - 5$. No; for example, $8 \div 4 \neq 4 \div 8$.

5. *Example:* $(18 \div 6) \div 3 \overset{?}{=} 18 \div (6 \div 3), 3 \div 3 \overset{?}{=} 18 \div 2, 1 \neq 9$. No.

Exercises 5.10

1.

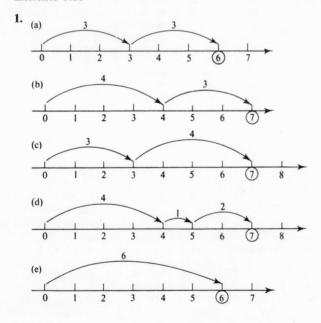

(f)
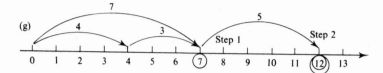

(g)
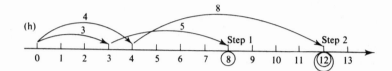

(h)
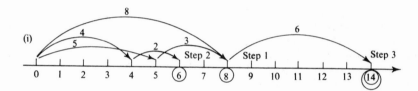

(i)

(j)
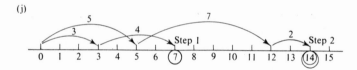

(k)
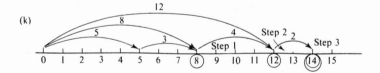

(l)
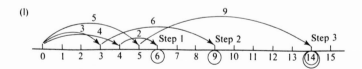

Exercises 5.11

1. (a) >   (b) <   (c) =   (d) <   (e) =   (f) <
   (g) >   (h) ≦   (i) ≦   (j) <   (k) =, >   (l) <, >

Exercises 5.12

1. (a) Reflexive, symmetric, transitive. (b) Reflexive, symmetric.
(c) Transitive. (d) Reflexive, transitive.
(e) Transitive. (f) Reflexive, transitive.
(g) Symmetric. (h) Transitive.
(i) Transitive. (j) None.
(k) Reflexive, symmetric, transitive. (l) Transitive.

3. (a) Commutative property of multiplication.
(b) Distributive property of multiplication over addition.
(c) Unique property of zero.
(d) Definition of order for whole numbers (or ordering property).
(e) Unique property of zero (or definition of multiplication).
(f) Ordering property.
(g) Associative property of multiplication.
(h) Identity element for multiplication.
(i) Identity element for addition.
(j) Symmetric property of equality.
(k) Closure under addition.
(l) Closure under multiplication.
(m) Distributive property of multiplication over addition.
(n) Multiplication property of equality.
(o) Commutative property of addition.
(p) Addition property of equality.
(q) Transitive property of equality.
(r) Reflexive property of equality.
(s) Associative property of addition.

CHAPTER 6

Exercises 6.3

1. (a) $^-15$ (b) $^-1$ (c) 0 (d) 1 (e) 15 (f) 9 (g) $^-9$ (h) 9

3. (a) 8 (b) $^-4$ (c) 4 (d) $^-8$ (e) $^-1$ (f) 6 (g) $^-7$ (h) $^-7$

Exercises 6.4

1. (a) 6 (b) 6 (c) 25 (d) 7 (e) 13 (f) 13 (g) 1 (h) 1
(i) 0 (j) 25 (k) 25 (l) 25

Exercises 6.5

1.
(a)

(b)

(c)

(d)

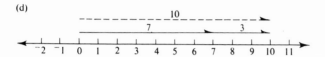

(e)

(f)

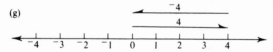

(g)

The result is represented by the zero vector $\overrightarrow{P_0P_0}$.

(h)

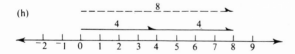

(i)

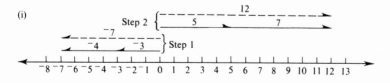

(j)

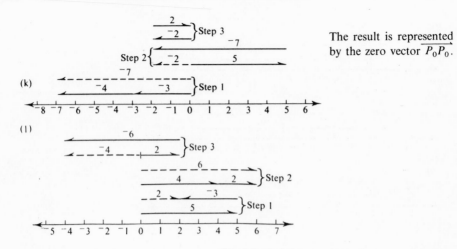

The result is represented by the zero vector $\overrightarrow{P_0 P_0}$.

(k)

(l)

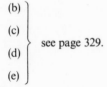

3. Examples: $7 + {}^-3 = 4, {}^-6 + {}^-7 = {}^-13, {}^-7 + 6 = {}^-1$, etc. Since subtraction of integers is expressed in terms of addition of integers, the closure property of integers under subtraction is a consequence of the closure property of integers under addition.

Exercises 6.6

1. (a)

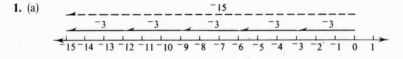

(b)
(c)
(d) } see page 329.
(e)

3. (a) 1   (b) ${}^-3$ (c) 3   (d) 3   (e) ${}^-2$ (f) 0   (g) ${}^-2$ (h) ${}^-2$
(i) 75   (j) 1   (k) ${}^-12$

Exercises 6.7

1. (a) odd, $2 \cdot 0 + 1$ (b) even, $2 \cdot 1$ (c) even, $2 \cdot {}^-4$ (d) even, $2 \cdot 26$
(e) odd, $2 \cdot {}^-26 + 1$ (f) even, $2 \cdot 0$ (g) odd, $2 \cdot 8 + 1$ (h) odd, $2 \cdot {}^-1 + 1$
(i) even, $2 \cdot {}^-26$ (j) odd, $2 \cdot 50 + 1$ (k) even, $2 \cdot 75$ (l) even, $2 \cdot {}^-36$

3. $3 + 5 = 8$; since 8 is even, the set of odd integers is not closed under addition.

$$(2m + 1) \cdot (2n + 1) = (2m + 1) \cdot 2n + (2m + 1) \cdot 1$$

$$= 4mn + 2n + 2m + 1$$

$$= 2(2mn + m + n) + 1$$

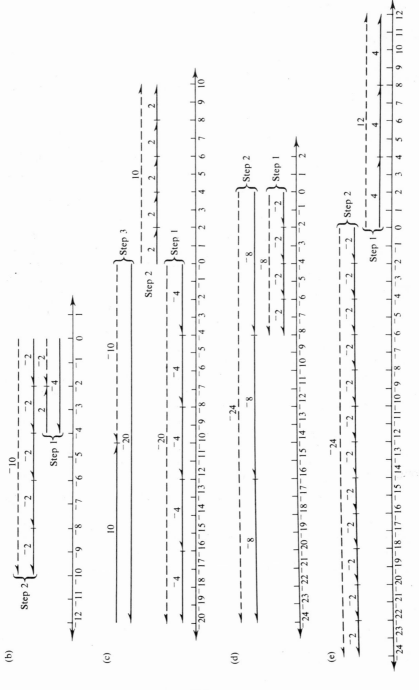

329

Since $mn + m + n$ is an integer, $2(2mn + m + n) + 1$ is an odd integer; thus the set of odd numbers is closed under multiplication.

Exercises 6.8

1. (a) neither   (b) prime   (c) prime   (d) composite
(e) prime   (f) composite   (g) composite   (h) composite
(i) prime   (j) composite   (k) prime   (l) composite
(m)composite   (n) composite   (o) prime   (p) prime
(q) composite   (r) prime

3. (a) 2^4 (b) $2 \cdot 3^2$ (c) prime   (d) $2^2 \cdot 7$ (e) $2 \cdot 3 \cdot 5$ (f) 2^5
(g) $3 \cdot 13$ (h) cannot be written as a product of prime numbers   (i) $3 \cdot 17$
(j) $2^2 \cdot 13$ (k) $3 \cdot 19$ (l) prime   (m)prime   (n) prime   (o) $2 \cdot 5 \cdot 13$
(p) $3^2 \cdot 5^2$ (q) 2^8 (r) $2^4 \cdot 3 \cdot 5$ (s) $2^4 \cdot 5^2$ (t) $2^3 \cdot 3 \cdot 29$ (u) $2^3 \cdot 5^3$

5.

1̸	2	3	4̸	5	6̸	7	8̸	9̸	1̸0̸	11	1̸2̸	13	1̸4̸	1̸5̸
1̸6̸	17	1̸8̸	19	2̸0̸	2̸1̸	2̸2̸	23	2̸4̸	2̸5̸	2̸6̸	2̸7̸	2̸8̸	29	3̸0̸
31	3̸2̸	3̸3̸	3̸4̸	3̸5̸	3̸6̸	37	3̸8̸	3̸9̸	4̸0̸	41	4̸2̸	43	4̸4̸	4̸5̸
4̸6̸	47	4̸8̸	4̸9̸	5̸0̸	5̸1̸	5̸2̸	53	5̸4̸	5̸5̸	5̸6̸	5̸7̸	5̸8̸	59	6̸0̸
61	6̸2̸	6̸3̸	6̸4̸	6̸5̸	6̸6̸	67	6̸8̸	6̸9̸	7̸0̸	71	7̸2̸	73	7̸4̸	7̸5̸
7̸6̸	7̸7̸	7̸8̸	79	8̸0̸	8̸1̸	8̸2̸	83	8̸4̸	8̸5̸	8̸6̸	8̸7̸	8̸8̸	89	9̸0̸
9̸1̸	92	9̸3̸	94	9̸5̸	9̸6̸	97	98	99	1̸0̸0̸					

Prime numbers: 2, 3, 5, 7, 11, 13, 17, 19, 23, 29, 31, 37, 41, 43, 47, 53, 59, 61, 67, 71, 73, 79, 83, 89, 97.

7. (1) Commutative property of addition for integers. $4 + 3 = 3 + 4 = 7$
(2) Commutative property of multiplication for integers. $4 \cdot 3 = 3 \cdot 4 = 12$
(3) Associative property of addition for integers. $(2 + 3) + 4 = 2 + (3 + 4) = 9$
(4) Associative property of multiplication for integers. $(2 \cdot 3) \cdot 4 = 2 \cdot (3 \cdot 4) = 24$
(5) Identity element for addition of integers. $5 + 0 = 5$
(6) Identity element for multiplication of integers. $5 \cdot 1 = 5$
(7) Closure property of integers under addition. $5 + 6$ is an integer.
(8) Closure property of integers under multiplication. $5 \cdot 6$ is an integer.
(9) Closure property of integers under subtraction. $5 - 6$ is an integer.
(10) Additive inverse property of integers. $5 + {}^-5 = 0$
(11) Distributive property of multiplication over addition for integers. $5 \cdot (6 + 7) = 5 \cdot 6 + 5 \cdot 7 = 65$
(12) Unique property of zero under multiplication: $p \cdot 0 = 0$ (p an integer). $5 \cdot 0 = 0$
(13) Unique property of zero under multiplication: If $p \cdot q = 0$, then $p = 0$ or $q = 0$ or both p and q equal 0 (p and q integers). If $5 \cdot x = 0$, then $x = 0$
(14) Ordering property for integers: $p \gtreqless q$ (p and q integers). $6 > 5, 6 = 6, 6 < 7$

CHAPTER 7

Exercises 7.2

1. (a) $\frac{3}{2}$ (b) $\frac{5}{8}$ (c) $\frac{8}{3}$ (d) $\frac{1}{3}$ (e) $\frac{1}{16}$ (f) $\frac{1}{10}$
(g) $\frac{1}{100}$ (h) 1   (i) $-\frac{1}{3}$ (j) $-\frac{3}{5}$ (k) ${}^-1$ (l) $-\frac{5}{3}$

3. (a) Definition of multiplication for rational numbers. Commutative property of multiplication for integers. Factoring of integers performed. Associative property of multiplication for integers. Definition of multiplication for rational numbers and multiplication of integers performed. Identity element for multiplication of rational numbers.

    (b) Theorem on division of rational numbers. Definition of multiplication for rational numbers. Commutative property of multiplication for integers. Factoring of integers performed. Associative property of multiplication for integers. Definition of multiplication for rational numbers. Identity element for multiplication and multiplication of integers performed.

5. $^-1$ and 1. No.

7. Yes. The theorem on the quotient of two rational numbers confirms this: $\dfrac{a}{b} \div \dfrac{c}{d} = \dfrac{ad}{bc}$.

Exercises 7.4

1. (a) $\dfrac{4}{6} = \dfrac{2 \cdot 2}{2 \cdot 3} = \dfrac{2}{2} \cdot \dfrac{2}{3} = \dfrac{2}{3}$

    (b) $\dfrac{3}{3} = 1$

    (c) $\dfrac{7}{91} = \dfrac{7 \cdot 1}{7 \cdot 13} = \dfrac{7}{7} \cdot \dfrac{1}{13} = \dfrac{1}{13}$

    (d) $\dfrac{12}{56} = \dfrac{2 \cdot 2 \cdot 3}{2 \cdot 2 \cdot 2 \cdot 7} = \dfrac{2 \cdot 2}{2 \cdot 2} \cdot \dfrac{3}{2 \cdot 7} = \dfrac{4}{4} \cdot \dfrac{3}{14} = \dfrac{3}{14}$

    (e) $\dfrac{^-16}{64} = \dfrac{^-2 \cdot 2 \cdot 2 \cdot 2}{2 \cdot 2 \cdot 2 \cdot 2 \cdot 2 \cdot 2} = \dfrac{^-2 \cdot 2 \cdot 2 \cdot 2}{2 \cdot 2 \cdot 2 \cdot 2} \cdot \dfrac{1}{2 \cdot 2} = \dfrac{16}{16} \cdot \dfrac{^-1}{4} = -\dfrac{1}{4}$

    (f) $\dfrac{15}{^-225} = \dfrac{3 \cdot 5}{^-3 \cdot 5 \cdot 3 \cdot 5} = \dfrac{3 \cdot 5}{3 \cdot 5} \cdot \dfrac{1}{^-3 \cdot 5} = \dfrac{15}{15} \cdot \dfrac{1}{^-15} = \dfrac{1}{^-15}$

    (g) $\dfrac{^-72}{144} = \dfrac{^-2 \cdot 2 \cdot 2 \cdot 3 \cdot 3}{2 \cdot 2 \cdot 2 \cdot 3 \cdot 3 \cdot 2} = \dfrac{2 \cdot 2 \cdot 2 \cdot 3 \cdot 3}{2 \cdot 2 \cdot 2 \cdot 3 \cdot 3} \cdot \dfrac{^-1}{2} = \dfrac{72}{72} \cdot \dfrac{^-1}{2} = \dfrac{^-1}{2}$

    (h) $\dfrac{^-63}{^-91} = \dfrac{^-7 \cdot 3 \cdot 3}{^-7 \cdot 13} = \dfrac{^-7}{^-7} \cdot \dfrac{3 \cdot 3}{13} = \dfrac{9}{13}$

    (i) $\dfrac{112}{48} = \dfrac{2 \cdot 2 \cdot 2 \cdot 2 \cdot 7}{2 \cdot 2 \cdot 2 \cdot 2 \cdot 3} = \dfrac{2 \cdot 2 \cdot 2 \cdot 2}{2 \cdot 2 \cdot 2 \cdot 2} \cdot \dfrac{7}{3} = \dfrac{16}{16} \cdot \dfrac{7}{3} = \dfrac{7}{3}$

3. PROOF

    (a) $\dfrac{a}{b} - \dfrac{c}{d} = \dfrac{a}{b} \cdot \dfrac{d}{d} - \dfrac{b}{b} \cdot \dfrac{c}{d}$ (a) Identity element for multiplication of rational numbers.

    (b) $\dfrac{a}{b} \cdot \dfrac{d}{d} - \dfrac{b}{b} \cdot \dfrac{c}{d} = \dfrac{ad}{bd} - \dfrac{bc}{bd}$ (b) Definition of multiplication for rational numbers.

    (c) $\dfrac{ad}{bd} - \dfrac{bc}{bd} = \dfrac{ad - bc}{bd}$ (c) Definition of subtraction for integers with like denominators.

(d) $\dfrac{a}{b} - \dfrac{c}{d} = \dfrac{ad - bc}{bd}$ (d) Transitive property of equality for rational numbers.

5. (a) Identity element for multiplication of rational numbers. Definition of multiplication for rational numbers. Multiplication of integers performed. Definition of addition for rational numbers. Factoring of integers performed. Definition of multiplication of rational numbers. Identity element for multiplication of rational numbers.

(b) Theorem on subtraction of rational numbers (Exercise 3). Multiplication of integers performed. Subtraction of integers performed.

7. (a) *Prove*: $\dfrac{a}{b} + \dfrac{c}{d} = \dfrac{c}{d} + \dfrac{a}{b}$

PROOF $\dfrac{a}{b} + \dfrac{c}{d} = \dfrac{(ad + bc)}{bd}$ Theorem: Addition of rational numbers with unlike denominators.

$\dfrac{(ad + bc)}{bd} = \dfrac{(bc + ad)}{bd}$ Commutative property of addition for integers.

$\dfrac{(bc + ad)}{bd} = \dfrac{bc}{bd} + \dfrac{ad}{bd}$ Definition of addition for rational numbers with like denominators.

$\dfrac{bc}{bd} + \dfrac{ad}{bd} = \dfrac{b}{b}\cdot\dfrac{c}{d} + \dfrac{a}{b}\cdot\dfrac{d}{d}$ Definition of multiplication for rational numbers.

$\dfrac{b}{b}\cdot\dfrac{c}{d} + \dfrac{a}{b}\cdot\dfrac{d}{d} = \dfrac{c}{d} + \dfrac{a}{b}$ Identity element for multiplication of rational numbers.

$\dfrac{a}{b} + \dfrac{c}{d} = \dfrac{c}{d} + \dfrac{a}{b}$ Transitive property of equality.

(b) *Prove*: $\dfrac{a}{b} + \left(\dfrac{c}{d} + \dfrac{e}{f}\right) = \left(\dfrac{a}{b} + \dfrac{c}{d}\right) + \dfrac{e}{f}$

PROOF $\dfrac{a}{b} + \left(\dfrac{c}{d} + \dfrac{e}{f}\right) = \dfrac{a}{b} + \dfrac{cf + de}{df}$ Theorem: Addition of rational numbers with unlike denominators.

$\dfrac{a}{b} + \dfrac{cf + de}{df} = \dfrac{adf + b(cf + de)}{bdf}$ Theorem: Addition of rational numbers with unlike denominators.

$\dfrac{adf + b(cf + de)}{bdf} = \dfrac{adf + bcf + bde}{bdf}$ Distributive property of multiplication over addition for integers.

$\dfrac{adf + bcf + bde}{bdf} = \dfrac{(ad + bc)f + bde}{bdf}$ Distributive property of multiplication over addition for integers.

$\dfrac{(ad + bc)f + bde}{bdf} = \dfrac{ad + bc}{bd} + \dfrac{e}{f}$ Definition of addition for rational numbers.

$\dfrac{ad + bc}{bd} + \dfrac{e}{f} = \left(\dfrac{a}{b} + \dfrac{c}{d}\right) + \dfrac{e}{f}$ Definition of addition for rational numbers.

$\dfrac{a}{b} + \left(\dfrac{c}{d} + \dfrac{e}{f}\right) = \left(\dfrac{a}{b} + \dfrac{c}{d}\right) + \dfrac{e}{f}$ Transitive property of equality.

Exercises 7.5

1. (a) $16 = 2^4, 52 = 2^2 \cdot 13, 60 = 2^2 \cdot 3 \cdot 5$　G.C.F. $= 2^2 = 4$
   (b) $14 = 2 \cdot 7, 35 = 5 \cdot 7, 98 = 2 \cdot 7^2$　G.C.F. $= 7$
   (c) $18 = 2 \cdot 3^2, {}^-36 = {}^-2^2 \cdot 3^2, 56 = 2^3 \cdot 7$　G.C.F. $= 2$
   (d) 17 is prime, 31 is prime, 47 is prime　G.C.F. $= 1$
   (e) ${}^-24 = {}^-2^3 \cdot 3, 36 = 2^2 \cdot 3^2, {}^-144 = {}^-2^4 \cdot 3^2$　G.C.F. $= 2^2 \cdot 3 = 12$
   (f) $30 = 2 \cdot 3 \cdot 5, 140 = 2^2 \cdot 5 \cdot 7, 420 = 2^2 \cdot 3 \cdot 5 \cdot 7$　G.C.F. $= 2 \cdot 5 = 10$
   (g) $165 = 3 \cdot 5 \cdot 11, 429 = 3 \cdot 11 \cdot 13, {}^-495 = {}^-3^2 \cdot 5 \cdot 11$　G.C.F. $= 3 \cdot 11 = 33$
   (h) $84 = 2^2 \cdot 3 \cdot 7, 294 = 2 \cdot 3 \cdot 7^2, 2002 = 2 \cdot 7 \cdot 11 \cdot 13$　G.C.F. $= 2 \cdot 7 = 14$

3. (a) $\dfrac{2}{3}$　　(b) 1　　(c) $\dfrac{1}{13}$　　(d) $\dfrac{{}^-3}{14}$　　(e) $-\dfrac{1}{4}$　　(f) $\dfrac{1}{15}$

   (g) $-\dfrac{1}{2}$　　(h) $\dfrac{9}{13}$　　(i) $\dfrac{7}{3}$　　(j) $\dfrac{5}{23}$　　(k) $\dfrac{7}{41}$　　(l) $\dfrac{4}{9}$

Exercises 7.6

1. (a) $6 = 2 \cdot 3, 10 = 2 \cdot 5, 3$ is prime　L.C.M. $= 2 \cdot 3 \cdot 5 = 30$
   (b) $9 = 3^2, 15 = 3 \cdot 5, 21 = 3 \cdot 7$　L.C.M. $= 3^2 \cdot 5 \cdot 7 = 315$
   (c) $4 = 2^2, 12 = 2^2 \cdot 3, 21 = 3 \cdot 7$　L.C.M. $= 2^2 \cdot 3 \cdot 7 = 84$
   (d) 5 is prime, $20 = 2^2 \cdot 5, 44 = 2^2 \cdot 11$　L.C.M. $= 2^2 \cdot 5 \cdot 11 = 220$
   (e) 3, 5, and 47 are each prime　L.C.M. $= 3 \cdot 5 \cdot 47 = 705$
   (f) 17, 31, and 47 are each prime　L.C.M. $= 17 \cdot 31 \cdot 47 = 24{,}769$
   (g) $30 = 2 \cdot 3 \cdot 5, 140 = 2^2 \cdot 5 \cdot 7, 420 = 2^2 \cdot 3 \cdot 5 \cdot 7$　L.C.M. $= 2^2 \cdot 3 \cdot 5 \cdot 7 = 420$
   (h) $165 = 3 \cdot 5 \cdot 11, 429 = 3 \cdot 11 \cdot 13, 495 = 3^2 \cdot 5 \cdot 11$　L.C.M. $= 3^2 \cdot 5 \cdot 11 \cdot 13 = 6435$

3. (a) $\dfrac{15}{18} + \dfrac{{}^-7}{18} = \dfrac{8}{18} = \dfrac{4}{9}$

   (b) $\dfrac{25}{60} + \dfrac{28}{60} = \dfrac{53}{60}$

   (c) $\dfrac{56}{160} + \dfrac{55}{160} = \dfrac{111}{160}$

   (d) $\dfrac{{}^-13}{27} + \dfrac{2}{21} = \dfrac{{}^-91}{189} + \dfrac{18}{189} = \dfrac{{}^-73}{189}$

   (e) $\dfrac{{}^-140}{315} + \dfrac{147}{315} + \dfrac{120}{315} = \dfrac{127}{315}$

   (f) $\dfrac{1}{4} + \dfrac{5}{12} + \dfrac{{}^-4}{21} = \dfrac{21}{84} + \dfrac{35}{84} - \dfrac{16}{84} = \dfrac{40}{84} = \dfrac{10}{21}$

   (g) $\dfrac{3}{5} + \dfrac{9}{20} + \dfrac{27}{44} = \dfrac{132}{220} + \dfrac{99}{220} + \dfrac{135}{220} = \dfrac{183}{110}$

   (h) $\dfrac{36}{153} + \dfrac{{}^-78}{153} + \dfrac{85}{153} = \dfrac{43}{153}$

(i) $\dfrac{50}{60} + \dfrac{35}{60} + \dfrac{22}{60} = \dfrac{107}{60}$

(j) $\dfrac{^-13}{27} + \dfrac{5}{12} + \dfrac{^-13}{24} = \dfrac{^-104}{216} + \dfrac{90}{216} + \dfrac{^-117}{216} = \dfrac{^-131}{216}$

Exercises 7.7

1. No.

3. $\left\{ \dfrac{^-64}{13}, \dfrac{^-7}{12}, \dfrac{^-22}{45}, \dfrac{^-3}{7}, \dfrac{^-1}{4}, 0, \dfrac{35}{132}, \dfrac{1}{2} = \dfrac{16}{32}, \dfrac{14}{15}, \dfrac{77}{79}, \dfrac{369}{371}, \dfrac{65}{63}, \dfrac{7}{5}, \dfrac{45}{22}, 3, \dfrac{16}{3} \right\}$

Exercises 7.8

1. (a) $0.2\bar{0}$ (b) $0.6\bar{0}$ (c) $1.2\bar{0}$ (d) $0.\bar{1}$

   (e) $0.\bar{3}$ (f) $0.\overline{142857}$ (g) $0.\overline{285714}$ (h) $0.\overline{428571}$

   (i) $0.\overline{571428}$ (j) $0.\overline{0588235294117647}$ (k) $0.0625\bar{0}$

   (l) $0.0\bar{5}$ (m) $0.04\bar{0}$ (n) $0.0\overline{285714}$

Exercises 7.9

1. (a) 4^3 (b) 3^5 (c) 2^7 (d) 2 (e) 4^4

   (f) 4^{-3} (g) 4^{-3} (h) 3^{-1} (i) 4 (j) 4^8

   (k) 3^{11} (l) 4^{-7} (m) 3 (n) 3^3 (o) 3^3

CHAPTER 8

Exercises 8.3

1. (a) $1^2 = 1 < 3 < 4 = 2^2$; thus $1 < \sqrt{3} < 2$.

   $(1.7)^2 = 2.89 < 3 < 3.24 = (1.8)^2$; thus $1.7 < \sqrt{3} < 1.8$

   $(1.73)^2 = 2.9929 < 3 < 3.0276 = (1.74)^2$; thus $1.73 < \sqrt{3} < 1.74$

   $(1.732)^2 = 2.999824 < 3 < 3.003289 = (1.733)^2$; thus $1.732 < \sqrt{3} < 1.733$

   $(1.7320)^2 = 2.99982400 < 3 < 3.00017041 = (1.7321)^2$; thus $1.7320 < \sqrt{3} < 1.7321$

  (b) $2^2 = 4 < 5 < 9 = 3^2$; thus $2 < \sqrt{5} < 3$

   $(2.2)^2 = 4.84 < 5 < 5.29 = (2.3)^2$; thus $2.2 < \sqrt{5} < 2.3$

   $(2.23)^2 = 4.9729 < 5 < 5.0176 = (2.24)^2$; thus $2.23 < \sqrt{5} < 2.24$

   $(2.236)^2 = 4.999696 < 5 < 5.004169 = (2.237)^2$; thus $2.236 < \sqrt{5} < 2.237$

   $(2.2360)^2 = 4.99969600 < 5 < 5.00014321 = (2.2361)^2$; thus $2.2360 < \sqrt{5} < 2.2361$

  (c) $2^2 = 4 < 6 < 9 = 3^2$; thus $2 < \sqrt{6} < 3$

   $(2.4)^2 = 5.76 < 6 < 6.25 = (2.5)^2$; thus $2.4 < \sqrt{6} < 2.5$

   $(2.44)^2 = 5.9536 < 6 < 6.0025 = (2.45)^2$; thus $2.44 < \sqrt{6} < 2.45$

   $(2.449)^2 = 5.997601 < 6 < 6.002500 = (2.450)^2$; thus $2.449 < \sqrt{6} < 2.450$

   $(2.4494)^2 = 5.99956036 < 6 < 6.00005025 = 2.4495$; thus $2.4494 < \sqrt{6} < 2.4495$

  (d) $2^2 = 4 < 7 < 9 = 3^2$; thus $2 < \sqrt{7} < 3$

   $(2.6)^2 = 6.76 < 7 < 7.29 = (2.7)^2$; thus $2.6 < \sqrt{7} < 2.7$

   $(2.64)^2 = 6.9696 < 7 < 7.0225 = (2.65)^2$; thus $2.64 < \sqrt{7} < 2.65$

$(2.645)^2 = 6.996025 < 7 < 7.001316 = (2.646)^2$; thus $2.645 < \sqrt{7} < 2.646$

$(2.6457)^2 = 6.99972849 < 7 < 7.00025764 = (2.6458)^2$; thus $2.6457 < \sqrt{7} < 2.6458$

3. We know that $\sqrt{9}$, in fact, is a rational number: $\sqrt{9} = 3$. If we attempt an indirect proof that it is not a rational number, the proof fails. The assumption that $\sqrt{9}$ is a rational number does not lead to a contradiction. We demonstrate this: Assume that $\sqrt{9}$ is a rational number. Then $\sqrt{9} = a/b$ where a and b are mutually prime.

$$9 = \frac{a^2}{b^2}$$

$$a^2 = 9b^2$$

Then

$$a = 3b$$

Thus a is divisible by 3. Then $a = 3n$, and $a^2 = 9n^2$. Substituting:

$$9b^2 = 9n^2$$

or

$$b^2 = n^2$$

and

$$b = n$$

This approach has produced no common factor of a and b and does not contradict the initial hypothesis.

Exercises 8.5

1. (a) irrational   (b) irrational   (c) rational   (d) irrational
(e) rational   (f) rational   (g) irrational   (h) irrational
(i) irrational   (j) rational   (k) rational   (l) irrational

3. All are integers:
(a) $\log_{10} 10 = 1$, since $10^1 = 10$ (b) $\log_{10} 1 = 0$, since $10^0 = 1$

(c) $\log_{10} 1000 = 3$, since $10^3 = 1000$ (d) $\log_{10} \frac{1}{10} = {}^-1$, since $10^{-1} = \frac{1}{10}$

(e) $\log_{10} \frac{1}{10,000} = {}^-4$, since $10^{-4} = \frac{1}{10,000}$ (f) $\log_{10} \frac{1}{100} = {}^-2$, since $10^{-2} = \frac{1}{100}$

CHAPTER 9

Exercises 9.2

1. (e) The three perpendicular bisectors always meet in a single point.
3. (d) The three medians always meet in a single point.
5. (d) They are equal in length.

7. (e) They are equal in length.
9. (e) It is a parallelogram.
11. (g) They are equal in length.
13. (e) The circle contains points A, B, and C.

Exercises 9.7

1. Extremity, finite, adjacent angles, standing on, set up on, falling on, line, center, etc.

3. (a) An accurate drawing would show that $\overline{CD}$ intersects both circles at B; thus $\triangle AFE$ would not exist.

(b) An accurate drawing would show that $\overline{PE}$ does not intersect the rectangle $ABCD$, but lies outside it. Thus $\angle BCE \cup \angle y$ is not an angle of $\triangle PCE$. Therefore, a proof that $\triangle ABP \cong \triangle ECP$ does not lead to the conclusion that $\angle ABC \cong \angle BCE$.

CHAPTER 10

Exercises 10.2

1. (a) One figure   (b) three figures   (c) seven figures   (d) fifteen figures
3. (f) (ABD); (ABE); (FAB); (FDE); (BDC); (FBE);   (g) No.
5. (a) Infinitely many.   (b) One.

Exercises 10.3

1. Suppose two distinct lines meet in more than one point. This possibility may be considered because, by Assumption 10.1, there exist at least two distinct points. Then any two of these points are contained in two lines, which contradicts Assumption 10.2. Therefore two distinct lines intersect in at most one point.

3. (c) No.

(d) The betweenness relationship for three points relates only to points that are in the same line according to Assumption 10.3. At least two points exist according to Assumption 10.1. Between these two points there exists at least one additional point according to Assumption 10.4. Let two points be the endpoints of segment $\overline{AB}$; then C_1 exists such that (AC_1B), C_2 exists such that (AC_2C_1), C_3 exists such that (AC_3C_2), etc., by Assumption 10.3. Also, by Assumption 10.3, we could continue this process indefinitely. Thus $\overline{AB}$ is an infinite set of points.

5. (a) P, A_1, and A_2 are in the same plane by Assumption 10.10.

(b) Yes. By Assumption 10.11.

(c) Yes.

(d) No. By Theorem 10.3, l is an infinite set of points A_n where n is any real number. Thus there is no limit to the number of lines $\overleftrightarrow{PA_n}$ that can be drawn.

(e) The plane contains an infinite number of lines as shown by parts a, b, c, and d. (We were justified, in part a, in locating a point P not on l by Assumption 10.9.)

7. Proof of Theorem 10.5: For any line l and any point P not on l, there is one and only one plane containing P and l.

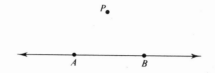

PROOF

Statements	*Reasons*
1. Locate points A and B on l.	1. Theorem 10.3.
2. A, B, and P are in a single plane, α.	2. Assumption 10.10.
3. l is in this plane α.	3. Assumption 10.11.
4. Thus, P and l are in the same plane.	4. Statements 2 and 3.

9. (a) Theorem 10.6.

   (b) Infinitely many.

   (c) By Assumption 10.12, we know there is at least one point P not on α. By the method of part (a), justified by Theorem 10.6, we can determine infinitely many planes in space.

Exercises 10.4

1. A line may be contained in a plane ($l \subset \alpha$); a line may intersect a plane in exactly one point ($l \cap \alpha = P$); a line may be parallel to a plane ($l \cap \alpha = \varnothing$).

3. (a) Theorem 10.6.   (b) Assumption 10.13 (parallel postulate).

   (c) No.   (d) No.

   (e) Theorem 10.8.   (f) Definition of parallel lines.

   (g) Assumption 10.11.   (h) Assumption 10.11.

   (i) Theorem 10.1.

5. A point, a single line, two lines, three lines.

Exercises 10.5

1. (a)

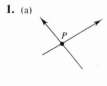

(b)

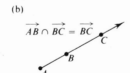

$$\overrightarrow{AB} \cap \overrightarrow{BC} = \overrightarrow{BC}$$

(c)

$$\overrightarrow{AB} \cap \overrightarrow{BA} = \overline{AB}$$

(d)

(e)   Cannot be a line.

3. a and g are convex sets; in b, c, d, and e the curve with its interior (shaded area) is a convex set; in f the curve with its interior is not a convex set; in h, the ball considered as a solid is a convex set.

5. (c) The double-shaded area; we call this the interior of $\angle ABC$.

7. Any points P and Q that are in $A \cap B$ are both in A and also both are in B, by definition of intersection of sets. If A is convex, then $\overline{PQ}$ is entirely contained in A, by definition of convex set; by the same definition, if B is convex, then $\overline{PQ}$ is entirely contained in B. Since $A \cap B$ contains all points in A that are also in B and no others, $\overline{PQ}$ is in $A \cap B$. Thus $A \cap B$ is a convex set.

9. (a) By definition of interior of an angle and Theorem 10.12.
  (b) By definition of interior of an angle a point in the interior must be in $\overleftrightarrow{AB}|C$; P is not in $\overleftrightarrow{AB}|C$ nor is it on $\angle ABC$.
  (c) By definition of interior of an angle a point in the interior must be in $\overleftrightarrow{BC}|A$. Q is not in $\overleftrightarrow{BC}|A$, nor is it on $\angle ABC$.
  (d) Given.
  (e) Definition of exterior of an angle.

Exercises 10.7

1. (a) point, ray, line, segment, angle, triangle
  (b) point, triangle
  (c) point, ray, line, segment, angle, triangle
  (d) point, triangle

5. (a) No.
  (b) Yes; $\angle ABC$.
  (c) The interior of $\triangle ABC$ is the intersection of the interiors of $\angle A$, $\angle B$, and $\angle C$.

Exercises 10.8

1. (a) curve, simple curve, closed curve, polygon, convex polygon.
  (b) curve, closed curve.
  (c) curve, simple curve, closed curve, polygon, convex polygon.
  (d) curve, simple curve, closed curve, polygon, convex polygon.
  (e) curve, simple curve, closed curve, polygon, concave polygon.
  (f) curve, simple curve, closed curve, polygon, concave polygon.
  (g) curve, simple curve, closed curve, polygon, concave polygon.
  (h) curve, simple curve, closed curve, polygon, concave polygon.
  (i) curve, closed curve.
  (j) curve, closed curve.
  (k) curve.
  (l) curve, simple curve, closed curve, polygon, convex polygon.
  (m) curve.
  (n) curve.

3. (a) No. (b) Yes. (c) Yes.

5. The line, angle, simple closed curve, triangle, and polygon (both concave and convex) separate the plane into three disjoint sets. For each of these figures, except the line, one of the sets is the *interior* of the figure and one the *exterior*. The simple closed curve and the polygon may or may not have interiors which are convex sets. The interiors of the angle, triangle and convex polygon are always convex sets. The interior of a concave polygon is not a convex set. A line divides a plane into three disjoint, convex sets. A non-simple closed curve separates the plane into four or more disjoint sets.

CHAPTER 11

Exercises 11.2

1. (c) Yes, by Assumption 11.1 and definition of the length of a line segment. (d) Yes, by Theorem 11.1.

   (e) Yes.   No, it is not unique; by Theorem 11.1, N may be contained in either $\overrightarrow{MQ}$ or in $\overrightarrow{AP}$.

3. (a) Q: 4, R: 6
   (b) $m(\overline{QR}) = 6 - 4 = 2$. Coordinates of Q and R above the line: $^-3$, $^-1$. Using these coordinates, $m(\overline{QR}) = \ ^-1 - \ ^-3 = 2$.
   (d) $m(\overline{PQ}) = 4$; $m(\overline{PR}) = 6$.
   (e) $m(\overline{PQ}) = 4$; $m(\overline{PR}) = 6$.

5. (b) and (e)

Exercises 11.3

1. (a)

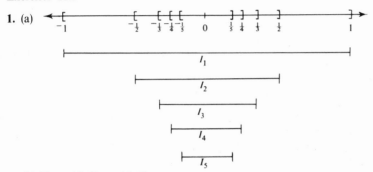

   (b) Yes.   (c) Yes.   (d) Zero.

3. (a) $(0.1111, 0.1112), (0.11111, 0.11112), (0.111111, 0.111112)$
   (b) $0.\overline{1}$ or $1/9$.

5. Coordinate of P: $7/6 = 1.1\overline{6}$
   $(1, 2), (1.1, 1.2), (1.16, 1.17), (1.166, 1.167), (1.1666, 1.1667)$

Exercises 11.5

1. Two; one on each of the rays determined by P.

3.

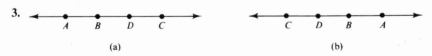

(a)   (b)

5. *Prove:* If (APQ), then $m(\overline{AP}) < m(\overline{AQ})$.
   Proof   If (APQ), then $m(\overline{AP}) + m(\overline{PQ}) = m(\overline{AQ})$ by the addition postulate. Then $m(\overline{AP}) < m(\overline{AQ})$ by definition of order for real numbers (Section 8.4).

7. *Prove:* If A, B and C lie on a line and if $m(\overline{AB}) + m(\overline{BC}) = m(\overline{AC})$, then (ABC).
   Proof   If A, B and C lie on a line, then (ABC), (ACB) or (BAC) by Assumption 10.6.

If (ACB) then $m(\overline{AC}) + m(\overline{CB}) = m(\overline{AB})$ by Assumption 11.2; then $m(\overline{AB}) - m(\overline{BC}) = m(\overline{AC})$ by the addition property of equality of real numbers. This is contrary to hypothesis. If (BAC) then $m(\overline{BA}) + m(\overline{AC}) = m(\overline{BC})$ by Assumption 11.2; then $m(\overline{BC}) - m(\overline{AB}) = m(\overline{AC})$ by the addition property of equality of real numbers. This is also contrary to hypothesis. Thus we must conclude that (ABC).

Exercises 11.7

1. (a) 19   (b) 180   (c) 24   (d) 84   (e) 97   (f) 36   (g) 41   (h) 137   (i) 83
   (j) 97   (k) 89   (l) 77   (m) 180   (n) 216   (o) 36   (p) 5   (q) 96   (r) 17

3. *Prove:* If $(\overrightarrow{BA}\ \overrightarrow{BP}\ \overrightarrow{BC})$ then $m(\angle ABP) < m(\angle ABC)$.
   PROOF   If $(\overrightarrow{BA}\ \overrightarrow{BP}\ \overrightarrow{BC})$, then $m(\angle ABP) + m(\angle PBC) = m(\angle ABC)$, then $m(\angle ABP) < m(\angle ABC)$ by definition of order for real numbers.

5. (a) Angle construction postulate.
   (b) Assumption 11.11.
   (c) Theorem 11.8.
   (d) Since $m(\angle DEP) = m(\angle ABC)$, if $m(\angle DEF) < m(\angle DEP)$, it follows that $m(DEF) < m(\angle ABC)$.
   (e) Since $m(\angle DEP) = m(\angle ABC)$, if $m(\angle DEF) = m(\angle DEP)$, it follows that $m(\angle DEF) = m(\angle ABC)$.

Exercises 11.8

1. (a) False   (b) True   (c) True   (d) True   (e) True   (f) False   (g) False
   (h) True   (i) True   (j) False   (k) False   (l) True   (m) False   (n) False
   (o) True   (p) False   (q) True   (r) False   (s) True   (t) False   (u) True
   (v) True   (w) False   (x) False   (y) True   (z) True

3. Two pairs. Measures of other angles: $x, 180 - x, 180 - x$.

5. *Prove:* Supplements of congruent angles are congruent.
   PROOF   Let $\angle A \cong \angle B$. Suppose $m(\angle A) + m(\angle C) = 180$ and $m(\angle B) + m(\angle D) = 180$ (that is, $\angle C$ and $\angle D$ are supplements of $\angle A$ and $\angle B$ respectively). Since $m(\angle A) = m(\angle B)$ (by definition of congruent angles), $m(\angle A) + m(\angle C) = m(\angle A) + m(\angle D)$. By the addition property for real numbers, $m(\angle C) = m(\angle D)$. By definition of congruent angles, $\angle C \cong \angle D$.

7. (a) Assumption 11.4 (the angle construction postulate).
   (b) Alternate definition of right angle (Section 11.8).
   (c) Assumption 11.4.
   (d) Assumption 10.2. (The distinct points P and R that determine the ray $\overrightarrow{PR}$ are contained in one and only one line.)

Exercises 11.9

1. (a), (d), and (f) are the same; (b), (e), and (g) are the same; (c) and (h) are the same.
3. (a) $ABC \leftrightarrow DEF$ (b) $PQR \leftrightarrow ZXY$ (c) $MNO \leftrightarrow BCA$ (d) $RST \leftrightarrow QPO$
   (e) $ABC \leftrightarrow DEF$.
5. **Theorem 11.22:**   If two angles of a triangle are congruent, then the sides opposite these angles are congruent.

PROOF

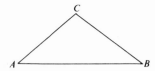

Given: $\triangle ABC$ where $\angle A \cong \angle B$.

Prove: $\overline{AC} \cong \overline{BC}$. Consider the correspondence of vertices $ABC \leftrightarrow BAC$ between $\triangle ABC$ and itself.

Statements	*Reasons*
1. $\angle A \cong \angle B$ and $\angle B \cong \angle A$	1. Given.
2. $\overline{AB} \cong \overline{BA}$	2. By identity.
3. $\triangle ABC \cong \triangle BAC$	3. Theorem 11.21 (ASA theorem).
4. $\overline{AC} \cong \overline{BC}$	4. Definition of congruent triangles.

7. Reason 1. Given.
   2. Assumption 11.4 (angle construction postulate).
   3. Theorem 11.6, part 4.
   5. Transitive property of congruence of angles.
   6. Assumption 11.15 (SAS postulate).
   7. Definition of congruent triangles.
   9. Transitive property of congruence of segments.
   10. Statement 9 and Theorem 11.23.
   11. Assumption 10.2, and definition of perpendicular bisector.
   12. Definition of a perpendicular bisector.

Exercises 11.10

1. (a) True   (b) True   (c) False   (d) True   (e) True   (f) False
   (g) True   (h) True   (i) True   (j) False   (k) False   (l) False
   (m) True

3. $\angle EAF$ and $\angle PDA$; $\angle EAD$ and $\angle PDQ$; $\angle FAB$ and $\angle ADC$; $\angle DAB$ and $\angle QDC$; $\angle GBH$ and $\angle BCS$; $\angle CBH$ and $\angle RCS$; $\angle ABG$ and $\angle DCB$; $\angle ABC$ and $\angle DCR$; $\angle PDQ$ and $\angle DCR$; $\angle QDC$ and $\angle RCS$; $\angle PDA$ and $\angle DCB$; $\angle ADC$ and $\angle BCS$; $\angle EAD$ and $\angle ABC$; $\angle DAB$ and $\angle CBH$; $\angle EAF$ and $\angle ABG$; $\angle FAB$ and $\angle GBH$.

5. (a) **Corollary (Theorem 11.29):** Given a correspondence between the vertices of two triangles, if two pairs of corresponding angles are congruent, then the third pair of corresponding angles are congruent.

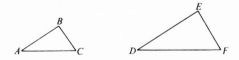

Given: $\triangle ABC$ and $\triangle DEF$. Under correspondence between vertices $ABC \leftrightarrow DEF$, $\angle A \cong \angle D$ and $\angle C \cong \angle F$.

Prove: $\angle B \cong \angle E$.

PROOF

Statements	Reasons
1. $\angle A \cong \angle D$; $\angle C \cong \angle F$.	1. Given.
2. $m(\angle A) = m(\angle D)$; $m(\angle C) = m(\angle F)$.	2. Definition of congruent angles.
3. $m(\angle A) + m(\angle C) = m(\angle D) + m(\angle F)$.	3. Addition property of equality of real numbers.
4. $m(\angle A) + m(\angle C) + m(\angle B) = 180$; $m(\angle D) + m(\angle F) + m(\angle E) = 180$.	4. Theorem 11.29.
5. $[m(\angle A) + m(\angle C)] + m(\angle B)$ $= [m(\angle D) + m(\angle F)] + m(\angle E)$.	5. Transitive property of equality for real numbers.
6. $m(\angle B) = m(\angle E)$.	6. Statements 3 and 5 and addition property of equality for real numbers.
7. $\angle B \cong \angle E$	7. Definition of congruent angles.

(b) *Prove:* For any triangle, the measure of an exterior angle is equal to the sum of the measures of the two remote interior angles.

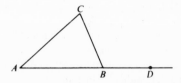

Given: $\triangle ABC$ with exterior angle CBD.
Prove: $m(\angle CBD) = m(\angle A) + m(\angle C)$.

PROOF

Statements	Reasons
1. $\angle CBD$ is an exterior angle of $\triangle ABC$.	1. Given.
2. $[m(\angle A) + m(\angle C)] + m(\angle ABC)$ $= 180$.	2. Theorem 11.25 (the exterior angle theorem).
3. $\angle ABC$ is supplementary to $\angle CBD$.	3. Definition of exterior angle.
4. $m(\angle CBD) + m(\angle ABC) = 180$.	4. Assumption 11.14 (the supplement postulate).
5. $[m(\angle A) + m(\angle C)] + m(\angle ABC)$ $= m(\angle CBD) + m(\angle ABC)$.	5. Transitive property of equality for real numbers.
6. $m(\angle CBD) = m(\angle A) + m(\angle C)$.	6. Addition property of equality for real numbers.

7. (a) Yes. (b) No.

9. Reasons 1. Given.
           2. Theorem 11.13 (vertical angles are congruent).
           3. Transitive property of congruence of angles.
           4. Theorem 11.27.

11. Theorem: If parallel lines *l* and *m* are cut by a transversal *n*, then a pair of corresponding angles are congruent.

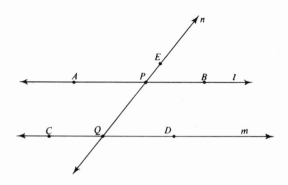

Given: Parallel lines *l* and *m* cut by transversal *n*. Let *P* and *Q* be the points of intersection of *n* with *l* and *m* respectively. Select points *A* and *B* on *l* so that (*APB*); points *C* and *D* on *m* so that (*CQD*) and *E* on *n* so that (*EPQ*). Points *A* and *C* are on the same side of $\overleftrightarrow{PQ}$. *Prove:* $\angle EPB \cong \angle PQD$.

PROOF

Statements	*Reasons*
1. $l \parallel m$; transversal *n* intersects *l* and *m* at *P* and *Q* respectively. Points *A* and *B* are on *l* and (*APB*). Points *C* and *D* are on *m* and (*CQD*). Point *E* is on *n* and (*QPE*).	1. Given.
2. $\angle APQ \cong \angle PQD$	2. Theorem 11.28.
3. $\angle APQ \cong \angle EPB$	3. Theorem 11.13.
4. $\angle EPB \cong \angle PQD$	4. Transitive property of congruence of angles.

13. PROOF

Statements	*Reasons*
1. *ABCD* is a parallelogram with diagonal $\overline{BD}$.	1. Given.
2. $\overline{DC} \parallel \overline{AB}$	2. Definition of a parallelogram.
3. $\angle BDC \cong \angle ABD$	3. Theorem 11.28.
4. $\overline{AD} \parallel \overline{BC}$	4. Definition of a parallelogram.
5. $\angle ADB \cong \angle CBD$	5. Theorem 11.28.
6. $\overline{BD} \cong \overline{DB}$	6. By identity.
7. $\triangle ABD \cong \triangle BCD$	7. Theorem 11.21 (ASA Theorem).

15. Theorem: If a line *l* is parallel to a line *m*, then the distances of any points *P* and *Q* on *l* from *m* are equal.

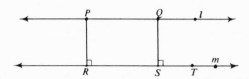

Given: Parallel lines *l* and *m* with points *P* and *Q* on *l*. *R* is the point where a line through *P* that is ⊥ to *m* meets *m*; *S* is the point where a line through *Q* that is ⊥ to *m* meets *m*. Let *T* be a point in *m* such that (*RST*).
Prove: $m(\overline{PR}) = m(\overline{QS})$.

Proof

Statements	*Reasons*
1. $l \parallel m$. Points *P* and *Q* are on *l*; $\overline{PR} \perp m$ and $\overline{QS} \perp m$. Point *T* is on *m* and (*RST*).	1. Given.
2. ∠*PRS* and ∠*QST* are right angles, and are congruent.	2. Definition of a right angle.
3. $\overline{PR} \parallel \overline{QS}$.	3. Theorem in Exercise 9: If two lines are cut by a transversal so that a pair of corresponding angles are congruent, then the lines are parallel.
4. *PQRS* is a parallelogram.	4. Definition of a parallelogram.
5. $\overline{PR} \cong \overline{QS}$.	5. Theorem in Exercise 14: In a parallelogram, any two opposite sides are congruent.
6. $m(\overline{PR}) = m(\overline{QS})$	6. Definition of congruent segments.

17. Reason 2. Theorem 11.28.
       3. By identity.
       4. Assumption 11.15 (SAS postulate).
       5. Definition of congruent triangles.
       6. Theorem 11.27.
       7. Definition of a parallelogram.

19. Reason 1. Given.
       2. Theorem in Exercise 16: In a parallelogram, any two opposite angles are congruent.
       3. Definition of a parallelogram.
       4. Theorem 28.
       5. Transitive property of congruence of angles.
       6. Definition of a linear pair and definition of supplementary angles.
       7. Any supplement of a right angle is a right angle.

21. Theorem: If the hypotenuse and an acute angle of a right triangle are congruent to the corresponding parts of a second right triangle, then the triangles are congruent.

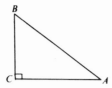 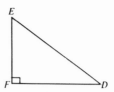

Given: $\triangle ABC$ and $\triangle DEF$ where $\angle C$ and $\angle F$ are right angles. $\overline{AB} \cong \overline{DE}$ and $\angle A \cong \angle D$.
Prove: $\triangle ABC \cong \triangle DEF$.

PROOF

Statements	*Reasons*
1. In $\triangle ABC$ and $\triangle DEF$, $\angle A \cong \angle D$ and $\angle C$ and $\angle F$ are right angles. $\overline{AB} \cong \overline{DE}$.	1. Given.
2. $\angle C$ and $\angle F$ are congruent.	2. Definition of right angles.
3. $\angle B \cong \angle E$.	3. Corollary to Theorem 11.29.
4. $\triangle ABC \cong \triangle DEF$.	4. Theorem 11.21.

23. Reason 1. Given.
2. Definition of a right angle.
3. Corollary to Theorem 11.29. (The acute angles of a right triangle are complementary.)
4. Definition of complementary angles.
5. Statements 1, 2, and 4 and the addition property of equality for real numbers.
6. Theorem 11.30.
7. By definition of a midpoint, $m(\overline{BD}) = m(\overline{DA}) = \frac{1}{2}m(\overline{AB})$; thus $m(\overline{CD}) = m(BD)$ by the transitive property of equality of real numbers, and $\overline{CD} \cong \overline{BD}$ by definition of congruent segments.
8. Theorem 11.20.
9. Addition property of equality of real numbers.
10. Theorem 11.29 and the addition property of equality of real numbers.
11. Theorem 11.22.
12. Transitive property of equality of real numbers.

25.

Statements	*Reasons*
1. $ABCD$ is a parallelogram, with diagonals $\overline{AC}$ and $\overline{BD}$ intersecting at point E.	1. Given
2. $\overline{AB} \parallel \overline{DC}$; $\overline{AD} \parallel \overline{BC}$	2. Definition of a parallelogram.
3. $\angle CAD \cong \angle ACB$; $\angle ADB \cong \angle CBD$.	3. Theorem 11.28

4. $\overline{AD} \cong \overline{BC}$

5. $\triangle ADE \cong \triangle BCE$ (under correspondence $ADE \leftrightarrow CBE$)

6. $\overline{AE} \cong \overline{EC}$; $\overline{DE} \cong \overline{EB}$

4. Theorem in Exercise 14: Opposite sides of a parallelogram are congruent.

5. Theorem 11.20 (ASA theorem).

6. Definition of congruent triangles.

CHAPTER 12

Exercises 12.1

3. (a) Area of square in part (d) is 4 times the area of the square in part (a).
   (b) When a side of a square is doubled, its area is quadrupled.
   (c) Area of square in part (f) is $\frac{1}{16}$ times area of square in part (e).
   (d) When a side of a square is divided by 4, its area is divided by 16.
   (e) When a side of a square is multiplied by k, its area is multiplied by k^2.
   (f) When a side of a square is divided by k, its area is divided by k^2.

5. (a) The area is multiplied by 3.
   (b) The area is multiplied by 5.
   (c) The area is multiplied by 4.
   (d) The area would be multiplied by n.
   (e) The area would be multiplied by n^2.

7. 108.

Exercises 12.2

1. (a) 20   (b) $10\frac{1}{2}$ (c) 14   (d) 11   (e) $\dfrac{49\sqrt{3}}{4}$ (f) $12\frac{1}{2}$ (g) 6.25

3. $s^2 - 2hs = s(s - 2h)$

5. (a) $52\frac{1}{2}$ (b) $71\frac{1}{4}$ (c) 46   (d) 54

7. **Theorem:**   The area of a rhombus is equal to one-half the product of the lengths of its diagonals.

Given: Rhombus $ABCD$ with diagonals AC and BD intersecting at point E.
Prove: Area($ABCD$) = $\frac{1}{2}m(\overline{AC}) \cdot m(\overline{BD})$.

PROOF

Statements	*Reasons*
1. $ABCD$ is a rhombus.	1. Given.
2. $\triangle ACD \cong \triangle ABC$	2. Theorem, Exercise 13, Exercises 11.10: A diagonal of a parallelogram forms 2 congruent triangles.
3. Area($\triangle ACD$) + Area($\triangle ABC$) = Area($ABCD$).	3. Assumption 12.3.
4. $\overline{AC} \perp \overline{BD}$.	4. Theorem, Exercise 20, Exercises 11.10: The diagonals of a rhombus are perpendicular to each other.

5. With AC as base, $\overline{DE}$ is the corresponding altitude of $\triangle ACD$.

5. Definition of altitude of a triangle.

6. Area($\triangle ACD$) = $\frac{1}{2}m(AC) \cdot m(\overline{DE})$

6. Theorem 12.3.

7. Area($ABCD$) = $2[\frac{1}{2}m(\overline{AC}) \cdot m(\overline{DE})]$
$= m(\overline{AC}) \cdot m(\overline{DE})$

7. Statements 2, 3 and 6.

8. $m(DE) = \frac{1}{2}m(BD)$

8. Theorem, Exercise 25, Exercises 11.10: The diagonals of a parallelogram bisect each other.

9. Area($ABCD$) = $\frac{1}{2}m(\overline{AC}) \cdot m(\overline{BD})$

9. Statements 7 and 8.

9. Reason 1. Given.
   2. Theorem, Exercise 13, Exercises 11.10: A diagonal of a parallelogram forms 2 congruent triangles.
   3. Assumption 12.2.
   4. Theorem 12.3.
   5. Statement 3.
   6. Assumption 12.3.
   7. Statements 4, 5, and 6.
   8. Addition of real numbers performed.

Exercises 12.3

1. (a) $\sqrt{61}$ (b) 4   (c) 12   (d) $2\sqrt{5}$ (e) $\sqrt{21}$ (f) $\sqrt{85}$

3. (a) $\sqrt{2}$ (b) $\sqrt{3}$ (c) $\sqrt{4} = 2$ (d) $\sqrt{6}$

5. $2\sqrt{39}$

7. 15; $\dfrac{108}{15} = \dfrac{36}{5}$

CHAPTER 13

Exercises 13.2

1. (a) 6   (b) 1   (c) 20   (d) 2

3. Any pair of (b), (e), (f), and (j); (a) and (c); Any pair of (d), (g), and (h).

Exercises 13.3

1. **Theorem:**   The triangle whose vertices are the midpoints of a given triangle is similar to the given triangle.

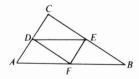

Given: $\triangle ABC$ and $\triangle DEF$ with D, E and F midpoints of $\overline{AC}$, $\overline{BC}$ and $\overline{AB}$ respectively.
Prove: $\triangle ABC \sim \triangle DEF$.

PROOF

Statements	Reasons
1. Vertices D, E, and F of $\triangle DEF$ are midpoints of sides $\overline{AC}$, $\overline{BC}$, and $\overline{AB}$ of $\triangle ABC$ respectively.	1. Given.
2. $m(\overline{DE}) = \frac{1}{2}m(\overline{AB})$; $m(\overline{DF}) = \frac{1}{2}m(\overline{BC})$; $m(\overline{EF}) = \frac{1}{2}m(\overline{AC})$.	2. Theorem, Exercise 18, Exercises 11.10.
3. $\dfrac{m(\overline{DE})}{m(\overline{AB})} = \dfrac{1}{2}$; $\dfrac{m(\overline{DF})}{m(\overline{BC})} = \dfrac{1}{2}$; $\dfrac{m(\overline{EF})}{m(\overline{AC})} = \dfrac{1}{2}$.	3. Multiplication property of equality of real numbers.
4. $\dfrac{m(\overline{DE})}{m(\overline{AB})} = \dfrac{m(\overline{DF})}{m(\overline{BC})} = \dfrac{m(\overline{EF})}{m(\overline{AC})}$	4. Transitive property of equality of real numbers.

Thus the corresponding sides of $\triangle ABC$ and $\triangle DEF$ are proportional.

Statements	Reasons
5. $\overline{DE} \parallel \overline{AB}$, $\overline{DF} \parallel \overline{BC}$, $\overline{EF} \parallel \overline{AC}$.	5. Theorem, Exercise 18, Exercises 11.10.

In $\triangle ABC$ and $\triangle AFD$:

Statements	Reasons
6. $\angle A \cong \angle A$	6. By identity.
7. $\angle C \cong \angle ADF$; $\angle B \cong \angle AFD$	7. Theorem, Exercise 11, Exercises 11.10.
8. $ADEF$ is a parallelogram.	8. Definition of a parallelogram.
9. $\triangle ADF \cong \triangle EFD$	9. Theorem, Exercise 13, Exercises 11.10.
10. $\angle A \cong \angle DEF$; $\angle AFD \cong \angle EDF$; $\angle ADF \cong \angle DFE$	10. Definition of congruent triangles.
11. $\angle A \cong \angle DEF$; $\angle B \cong \angle EDF$; $\angle C \cong \angle DFE$	11. Transitive property of congruence of angles.
12. $\triangle ABC \sim \triangle DEF$	12. Definition of similar triangles.

3. (a) and (c); any pair of (b), (e), (f), and (j); any pair of (d), (g), and (h).

5. (c), (d), (e), (f).

Exercises 13.4

1. SAS and SSS

3. (a) $\angle A$ and $\angle B$; $\angle A$ and $\angle ACD$; $\angle B$ and $\angle BCD$; $\angle ACD$ and $\angle BCD$.
(b) $\angle B$
(c) $\angle A$
(d) $\angle ADC$ and $\angle BDC$
(e) $\triangle ABC$ and $\triangle ACD$; $\triangle ABC$ and $\triangle BCD$; $\triangle ACD$ and $\triangle BCD$.

(f) In $\triangle ACD$ and $\triangle BAC$, $\dfrac{m(\overline{AD})}{m(\overline{AC})} = \dfrac{m(\overline{DC})}{m(\overline{BC})} = \dfrac{m(\overline{AC})}{m(\overline{AB})}$;

in $\triangle BCD$ and $\triangle BAC$, $\dfrac{m(\overline{CD})}{m(\overline{AC})} = \dfrac{m(\overline{BD})}{m(\overline{BC})} = \dfrac{m(\overline{BC})}{m(\overline{AB})}$;

in $\triangle ACD$ and $\triangle CBD$, $\dfrac{m(\overline{AD})}{m(\overline{CD})} = \dfrac{m(\overline{CD})}{m(\overline{BD})} = \dfrac{m(\overline{AC})}{m(\overline{BC})}$.

5. (a) Yes, by the SAS similarity theorem.
  (b) Yes. Since $\angle BCA \cong \angle DCE$ (vertical angles are congruent), $\triangle ABC \sim \triangle ECD$ by the AA similarity corollary.
  (c) Yes, by the SAS similarity theorem.
  (d) Not necessarily; none of the similarity theorems apply here.

7. *Prove:*   If a line that is $\parallel$ to one side of a triangle intersects the other two sides in distinct points, then a second triangle is formed that is similar to the first.

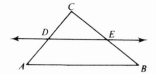

Given: $\triangle ABC$ with $\overleftrightarrow{DE}$ intersecting sides $\overline{AC}$ and $\overline{BC}$ in points D and E respectively; $\overleftrightarrow{DE} \parallel AB$.
Prove: $\triangle ABC \sim \triangle DEC$.

PROOF

Statements	*Reasons*
1. The line $\overleftrightarrow{DE}$ intersects sides $\overline{AC}$ and $\overline{BC}$ of $\triangle ABC$ in points D and E respectively. $\overleftrightarrow{DE} \parallel \overline{AB}$.	1. Given.
2. $\angle CDE \cong \angle A$ and $\angle CED \cong \angle B$.	2. Exercise 11, Exercises 11.10.
3. $\triangle DEC \sim \triangle ABC$.	3. AA similarity corollary.

9. Reason   1. Given.
          2. Theorem 11.4.
          3. Statements 1 and 2.
          4. Theorem 13.2.
          5. Exercise 11, Exercises 11.10.
          6. AA similarity theorem.
          7. Definition of similar triangles.
          8. Statements 2 and 7.
         10. Multiplication property for equality of real numbers.
         11. Given.
         12. Multiplication property for equality of real numbers.
         13. Transitive property for equality of real numbers.
         14. SSS congruence theorem.
         15. Definition of similar triangles.
         16. SAS similarity theorem.

Exercises 13.5

1. $\sin 60° = \sqrt{3}/2$; $\cos 60° = 1/2$; $\tan 60° = \sqrt{3}$.

3. (a) $\sin x° = 8/10$ or $4/5$; $\cos x° = 6/10$ or $3/5$; $\tan x° = 8/6$ or $4/3$
  (b) $\sin x° = 5/13$; $\cos x° = 12/13$; $\tan x° = 5/12$
  (c) $\sin x° = 9/41$; $\cos x° = 40/41$; $\tan x° = 9/40$
  (d) $\sin x° = 20/25$ or $4/5$; $\cos x° = 15/25$ or $3/5$; $\tan x° = 20/15$ or $4/3$
  (e) $\sin x° = 4/8$ or $1/2$; $\cos x° = 4\sqrt{3}/8$ or $\sqrt{3}/2$; $\tan x° = 4/4\sqrt{3}$ or $1/\sqrt{3} = \sqrt{3}/3$
  (f) $\sin x° = \sqrt{5}/2\sqrt{2}$; $\cos x° = \sqrt{3}/2\sqrt{2}$; $\tan x° = \sqrt{5}/\sqrt{3}$

5. (a) 0.951   (b) 0.231   (c) 0.423   (d) 2.904   (e) 0.999   (f) 0.731

7. $a = 4.770$ or approximately 5; $b = 7.632$ or approximately 8. $\angle B = 58°$.

9. Area $= 22.016$ or approximately 22.

Index